The New Economy of the Product Life Cycle

Andrey Tyulin • Alexander Chursin

The New Economy of the Product Life Cycle

Innovation and Design in the Digital Era

Springer

Andrey Tyulin
JSC Russian Space Systems
RUDN University
Moscow, Russia

Alexander Chursin
JSC Russian Space Systems
RUDN University
Moscow, Russia

ISBN 978-3-030-37813-4 ISBN 978-3-030-37814-1 (eBook)
https://doi.org/10.1007/978-3-030-37814-1

This Springer imprint is published by the registered company Springer Nature Switzerland AG.
The registered company address is: Gewerbestrasse 11, 6330 Cham, Switzerland

Preface

Over the past few years, a pronounced tendency has been observed in the economy to reduce the product life cycle due to the rapid innovation development, global competition, deepening of the processes of informatization and digitalization, and the rapid development of science, engineering, and technology, which entails the accumulation of the intellectual potential of companies and society. It is related to the formation of new management approaches arising from the development of the digital economy and the digital industry based on the implementation of Industry 4.0 concept, modern technologies for collecting, processing, and analyzing of big data, artificial intelligence, the Internet of Things, deepening of production automation, and building adaptive production lines and the creation of robotic complexes and entire plants capable of producing robots with minimal human involvement. This entails the transformation of organizational and economic relations in the field of development and production management of promising products, which in the future could capture a high market share or create a new consumer segment.

Consequently, an important competitive advantage of each industrial organization is the ability to organize the management of the product life cycle so that it is designed and manufactured in the shortest possible time at the optimal level of costs, providing the best technical and functional characteristics in the market in order to achieve global leadership.

Global leadership allows us to increase profit from sales in short terms and focus it in larger quantities compared to competitors on solving fundamental and applied problems of creating products with exceptional new value to society and satisfy needs in the short term.

Thus, there are conditions exponentially increasing the dynamism of the processes of economic development, which, in our opinion, today necessitates a revision of some economic provisions, for example, in the theory of large life cycles of N. D. Kondratiev, when their duration is reduced, and the phases of growth and decline have the form of a sharply increasing and decreasing curve.

The principles outlined above and the economic processes studied in the monograph have allowed us to formulate the economic law of rapid development. This law is the basis for the creation of axiomatic foundations for managing an organization's rapid development, establishing principles for quantifying the processes, determining the dynamics of changes in indicators and general principles for building models of the rapid development process, clarifying the effect of the law in conditions of imperfect competition, and linking rapid development with the formation of unique competencies of an organization.

On the basis of the economic law of rapid development, the main provisions of the new economy of the product life cycle are formed, given the dynamism of developing and implementing new technologies, accelerating innovation development, such as artificial intelligence technologies, cloud computing, end-to-end, quantum, and supercomputing technology, identification technology, the blockchain, neural networks, technologies of creation of cyber-physical systems, 3D technology (printing), robotization (including the creation of robotic systems for the production of robots), additive technology, open production technologies, paperless, biometric technologies, brain-computer technologies, etc.

These provisions, founded on the tools and economic mechanisms described in this work, based on the use of intellectual methods in solving management tasks in the development and production of promising products, have led to the creation of a new methodology that contains the main methods, ways, and strategic approaches to managing the product life cycle which will enable industrial organizations to achieve faster development and global market superiority. The methodology is fully disclosed in the monograph.

The monograph will be of interest to scientists, theoretical economists involved in the problems of microeconomics, production practitioners, heads of industrial organizations, and researchers interested in the described issues.

Moscow, Russia

Andrey Tyulin
Alexander Chursin

Acknowledgment

The reported study was funded by RFBR, project number 19-29-07348.

Contents

1 The Fundamentals of Product Life Cycle Economics 1
1.1 Evolution of the Product Life Cycle Theory: The Value and Goal of Rapid Development 1
1.2 Modern Product Life Cycle Management Approaches Used at Enterprises Practicing Digitized Work 14
1.2.1 Ensuring Stable Economic Development of an Organization 18
1.2.2 Building Mechanisms for Management of All Corporate Resources 21
1.2.3 Introduction of Advanced Product Design, Preproduction and Production Methods 24
1.3 Life Cycle Cost Calculation Methods 31
1.4 Informational Support of Automated Life Cycle Management Processes with the Use of Digital Technologies 41
References 49

2 Rapid Development of an Organization 51
2.1 Rapid Development Management Law 51
2.2 Axiomatic Fundamentals of Management of Rapid Development 59
2.2.1 The Main Axioms of Rapid Development 59
2.2.2 Dynamic Mathematical Model 61
2.2.3 The Hysteresis Type Rapid Development Model 63
2.2.4 Management of Rapid Development in Imperfect Competition 64
2.2.5 General Management of Rapid Development in Imperfect Competition 65
2.2.6 A Relationship Between Rapid Development and Competence Management by Organizations 66

2.3 The Role of Competences in Management of Organizations' Rapid Development . . . 68
2.4 Strategic Rapid Development Approaches and Instruments for Science-Intensive Companies and Branches . . . 77

3 Personification of Needs as a Landmark for Creating Future Goods . . . 89
3.1 Development of Technologies and Public Needs with Technological Waves and Information Technologies Enhancing in All Spheres of Public Life . . . 89
3.2 Big Data Analysis-Based Methods of Assessment of the Effectiveness of Next-Generation Product Creation Processes Aimed at Satisfying Future Needs . . . 105
3.3 The Economy–Production–Economy Cycle as a Form of Creating Competitive Goods . . . 119

4 Evaluation of an Organization's Ability to Tailoring Production to Set Parameters . . . 137
4.1 The Model and Dynamic Evaluation of Innovative Potential with Rapidly Growing Competitive Innovative Solutions and Expanding Informational Space . . . 137
4.2 Modeling, Evaluation, and Prognosis of the Development of Unique Competences to Satisfy Prospective Needs and Ensure Their Transfer . . . 146
4.3 Resource Provision Models for Future Products . . . 159
References . . . 170

5 A Product's Image as a Basis of Its Competitiveness . . . 171
5.1 Assessment of a Corporate Microenvironment and Its Role in Creating a Brand Image of a New Competitive Product . . . 171
5.2 Use of Communication Methods and Technologies in Market and Consumer Demand Analysis When Shaping a Product's Image . . . 177
5.3 Competitiveness Management of Science-Intensive Products When Shaping Its Technical and Economic Image . . . 187
Reference . . . 193

6 Economic Aspects of Developing Science-Intensive Products . . . 195
6.1 Principles of Building an Intelligent Automated Product Life Cycle Management System . . . 195
6.2 Advanced Digital Design, Modeling, and Production Methods . . . 208
6.3 The Process of Tailoring Products to a Fixed Prime Cost and Competitiveness with the Help of Intelligent Automated Systems . . . 220
Reference . . . 230

7 Preproduction of Advanced Products with High Technical and Economic Characteristics 231
7.1 Preproduction of Products with High Technical and Economic Characteristics Based on New Physical Principles 231
7.2 Prime Cost Optimization Through Effective Preproduction 246
7.3 Building an Intelligent Automated Preproduction Management Systems 262

8 Modern Manufacturing Process Management Methods 275
8.1 Setting Up Flexible Automated Manufacture Processes Relying on Digital Technologies' Advantages 275
8.2 Allocation of Production Costs Between Parent and Cooperating Enterprises 289
8.3 Building Advanced Production Management Systems 302

9 Product Life Cycle Management 321
9.1 Integrated Digital Platform Supporting Effective Managerial Decision-Making in Product Life Cycle Management 321
9.2 Information Support of a Corporate Life Cycle Management System 332
9.3 The Theoretical Basis of Creating Future Goods, Which Put a Business onto the Path of Rapid Development 343

10 Rapid Development of an Organization on the Basis of Product Life Cycle Management 351
10.1 Basic Management Tools for the Development and Production of Future Products in Order to Ensure an Organization's Rapid Development 351
10.2 Mechanisms for the Transition of Organizations to the Rapid Development Path 368
10.3 Rapid Development of an Organization: Key to Improving Sustainability and a Revolutionary Transition to a New Technological Order 378
References 387

11 Conclusions 389

Bibliography 393

Chapter 1
The Fundamentals of Product Life Cycle Economics

1.1 Evolution of the Product Life Cycle Theory: The Value and Goal of Rapid Development

In the current global economic system, which is characterized by globalization and a highly volatile environment for the functioning economic entities, competitive growth is a key element of stable progress for organizations, industries, and national economy in general.

The pluses of globalization are predicated by the economic benefits of the use of advanced scientific, technical, technological, and quality potential of economies, which are leaders in respective fields, in other economies. In such cases, new solutions can be introduced quickly and cost effectively.

Globalization stimulates competition and, consequently, further progress and distribution of new technologies; in a globalized world, direct investment grows much faster than global trade. It is key part of the transfer of industrial technologies, formation of transnational organizations, and it directly influences national economies. Globalization boosts labor efficiency, and as the manufacturing process is being optimized globally, it contributes to the distribution of advanced technologies and stimulates competitive pressure that ensures nonstop worldwide introduction of innovative solutions. At the same time, globalization allows nations to mobilize hefty financial resources, as investors can use broader financial toolkits, and the number of available markets has increased.

Meanwhile, globalization results in tougher international competition and creates an environment that forces companies to speed up the development and implementation process and to focus on unique types of expertise, which ensure leadership in new competitive fields.

Crucial for high competitiveness in a globalized world are industrial modernization and upgrades of innovative companies' scientific and technical infrastructure, as these companies focus on commercialization of expertise, and providing high-quality intellectual resources. Therefore, in the postindustrial economy, which is a

A. Tyulin, A. Chursin, *The New Economy of the Product Life Cycle*,
https://doi.org/10.1007/978-3-030-37814-1_1

knowledge-based economy, the biggest competitive advantages result from the use of the intellectual potential. To implement this potential effectively, companies, along with industries and regional authorities, should step up innovative work. In a highly competitive environment, this kind of activity helps to improve technical and economical characteristics of commercial projects, which could outrun top national and foreign design solutions. This allows companies to greatly expand in science-intensive product markets and create new ones through rapid development, which should be based on retaining leadership in making advanced products.

There are signs that a new kind of theory is forming as part of the economic science—the rapid development theory. Its formation and evolution is based on some fundamental strands of research, such as management of products' and organizations' competitiveness, corporate governance, innovation management, and product life cycle management.

The group of scientists, who have made the greatest contribution to the theory of competitiveness management, includes those who initially studied the "competition" concept. These are J. Schumpeter, who viewed entrepreneurship as the driver of competitiveness in Capitalism, Socialism and Democracy (1942) [1]; American economists A. Sloan and P. Drucker, who described the concept of management as a key element of competitiveness (see works by A. Sloan—"My Years with General Motors" (1963) [2] and P. Drucker's "The Age of Discontinuity" (1969) [3]); R. Solow, who studied the role of education, technological innovation, and expertise factors as the fundamentals of the high competitiveness of the USA, which was observed over the period from 1948 until 1982, in "Technical change and aggregate production function" (1957) [4]; M. Porter, whose research resulted in the Porter's Five Forces theory, which highlights the determinants that exert maximal influence on economic units in the competitive environment [5].

Scientists, who pioneered in studying the problem of management of competitiveness, looked at it as a set criterion predicated by various internal and external factors [6].

Later, as scientists' theories progressed, they began to consider the possibility and necessity of intentional management of competitiveness. That generated varied approaches, which enabled them to outline plans for reaching desired levels of competitiveness.

The approaches to management of competitiveness, as the global perception of the problem progressed and evolved, have changed greatly by now (Table 1.1).

In the late 1960s, there was a shift in priorities toward internal factors of corporate growth based on more rational use of available resources. Optimization of internal resources leveraged the majority of modern organization development concepts.

As follows from Table 1.1, scientists from around the world in the 1900s noted in their works that management of an organization's competitiveness largely depended on effective management of corporate expertise with account of the specifics of an industry.

According to A. Thompson and A. Strickland, factors of competitiveness include: products' quality and technical characteristics, reputation of organizations within an industry, production capacity and innovative potential within an industry,

Table 1.1 Evolution of corporate competitiveness management concepts

No.	Concept	Year	Authors (country)	Subject matter
1.	Complex quality management	1951	A. Feigenbaum [7], E. Deming [8] (the USA)	An organization's competitiveness and effectiveness depend on the quality of its product. Therefore, to ensure high output quality, the production process should be organized in such a way as to model quality at the main functional nodes and processes with direct participation of duly qualified, trained, and responsible operators, instead of mere statement of output quality.
2.	Expertise and organizational learning potential management	1966	P. Drucker [9, 10], P. Senge [11] (the USA)	Most changes in the society follow informational development, and expertise becomes a key source of information, which does not have a geographic location.
3.	Organizational culture	1970	E. Shein [12] (Switzerland), W. Ouchi [13] (the USA)	Organizational culture is a set of personnel behavior rules, regulations, standards, and guidelines, which help an organization to build a unified team, achieve a competitive status in the market and adapt to changes in the outer environment.
4.	Resource concept	1984	B. Wernelfelt (Denmark) [14], D. Collis, S. Montgomery [15] (the USA)	An organization's resources include physical (estate property, raw materials, etc.) and nonmaterial (the know-how, reputation, patents, etc.) assets, as well as managerial capabilities, which can provide a number of competitive advantages.
5.	Reengineering of business processes	1990	M. Hammer, J. Champy [16] (the USA)	In a postindustrial society, it is not division of labor that is the basis for organizations' formation, development, and competitive advantages, but reintegration of separate operations into single business processes.
6.	Organization's dynamic potential	1997	D. J. Teece [17] (the UK)	An organization's potential consists in dynamic adaptability, generation, and readjustment of spheres in a rapidly changing environment with the achievement of innovative competitive advantages.

incorporation of innovative technologies in industries, output product distribution potential (distribution networks), budget resources, and quality of customer service.

D. Cravens, on the contrary, used an abstract approach to the category of the competitiveness of an industry's and enterprises within the scope and described as the most important factors of competitiveness the following competitive advantages: versatility (competitive advantage in different situations) and low duplicability of technologies developed within an industry [18].

As the competitiveness management theory progressed, a competitiveness management law was formulated [19], which is based mainly on an idea that high competitiveness should be the first step toward higher financial and economic indicators. The law has produced a number of postulates, which can be used when solving applied management tasks. For example, there is a mathematical explanation to the necessity of stepped-up management of competitiveness during economic depressions and stagnation, to stimulate production and sales. Second, production of innovative science-intensive high-value-added goods plays a crucial role, as it is achieved through implementing innovative solutions and utilizing organization's key spheres while making one-of-a-kind products, which are of exceptional value for consumers.

The competitiveness management theory continues to evolve rapidly with new techniques and methods of organizing economic, manufacturing, financial, and other activities developing. The global digitization of all spheres of life and new technologies, such as 3D, artificial intelligence and machine learning, big data processing and analysis, digital design and production, stimulate the development of digital economics as a new kind of philosophy of organizing economic activity, which rests on a key resource that is digital data (as the data may originate from organizations' retrospective experience, industrial and governmental databases, and from the global information space).

According to S. Scantleberry, D. Ross, V. Bauridel, "digital" businesses design ecosystems, which embrace a variety of participants: customers/clients, suppliers, partners, researchers, etc. Also, they actively use social media and communities for different purposes, for example—getting customers to share feedback about new goods and services.

"Digital" organizations are less eager to rely on formal planning, control, and discussion mechanisms while developing common strategies, setting goals and objectives, and coordinating decisions made at all corporate levels with the strategies and each other. Speed is the most important and must-note factor. A digital organization works fast and therefore has a big competitive advantage in terms of deadlines and the production cost. Lower dependence on the hierarchical order and outlining of a hierarchy is impossible without cooperation. For many companies, a transition to cooperation and horizontal organizational structure, contrary to a pyramid organizational structure with three or four top levels acting as a network, is a huge organizational and cultural shift, and its implementation poses great difficulties. Organizations, which do everything correctly, develop guidelines at the upper

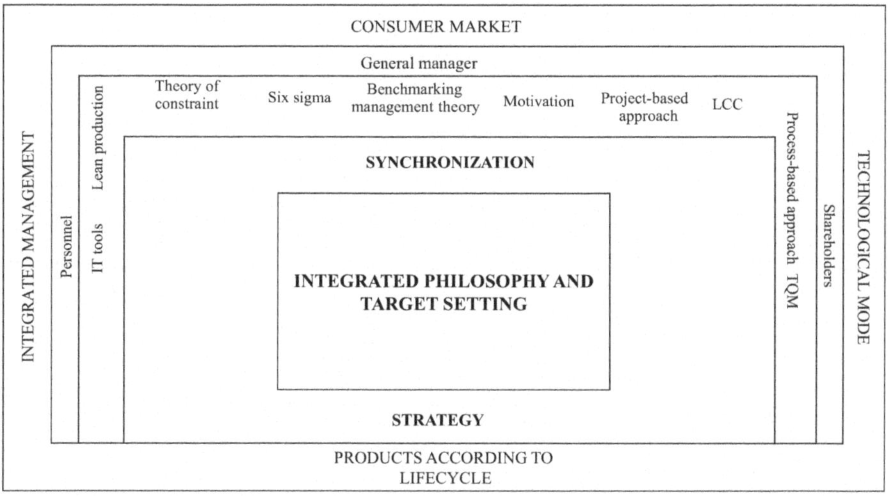

Fig. 1.1 Factors influencing the effectiveness of a management model's functioning

level. The guidelines set boundaries, within which authorized employees can effectively cooperate and achieve impressive results.[1]

Generally, when forming a digital management model, it is necessary to note the most significant factors, which influence its effectiveness (Fig. 1.1).

The factors shown in Fig. 1.1 depict integration between the competitiveness management theory and the corporate governance theory. Management of competitiveness is part of a whole corporate governance philosophy, which contributes to effective management of all business processes within an enterprise to achieve the best possible economic results.

The corporate governance theory originated recently—around the mid-1980s. It sprung from the agency theory—a dominating corporate governance theory developed by American economists M. Jensen and W. Meckling, in 1976. This theory views relationships within a company through the lens of agency costs. At the same time, at the turn of the 1990s, research works of separate applied aspects of corporate management emerged: market for corporate control, functional characteristics of the board of directors, motivation systems for managerial staff, etc. Rafael La Port's revolutionary work highlights the judicial and legal basis of corporate management. In the 1990s, first corporate governance codes appeared, and the Code of Corporate Governance of the Organization of Economic Cooperation and Development (OECD), which has given impetus to a broad range of national corporate governance projects, appeared in 1999.

Corporate governance views an organization as a system, management of which should provide economic and technical (scientific and technological) benefits, as shown in Fig. 1.2.

[1] https://www.bcg.com/ru-ru/about/bcg-review/design-of-digital-organizations.aspx

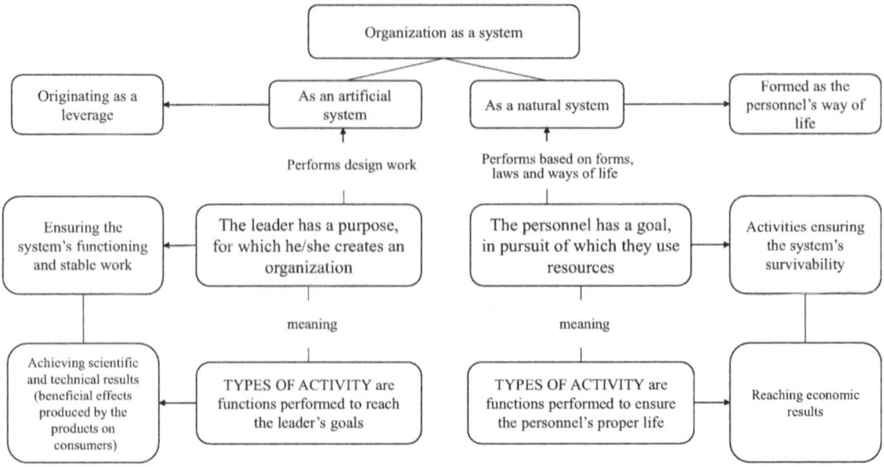

Fig. 1.2 A corporate governance system

Corporate governance has the following objectives:

- Defining a development strategy for an organization, its behavior in the market, and internal corporate governance policy.
- Countering unfriendly merger and acquisition attempts.
- Outlining investment and dividend policies.
- Modeling an organizational structure and business processes.
- Developing a proper motivational policy to ensure maximal labor efficiency, employee, mid-, and top-ranking manager performance.
- Building and utilizing an effective managerial decision-making support system.
- Building an integrated operation control system for manufacturers, which is intended for making and releasing highly competitive high-demand products with ultimately new technical and economical characteristics and capable of expanding to existing distribution areas and creating new ones.
- Ensuring client, consumer, customer, supplier, partner, and government loyalty.

Solving these tasks is largely related to effective management and governance of all business processes within an organization as shown in Fig. 1.3.

Successful corporate governance necessitates a search for and creation of new markets that should define the prospects of rapid growth, distribute financial investments between new projects to prevent a complete loss of capital and provide income, create a basis for producing new types of goods, services, and technologies that meet modern digital economy and production standards. Besides, modern responsibility management methods need to be used when forming key responsibility centers that would be more technologically advanced than their competitors, could solve focused tasks in their sphere, and generate unique technical solutions to produce highly competitive goods.

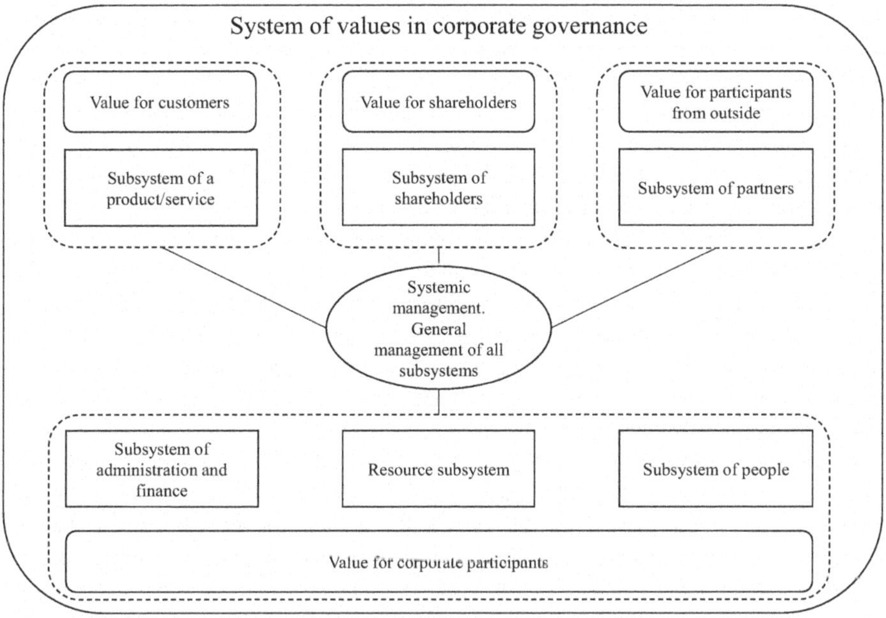

Fig. 1.3 Systemic governance of business processes within an organization

However, corporate governance is only effective with minimal influence of the human factor on the managerial decision-making process and with extensive use of expert systems that analyze big and diversified data and offer an optimal solution to each specific task.

As long as all financial operations (especially those involving public funds) are under tight anti-corruption and effectiveness control, decision makers must have good reasons for their decisions. The use of an expert system is a good opportunity to do so, because it provides objective retrospective information on the one hand and a decision-making protocol on the other hand.

The key point is that traditional methods of analysis and processing of original data, which needs to be used when making a decision, are supplemented by machine learning and artificial intelligence that should stimulate rapid development of an organization with carefully scheduled control of its work.

Meanwhile, rapid development is impossible without implementing an organization's key focuses and advanced technical and technological solutions in its output product. Therefore, the rapid development management theory should include the innovation management theory.

The concept of innovation as an economic category was introduced in the early 1900s by I. Schumpeter, an Austrian economy expert, in his "Theory of Economic Development" 1911, where he described the process of innovation from an economic standpoint.

To boost the innovation management theory, J. A. Hobson, a British economist, decided to add talent to the three existing factors of production—land, labor, and capital. He introduced the term "progressive industry sphere." Hobson understood this term as a sphere of economy that produced new types of goods, embraced new markets, and used new technologies. It was this sphere, Hobson believed, that would justify generation of capitalist profit. Actually, it would imply innovative economy, which could stimulate maximal fulfillment of the business class's potential.

Major research of innovations and their role in the economic development process were preconditioned by N. D. Kondratiev's works (1892–1938). Kondratiev himself did not analyze any innovation-related questions, although the huge business cycles, which he had studied, laid ground for further research of their origins and duration. Innovation was recognized to be the most important origin. Kondratiev revealed long cyclic waves that pervade the economic development process and described its uneven cyclic nature.

In the early 1960s, influenced by a new wave of scientific and technological progress, E. Denison, an American economist (1915–1992), attempted to find sources of economic growth in the knowledge acquisition process that is part of the innovation process. He paid special attention to education and other factors, which would influence the training of work force. Denison would link the economic nature of an innovative process to the progressing scientific expertise—the main driver of production, which, given a higher quality of education and personnel training, boosts national output.

There is an economic model emphasizing a crucial role of innovation-related technological changes and boasting maximal novelty—a theory by M. Kaletski (1899–1970), an American economist. He discovered a trend's ability to become exponential in the long run. Kaletski compared new technical solutions with cyclic shocks, most of which have a stimulating effect on economy. Over the long-term, these shocks shorten downfalls and prolong growth periods.

Therefore, a complete change in technological methods of production becomes an indicator of economic development. Innovative transformations become the core of socioeconomic processes. According to innovative concepts, every next generation of innovations within techniques and technologies extends its scope of influence on social life.

The understanding of human and intellectual resources as eternal values is a fundamental criterion of economic development. Given the role of resources in economic growth, the interlinking between intellectual potential and scientific and technical standards is paramount.

Russian researchers (A. E. Tyulin et al.) note in their works that management of an organization's intellectual potential and experience facilitates their use in business projects of all sizes, aligns data presentation process and standardizes routine algorithms. This helps to save a lot of time for the search and processing of information.

Also, knowledge management facilitates the formation and development of awareness about customers through building of respective databases, consumer profiles, and sales support systems. It helps to implement an organization's

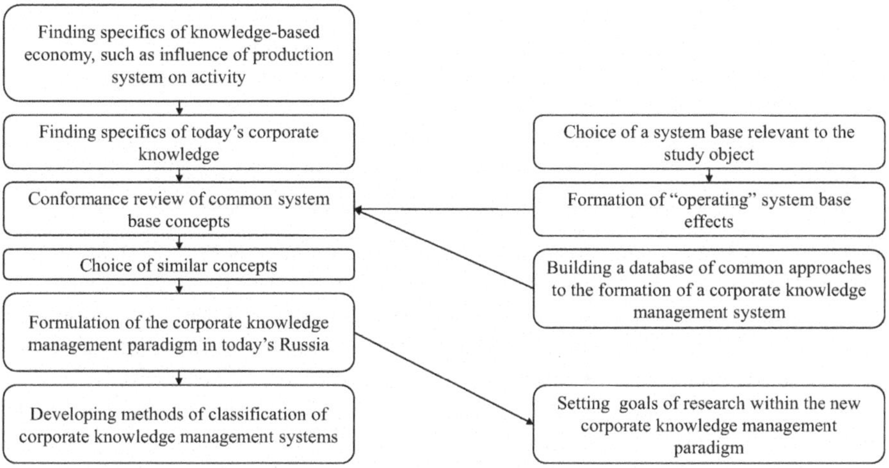

Fig. 1.4 Formation of the corporate knowledge management paradigm

intellectual potential (human, managerial, consumer, etc.), increases the output of existing nonmaterial assets and carries the effects of R&D over to baseline projects.

Also, it creates an environment that forms an innovative climate and stimulates specific innovative projects. Therefore, correct structuring of a corporate knowledge governance system is crucial.

A possible variation of this process [20] is shown in Fig. 1.4.

It is necessary to ensure that the knowledge management system solves the following tasks:

- Standardization of business processes, which should meet unified standards.
- Creating a database of feasible organizational, institutional, and managerial solutions, so it can be addressed in emergency situations.
- Creating a database of potentially interesting companies, innovative solutions, and their sources.
- Applying advanced technological standards, which are used within an organization, to suppliers' activities (small- and medium-sized businesses).

Part of a single knowledge management system are a group of competences that are linked with "breakthrough innovations" and organized into separate centers, which should distribute solutions around an entire hi-tech holding and speed up the company's development.

Rapid development of a company should be closely connected to the theory of life cycle management for an innovation (a product and project) concerning its development and manufacture.

The *Innovation Life Cycle Concept* plays a crucial role in scheduling the production of innovations and organizing the innovative process. This role consists in the following:

- The Innovation Life Cycle Concept forces a company leader to analyze economic activity from the perspective of the present day and future.
- The Innovation Life Cycle Concept explains the necessity of systemic planning of production and acquisition of innovations.
- The Innovation Life Cycle Concept is the basis for innovation analysis and planning. Analysis can help to define an innovation's current life cycle stage, its near-term prospects, the time of downfall and end of existence.

Innovation life cycles differ according to types of innovations. First of all, the differences concern the whole length of the cycle, the length of each stage, the cycle's development characteristics, and number of stages. Types and number of stages depend on an innovation's specific features. However, each innovation has the core (basic) element of the life cycle with clearly defined stages (Fig. 1.5).

Product life cycle management (PLM) issues are increasingly acute. The use of digital data on a product and its components at all stages—from drafting to withdrawal from service—can speed up and reduce development costs, facilitate the mastery of production, and allow for more effective use of it.

Although complex products and systems have been around for a long while, the life cycle subject gained importance only 10 to 15 years ago. Today, it is a key factor, because proper use of digital technologies is impossible without product life cycle management; nor is life cycle management possible without using digital technologies.

Previous digital systems were not fully integrated: at each stage—from drafting, designing, and manufacturing through technical maintenance and repair, separate solutions had to be used. Today, all processes have changed greatly. For an organization and their products to stay competitive, the developer should define all technological processes at the feasibility study stage. After-sales maintenance should be taken seriously too, and its main characteristics must be defined at the initial stage.

Integration of all technological processes is carried out as part of a life cycle management system, and its functioning is ensured through the use of advanced digital technologies, including big data collection, processing, analysis, artificial intelligence, and machine learning technologies that are applied in expert systems.

From an economic standpoint, effective operation in today's market environment is impossible without incorporation of life cycle management systems.

Integration of the innovation life cycle management theory with the competitiveness management theory makes a good case that the basis for creating competitive advantages, outrunning competitors, and taking over a substantial share of market, should be laid at the project concept stage.

Apart from these theories, the concept of rapid development rests on existing economic laws, such as the law of diminishing marginal utility, the law of rising necessities, and the law of interacting capacity building and consumer markets.

The law of diminishing marginal utility states that with consumption of goods total utility grows, while marginal utility diminishes as the consumer's satisfaction (need satisfaction) grows with every next consumed item. The law of diminishing marginal utility demonstrates that a manufacturer always has a good reason to build

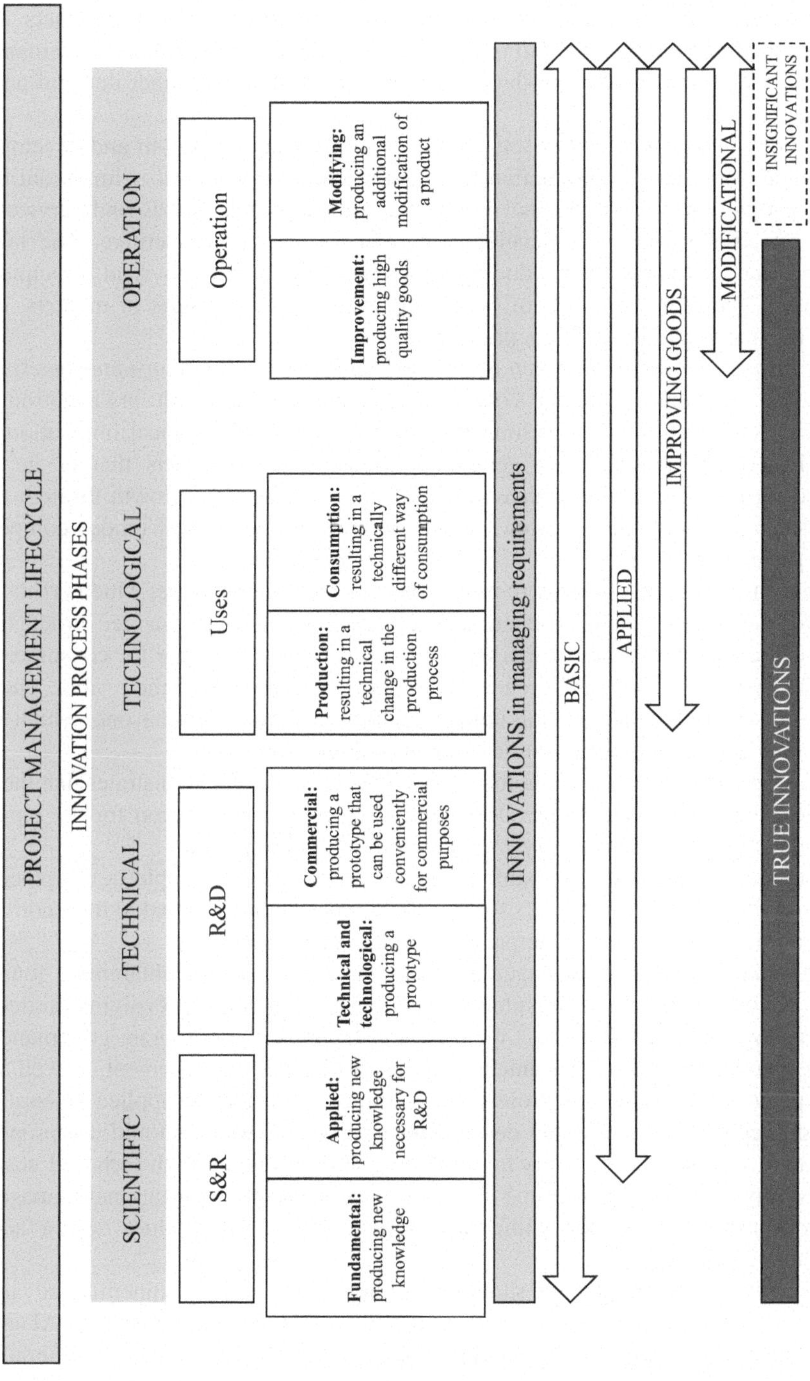

Fig. 1.5 Innovation life cycle

competitive advantages by *executing* rapid development through designing and implementing groundbreaking innovations that help to create new products or improve the functionality of existing ones. Meanwhile, not only should the demand keep pace. The newly made products should generate new consumer demand and new markets.

The law of rising necessities sets an objective necessity of growth and development in consumption, as production and culture improves. Also, it is important to switch people's needs to personal intellectual development. Goods and services, which effectively boost this development, will always be competitive. The law confirms that designing and producing goods that rely on new physical principles is going to generate new kinds of needs and values and new types of markets, in which the manufacturer will be a pioneer.

The law of *interaction between developing competences and consumer markets* states that creation of unique competences increases high-tech organizations' resources, which stimulate bunching growth of unique technological innovations. Their implementation helps to generate ultimately new products that create a demand for new benefits, stimulate new markets and economic growth through to rapid development. All this creates ample economic ground for new unique competences to develop.

Also, it triggers a cyclic growth of competences and innovations, which create a basis for development and production of competitive high-value-added products. These can create new types of markets and are extremely valuable for consumers. However, products' life cycle tends to be short, and they lose their value once competitors reproduce an existing unique solution or create a similar one, which is not second to the original version in terms of features and value.

Just a brief outline of the economic laws mentioned above demonstrates that they do reflect the basics of rapid development and provide a foundation for the rapid development management theory, which is looking to be created.

Based on the above, it is possible to set an algorithm for the rapid development theory and define the main vectors for research, which should be used to incorporate the algorithm into industrial enterprises' activity (Table 1.2).

The rapid development management theory is at the stage of obtaining a foundation, which should rely on the integration of existing and rapidly evolving product and organization competitiveness management theories, the corporate governance theory, innovation, and science-intensive product life cycle management, as well as development and production projects. It is time to develop an applied scientific methodology of managing rapid development processes, which should use proper economic tools and mechanisms for designing and producing high-tech and cost-effective goods. An optimal prime cost can be achieved through rational management of business processes within an organization at all product design and manufacturing stages.

Putting this theory in practice should result in gaining unique competitive advantages through the use of key competences and advanced technologies, such as AI and 3D technologies, as well as big data processing and analysis methods. This should

Table 1.2 Creation and development of the rapid development management theory

Outlining the essentials of the rapid development management theory:
Development of the theory of competitiveness of a product and its manufacturer
Development, description, explanation, and implementation of the competitiveness management law
Development of innovation management theories
Formation of the competence management theory as an organization's key resource within a functioning knowledge-based economy
Development and explanation of the economic law of competences' relationship with the market formation process
Development of the innovative project and output product life cycle management theory
Formulation of the main points of rapid development management with reliance on the product life cycle economy
Formulation of the rapid development law
The basics of rapid development management with account of the specifics of a product's life cycle:
Development of technologies and society's needs with changing technological paradigms and digitization pervading all spheres of public life
Development of the fundamental science, which enables creation of products based on new physical and technical principles
Development of an advanced high-tech economy
Creating a basis for high-tech and high-value-added *products of future*
Methodological basis of the rapid development theory:
Methods used to calculate the value of a product's life cycle
Adaptive methods used in creation of advanced products that are tailored to the "needs of future" based on big data analytics
Methods used to evaluate the innovative potential in the context of rapidly growing competitor innovations and expanding global information space
Methods used to create competitive advantages
Methods used to predict the development of unique competences that should satisfy future needs and ensure their transfer
Formation of resource supply models for new advanced projects
Methods of digital design, modeling, and product development
Mechanisms used to manage the creation of rapid development products:
A mechanism used to form a product's concept and technical/economical image
A mechanism used for complex assessment of resources that are needed for product development
A mechanism used to manage the development of production and technical facilities
A mechanism of development cost optimization, preproduction, and production
A mechanism used to assess the economic effectiveness of construction and production processes
Basics of rapid development product life cycle management:
A production management system tailored to a finished image of a product with a set value and competitiveness level
A system of preproduction principles for high-tech and cost-effective products that rely on new physical principles
An intelligent automated system for management of preproduction processes
Versatile automated production facilities utilizing the advantages of digital technologies
An digital platform supporting the process of making effective managerial decisions during product life cycle management

generate new markets and public needs, which only a rapidly developing and technologically advanced organization can satisfy.

In this context, the development of trends of implementing and using the rapid development theory in practice becomes critically important. These trends imply that an organization should achieve economic stability through formation of management mechanisms and optimal use of all resources at all stages of the life cycle of advanced, highly customer appealing, and competitive products.

1.2 Modern Product Life Cycle Management Approaches Used at Enterprises Practicing Digitized Work

The aforementioned scientific basics of the rapid development management theory demonstrate that an organization's stable work depends on its ability to design and produce highly competitive goods with unique characteristics, which are exceptionally valuable for customers. This poses a task of managing these goods' life cycle with account of the ongoing digital transformation of the production sector and society.

Production of new competitive goods requires specific conditions, and all organizations should adapt to them.

Most likely, these conditions are made up of closely connected key directions:

1. Formation of key priority areas for an organization and balanced management of all activities.
2. Ensuring stable economic corporate growth.
3. Formation of mechanisms for total corporate resource management.
4. Development of competences, informational solutions, and digital technologies to create advanced, uniquely customer-appealing, and highly competitive products.
5. Implementation of advanced product design and manufacturing methods (careful production, adaptive production facilities, versatile production lines, digital 3D-modeling, etc.).

These areas form a foundation for a major organizational and economic rapid development management system.

Outlining top-priority vectors for an organization to develop and balanced management of all activities.

Product life cycle management, which should contribute to rapid development of the organization producing it, should start with assessment of potentials of three main sectors, which shape the main activity areas for a high-tech organization:

- *The state customer's potential.* It is important for any high-tech organization to evaluate an opportunity to cooperate with state customers, because some of the state's needs are quite stable, and its role in stimulating innovative processes and digital technologies is globally important. The state sector has a greater influence on innovative activities than private buyers, suppliers, and competitors. Besides, to predict state customers' needs, it is necessary to monitor developing national programs, which finalize a nation's priority goals and generate new types of needs that result from changing technological setups and the growing public intellectual potential.

High-tech organizations should follow several directions as they cooperate with state customers. The most important ones include selling mass-produced goods, which do not require any further research or development; carrying out R&D state orders, which evolve into production; implementing R&D output products. This will hedge the risks arising out of selling products of different categories.

However, along with this segment, high-tech organizations should cooperate with individual customers as well (Fig. 1.6) through hedging the risks pertaining to areas lying within the goals management segment.

This creates a basis for the second prospective area, where high-tech companies can sell their products.

- *The potential of a commercial sector with account for market demand for innovative solutions and products with advanced technical features, which are exceptionally valuable for customers.* This suggests a necessity to set up a prospective market demand control system, the mechanism of which is shown in Fig. 1.7.

  The system not only defines, but also analyses prospective personified needs of non-state customers and uses big data processing technologies, ensures control over competitor products, monitors their vulnerability, and factors threatening their competitive supremacy.

  The analytical data generated by the market control system should highlight for an organization the most promising directions in creating advanced products, which should create the third customer segment.

- *Creation of new market niches and markets through developing and manufacturing all-new products.* This segment is the foundation for rapid development of an organization. The direction is represented by world high-tech whales like The IBM, Microsoft, Ford, Apple, and other renowned companies. It has helped them achieve stable economic growth, and actually they are setting the trend for the world we live in.

A new market can only be created when a new and unique product garners a critical amount of potential customers, who actually make up a market. However, this requires the right type of environment. On the one hand, this environment is formed by potential customers, who create the demand, and they should have some specific characteristics. In case whereby customers lack these characteristics, a market cannot be created. For instance, it is impossible to sell a computer program to a customer who does not have a PC. For this reason, assessment of potential public needs is crucial, and a company should produce goods, which customers are ready to buy.

When outlining vectors for an organization within different market segments, it is necessary to design a strategy to manage its development and to carry out a

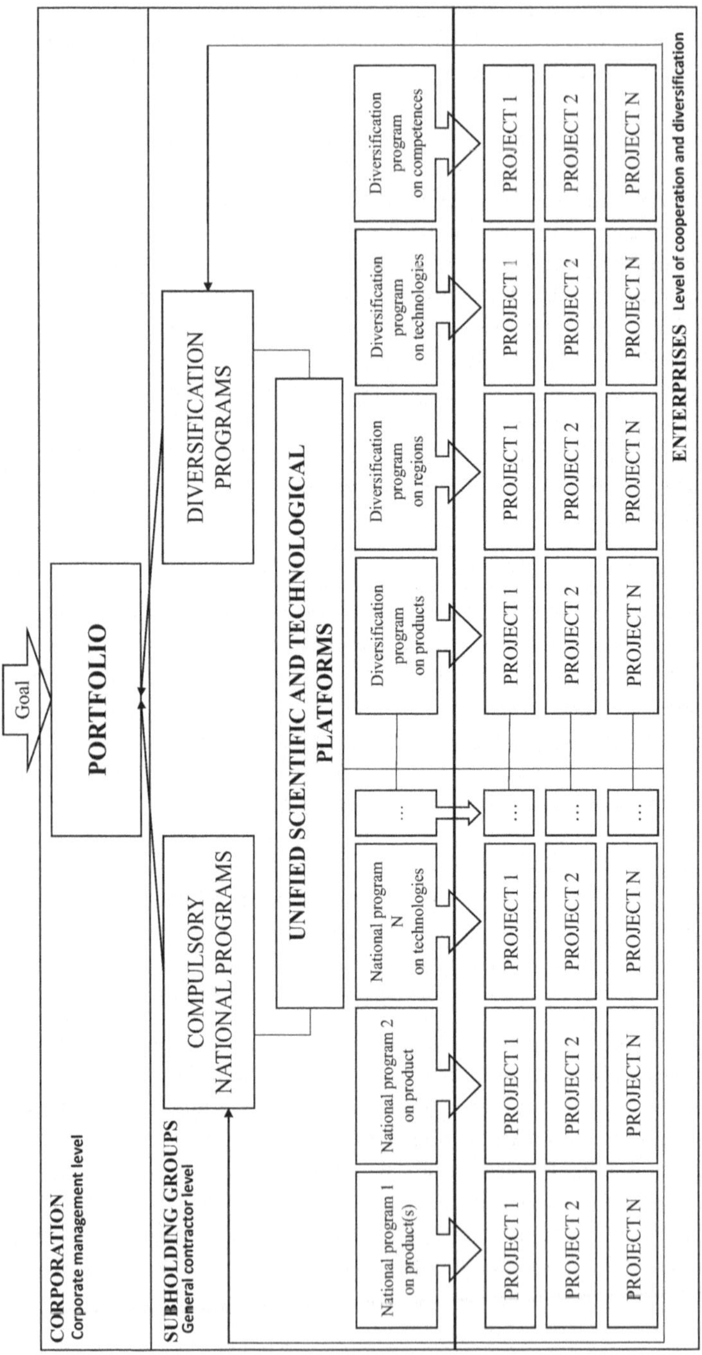

Fig. 1.6 Goals management

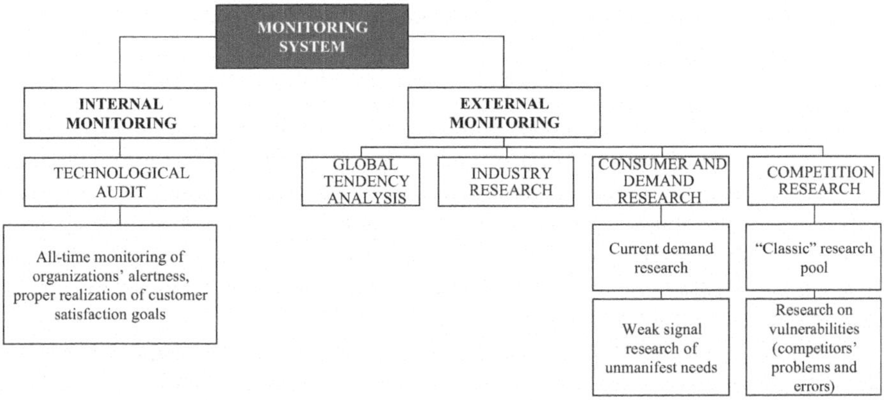

Fig. 1.7 A corporate market control system

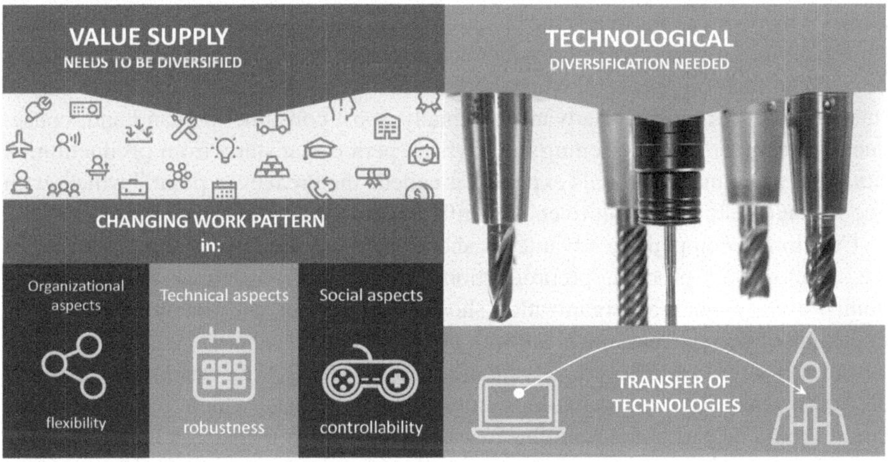

Fig. 1.8 Corporate product development system

modernization starting from the product design and engineering base down to existing management methods, which should include digital organization setup technologies.

It should be noted that product development should embrace all three segments, and it should rely on unification between separate product components and the production technology, also through exchange of technologies and competences used when creating products that are valuable for each of the segments (Fig. 1.8).

Manufacturing products of future for a variety of customers, who have different values, within a unified engineering framework, should enable a company to use its technological potential in an optimal way and ensure a stable economic growth. Achieving stable growth is the second vector, which forms the foundation for a major organizational and economic rapid growth management system.

1.2.1 Ensuring Stable Economic Development of an Organization

Stability characterizes a company's condition based on its reaction to external and internal influences. A stable company is one, which, notwithstanding any equal outer and inner shifts, is less likely to change and deviate from its current state. As noted above, economic stability can be achieved through diversification of a company's directions within different customer segments. It necessitates choosing an optimal diversification pattern based on the influence of the number and viability of activity types on the company's economic strength.

Stable economic condition is only possible when each product enjoys a demand thanks to its unique customer appeal and competitive advantages arising from their technical characteristics and price.

Therefore, it is necessary to predict the evolution of each product type, determine its lifespan, breakeven point, marginal utility, and fill rate, which will be followed by a sales slowdown as these advanced solutions invade the market.

The all-range monitoring of product development should start from the beginning of the manufacturing process, when the sales income is not yet big enough to ensure stable growth. Competitive advantages finally boost consumer demand and, consequently, income, also shortening the payback period that starts from production. A situation, when income equals expenses is called "the breakeven point," which, with due management of competitiveness, shifts toward smaller output.

Creation of competitive advantages should be more dynamic at the beginning of the development process, preproduction, and up until reaching the breakeven point—exactly when an organization should start getting sales income. Applicable to the innovation process, the breakeven point, we believe, should be calculated with consideration of competitive advantage creation costs, i.e., scientific, research, design, engineering, and technology implementation costs, which, given their specifics, should be part of the standing costs. However, their increase at the initial stage of R&D, when the new technology has not been introduced yet and when previous products with the previous productivity and price growth rate are still being released, causes the breakeven point to shift rightwards. This explains why many organizations do not want to invest in innovative projects.

However, once a new competitive item is launched into production, the income curve takes on the shape of a bent line (cash receipts and, consequently, income, will start to grow faster), which results from increased output and sales providing that the organization has created a highly competitive product enjoying a higher demand than its competitors, who are unable to quickly copy the technology or produce more competitive goods. In Fig. 1.9, the process is shown in a graphic form.

It is very important to predict the achievement of the breakeven point. A company wanting to reach it as soon as possible should design and develop advanced products faster. It is imperative that the breakeven point be reached before the product life cycle starts to decline. Once a product is close to the stage of maturity, which is

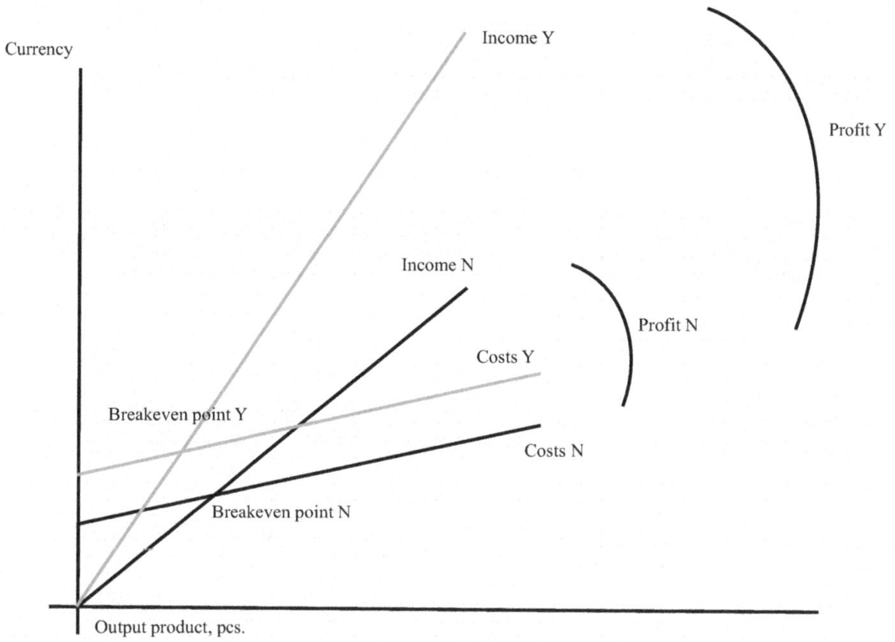

Fig. 1.9 A simplified organization development pattern, which demonstrates effects achieved by companies that do manage competitiveness (Y) and by ones that do not (N)

followed by a slowdown in sales, production should be upgraded with reference to predicted customer needs and values dynamics.

Upgrading should focus on a kind of modernization, which could ensure a takeover of a large share of the market or start a new market niche. A company should start an upgrade from creation and accumulation of a scientific and technological potential, its effective management, development of unique key competences and gaining cross-industry knowledge, technologies, and competences through their transfer.

The universal trajectory of success in executing product upgrade projects with a need for high competitiveness can look as shown in Fig. 1.10.

As shown in Fig. 1.10, successful implementation of any product upgrade project is largely predicated by the use of unique competences, which are a source of competitive advantages, and by providing a sufficient resource base for business processes relating to making products of future, which meet the current market demand.

Components of sustainable growth, on which an organization should rely in its work, include several categories that embrace all spheres of public life (Fig. 1.11).

However, sustainable growth is achieved primarily through management and building of innovative potential of a company and creating a resource base for implementing product upgrade projects and creating brand new products, which ensure rapid development.

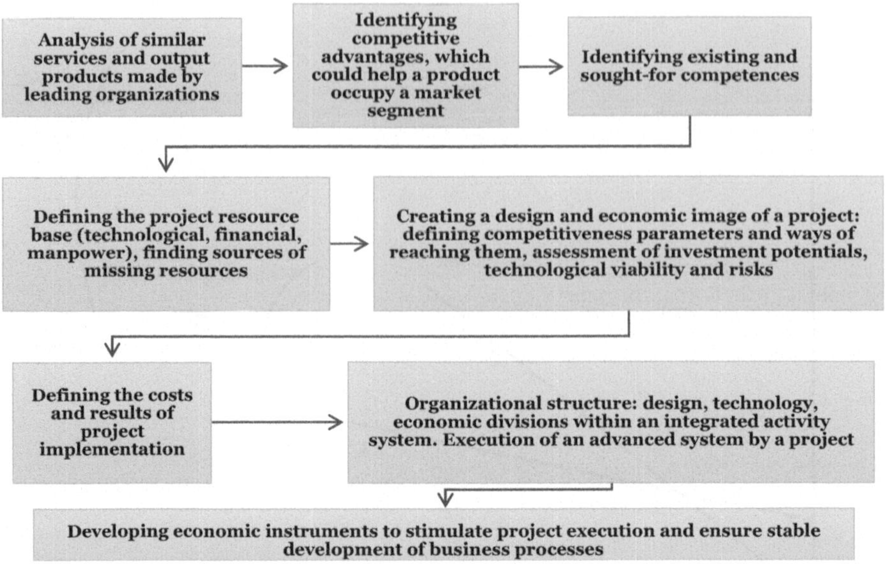

Fig. 1.10 Successful execution of a product improvement project looking for high competitiveness

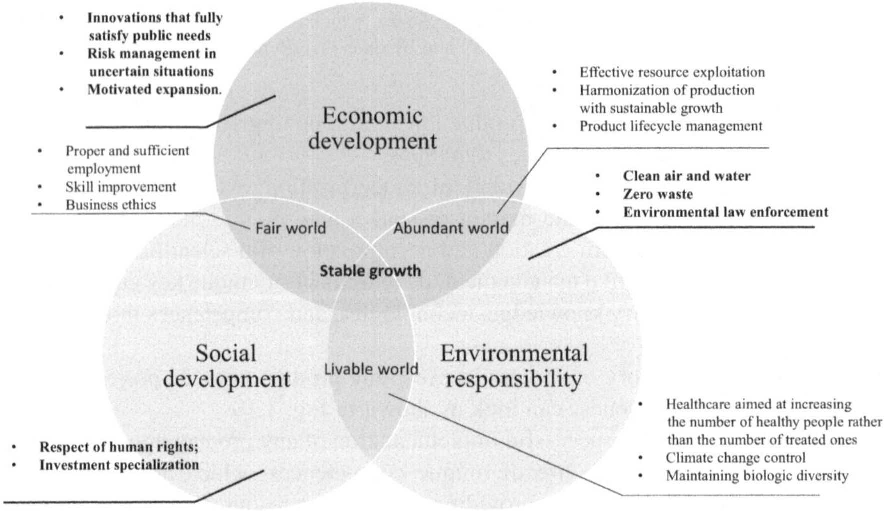

Fig. 1.11 Components of sustainable growth of an organization

1.2.2 Building Mechanisms for Management of All Corporate Resources

Traditionally, formation of a resource base for executing various activities and projects is treated as a key factor of an organization's and its products' competitiveness. Availability of cash flows ensures timely execution of activities and projects, which, in turn, improve a company's general competitiveness.

In this case, flows of funds that finance rapid development can depend on income resulting from selling competitive products, active use of stock market mechanisms, and venture funds to raise investment in innovations.

However, high investment does not necessarily generate highly effective innovations. On the other hand, a small investment can give a high yield providing that it is made at the right point. This may create a multiplicative effect throughout an organization (usually, this applies to improvement of management tools, talent management, use of basic technologies, etc.).

To set up effective production management mechanisms, an integration/logistic approach to management of strategic resources should be developed and used (Fig. 1.12), which should provide for high quality and due effectiveness of communication within the unified digital information space.

Because competences do make a special kind of resource, without which rapid development is not possible, management of these requires a closer look.

Development of competences, informational solutions, and digital technologies aimed at creating brand new, uniquely appealing, and highly competitive products.

Development and management of competences, informational solutions, and digital technologies within an organization aimed at rapid development should rely on the formation of a branch informational base for competence management technologies. Databases of technology have long been used on a global scale (NASA NAFCOM, SpinOff, etc.). Within these bases, technologies can be linked to key competences that are part of their creation and use. This can make the technology transfer process more effective. Obviously, it is organizations possessing appropriate competences who should be buyers of these technologies. This creates a potential for more effective use of innovations. Competences are indicators of adaptability to novelties, and using this potential enables one to quickly adopt someone else's experience and come up with projects that are similar to the most promising ones. This is vital for economies that experience economic sanctions and have to rely on import substitution programs.

In turn, competences should not be studied separately from the knowledge management system. Competences (abilities) are always related to resources, intellectual property assets, and other knowledge. Therefore, the knowledge management system and competences make up a category of informational resources.

Newly discovered knowledge is a kind of informational resource, which can be divided into three categories: data, information, and competences (expertise and abilities).

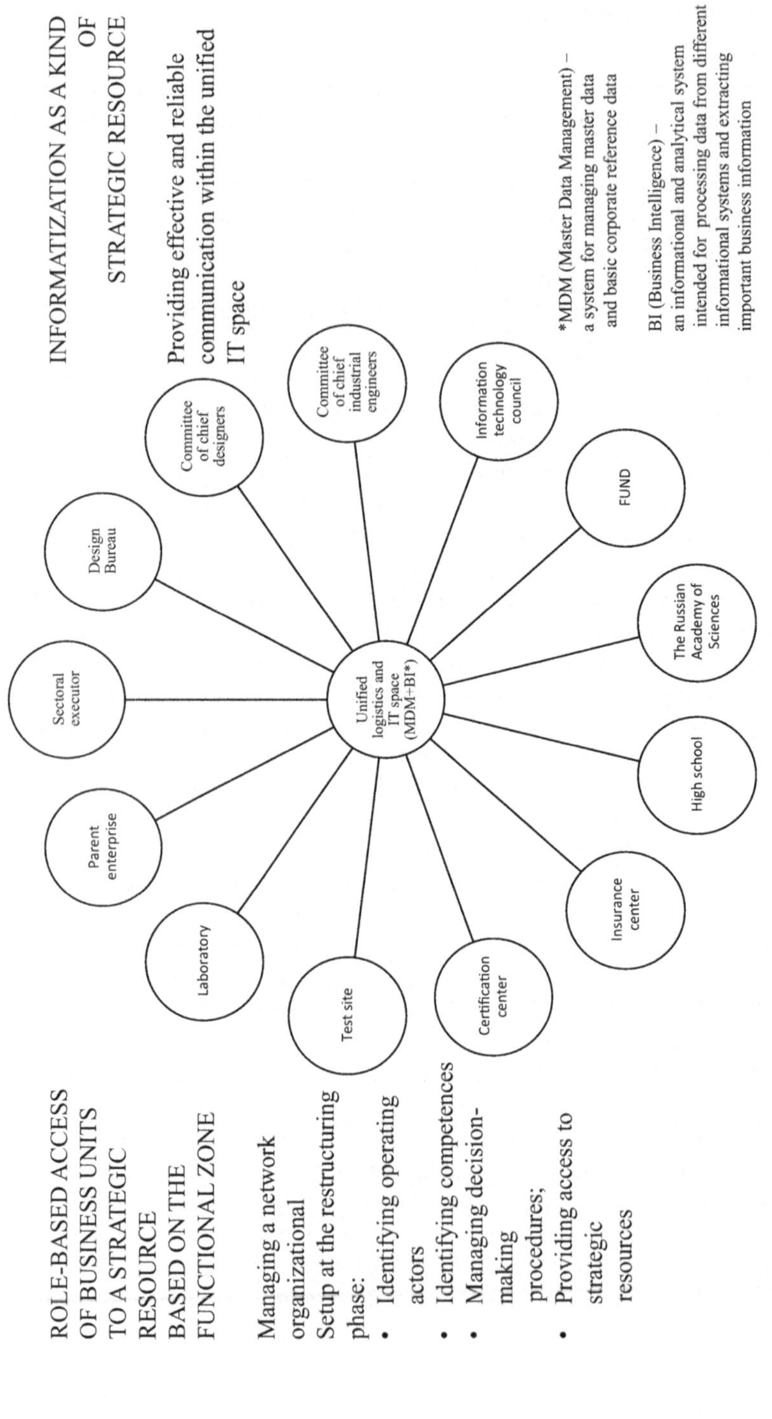

Fig. 1.12 Managing strategic resources within digital management structures

Data includes recorded facts, indicators, and values, which are unambiguously identified by a fixed set of measurements. For example, time and cost of a project's implementation, an operation's labor intensity, etc., are facts.

Information is a formalized text, which describes the properties and technical characteristics of objects being analyzed. Information is semi-structured data, possibly relating to a current task. For example, it can be a textual description of a technology being implemented with keywords. They can help to select a technology for carrying out a specific task. Information is description of properties and technical characteristics of materials, parts, other products, processes, as well as other types of information.

Competences make up a synergistic system of knowledge, rules, regulations, and methods that are used for reaching specific goals and solving specific problems. This knowledge enables an organization to solve specific issues and it acts as a prerequisite of effective work. Competences are the product of information processing; they have spheres of effective use and are long lived.

Data is the simplest type of information resources. A data warehouse is a hypercube, which aggregates data within a slice of organizational units, personnel, objects, processes, time, etc. Different facts (indicators) are stored in the warehouse as values (referring to the storage object): cost, labor intensiveness, resource intensiveness, production volume, number of employees, quality control data, current guidelines, standards, etc.

Data storage is also executed in the hypercube form, and it retains all basic measurements that are typical of the previous pattern, although the content is usually presented in a textual form.

Hypercube 3 is the most interesting one, as it presents competences in the form of documented procedures used in executing processes, methods, instructions, and problem-solving methods. The biggest issue is that the term "competence" is not a synonym for "method" or "skill"; it provides an effective solution to a broad spectrum of adjacent tasks and therefore embraces a number of processes, methods, etc. There cube cells store knowledge frames. The data pool structure is shown in Fig. 1.13.

In the Data hypercube, knowledge occupies its ascertained niche and is related to a respective process, division, resource, etc.

In the Information hypercube, one and the same piece of knowledge may belong to different categories (dimensions), i.e., it can be duplicated within cells many times.

Finally, competences cannot be straightforwardly linked to dimensions. This process becomes dynamic when a piece of knowledge, which was initially linked to particular dimensions (relating ex post to the main use), can be linked to other types of knowledge through putting respective data in slots. For instance, addressing this information with the help of the keyword "nanotechnology" implies that the competence is linked to all processes, equipment, and other assets, which are related to nanotechnologies. Today, Internet users are quite familiar with this approach as they continue to receive advertisements on topics coinciding with those of their earlier made search inquiries.

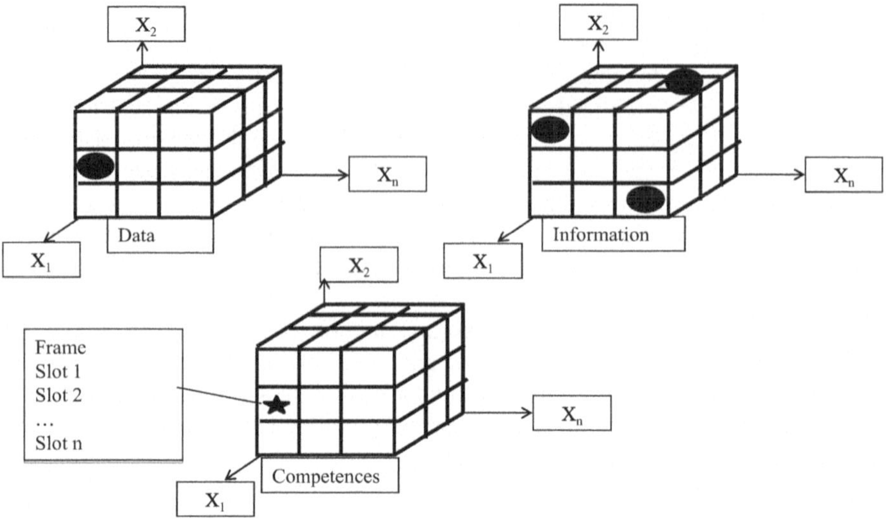

Fig. 1.13 Data pool structure

The multidimensional cube technology provides segments of data in all dimensions, as well as information on the data, knowledge, and competences.

An organization's innovative and competence potential should serve as a basis for effective life cycle management of science-intensive products, which create new markets relying on advanced digital product design management methods, preproduction, and production.

1.2.3 Introduction of Advanced Product Design, Preproduction and Production Methods

One of the most significant shifts taking place nowadays is manifested in the emergence of new types of automated and "intelligent" technology and management of production systems. This creates a basis for a manifold increase in labor efficiency and significant economical and, consequently, social changes. The concept of Industry Version 4.0 necessitates the use of new approaches to production management, and mass robotization within companies. Some estimates suggest that Industrial Revolution 4 will cause a full and complete change in traditional production management methods at the majority of enterprises, which design and implement innovative technologies.

Effective incorporation of innovative technologies created within a high-tech branch should be treated as a paramount segment of rapid development and boosting national goods' and services' competitiveness, which help them enter global markets, counter their key competitors and achieving leadership; finally, it stimulates the formation of new markets and buyers' needs.

Rapid development is inalienable part of switching to a more advanced level of combined production, building innovative scientific/industrial unions, and competence centers. This process is based on radically new digital production technologies. It necessitates adaptation of existing systems to the new industrial and economic environment and using a *program and adaptive* approach in management, which would reflect rapidly changing standards. A system of adaptive mechanisms that are used in strategic management is shown in Fig. 1.14. At the formation stage, a strategy should create a potential for changes, which, in turn, enable the system to react to the rapidly changing outer environment and provide optimal conditions for functioning.

The program and adaptive approach not only allows to smoothly switch between production lines; it also helps to monitor industrial performance and carry out a qualitative analysis of factors that cause deviations from target indicators.

For a major high-tech organization, it would be quite rational to set up a management center for analyzing information concerning the production system's current state and outlining a corporate policy to regulate the production of the main types and categories of products.

An important part of understanding, explaining, and stabilizing the adaptive system, is feedback. As the system develops, it deviates from set parameters and changes shape; this prompts managers to call entrepreneurship skills into service. On the one hand, automation of business processes that take place within a production system should balance the entire system and keep the process within set boundaries; on the other hand, feedback opens the potential for intentional development. Therefore, the balance in the system is still relative.

Under- and over-performance cause disturbances within the system. To bring it back to its initial state, a whole set of activities should be carried out. Not infrequently, the balancing process slows down. This is called a "relaxation." Length and intensity of disturbances largely depend on feedback's effectiveness. Because man is the key component of the system, feedback can be quite subjective. Insufficient information can make the manager feel unconfident and prompt him/her to do a random search in an effort to counter the uncertainty. The quality of search depends on the manager's qualification and experience, and it can affect the functioning of a whole division or department.

To directly manage the transition to the adaptive production management system, a unified and centralized system formation program should be developed, which would reflect the specifics of an organization's work and reflect all production stages (from drafting through implementation). This kind of program can integrate into the organization's innovative development program.

Introduction of the adaptive production management system allows managers to flexibly respond to disturbing deviations from current goals and/or strategies, speeds up the decision-making process and maintains the rapid development of a high-tech organization and industry.

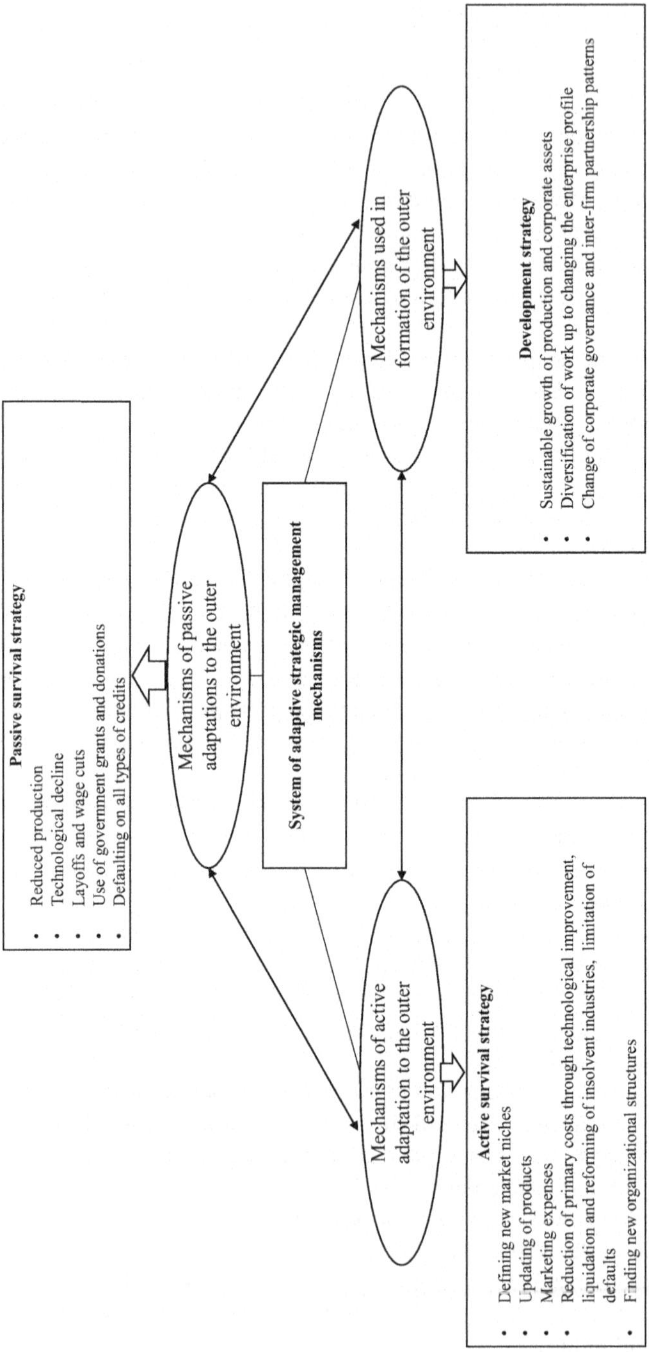

Fig. 1.14 Adaptive strategic management mechanisms

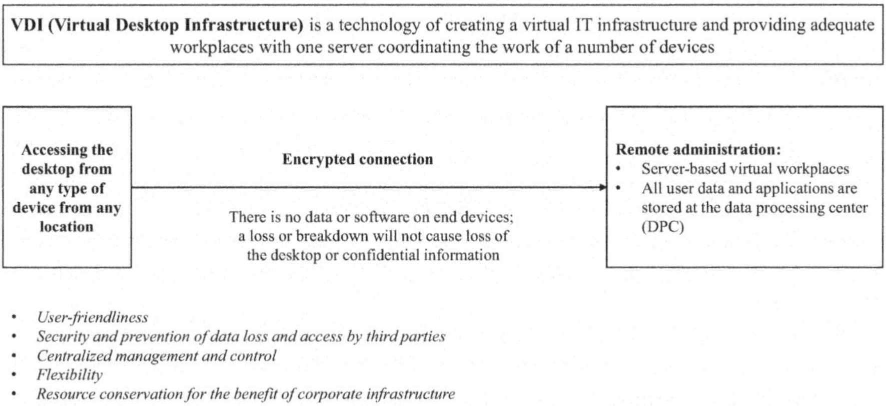

Fig. 1.15 A digital infrastructure for a unified corporate industrial platform

Along with adaptive production management systems, organizations should use mechanisms of broad cooperation, which ensure cost-effective development and production with the help of companies that are highly competent in specific fields, have a rich technological base and innovative potential. Also, it requires a deep awareness of design bureaus' and enterprises' potentials for mass production.

To work effectively and optimize business processes, high-tech industrial companies should have a kind of infrastructure that meets today's digitization standards and build a unified digital platform relying on information technologies (Fig. 1.15).

This kind of platform can be used as a tool for improving all business processes taking place within a modern company. It can provide a basis for mutual integration of all interconnected and balanced elements: functions, tasks, production organization and management tools, organizational structures, and competences of employees, who are focused on reaching strategic and operational goals. As it is known, partial automation produces a limited effect, and a maximum effect can be achieved through complex automation. On the other hand, over-automation and building fully automated enterprises may fail to produce the expected result. Therefore, when designing a virtual production platform, it is important to define the needed level of automation. It sets an economical task of defining man's role and influence on a production system at the current stage of social development.

Computerized integrated production, which relies on a virtual platform, can embrace the following business processes within a company:

1. Development of technological processes
2. Engineering analysis of structures and technologies
3. Engineering data management
4. Drafting instrumentation and preparing management programs
5. Operating CNC machines
6. Control over equipment used in engineering systems, networks, and terminals
7. Control over equipment and the product characteristic measurement process

Besides, a company's virtual platform should integrate into the global informational space, in which so called *competence stock markets* have just started to emerge. In some cases, they act as competence databases, which users (organizations) check when looking for potential partners possessing unique competences that can be applied while executing a mutually beneficial project (for example, the Internationalization for Joint Innovations in Photonics, www.competencedatabase-photonic.com). Also, those may be so called *competence selection and design group formation markets*, which are supposed to help organizations solve specific tasks (SmartSAP is a market pioneering in the use of "smartstaffing," Outstaffing AS, and other concepts).

Development of competence databases and markets offering competence selection and design group services can be an effective tool for opening markets for organizations and helping them reach high sectoral competitiveness. It will free organizations from lengthy searches for technical and technological solutions, which are not always properly described. Instead, companies will be able to address these virtual markets offering tried-and-true competences, which have already been applied in existing competitive products.

Eureka is just one of advanced venues for competence marketing, and it has functions that are typical of a stock market. This international platform embraces 40 countries and facilitates a search for and selection of competences for joint project implementation.

These trends are preconditioning the emergence of competence stock markets. A look back over the history of stock markets suggests that they have come a long way toward reaching their current level.

Therefore, it can be expected that respective competence markets will appear in the future. They will rely on effective theoretical mechanisms with their practical implementation resulting from the competence market trading experience gained thus far.

It should be noted that integration of a corporate virtual platform with such competence markets is going to be an important element (Fig. 1.16).

Integration of competence stock markets with a corporate platform can have the following benefits:

- Reduced amount of resources for coordination and performance quality control.
- New opportunities for penetration of international markets.
- Reduced total operational costs through transparent competition for each type of work.
- Optimization of competence development strategies through exclusion of uncompetitive directions and using resources in key competences.

Therefore, a unified virtual platform should help to build a broad cooperation network, distribute component development between organizations, and share production risks and costs.

The corporate platform development process is influenced by external and internal factors. This poses the question of the length and effectiveness of the platform development process.

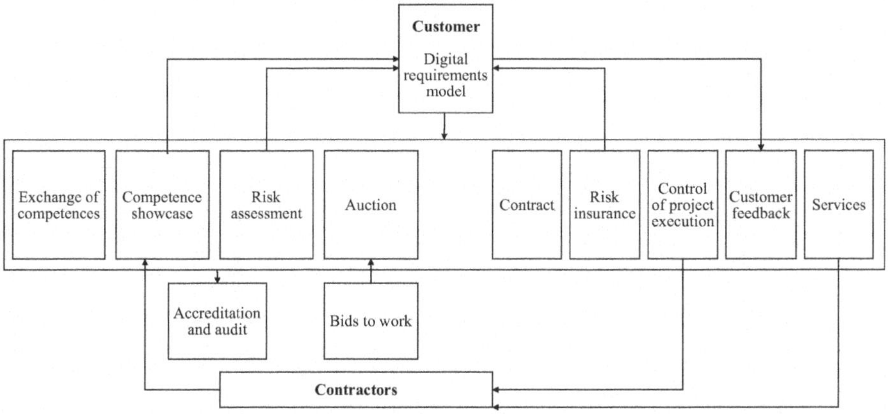

Fig. 1.16 Development of a virtual industrial platform

Prospects of virtual platform development are shown in Fig. 1.17.

Effective functioning of a virtual industrial platform depends on ultimately new management competences (internal factors), given the geographically diversified external economic relations. These relations strengthen exporters' positions in the system of microeconomic relationships (external factors).

All these conditions, which are musts for successful rapid development, should be taken into consideration in the process of building a digital company—one, which uses information technologies as a competitive advantage in all spheres: production, business process management, marketing, and company–customer relationships. This should ensure effective use of the information technology's potential in product life cycle management, the product's high competitiveness/quality, and low cost. A digital company is going to be a major organizational and economic system, which should manage the company's rapid development on the basis of product life cycle economy.

A digital organization should have a data exchange system, be integrated into a sectoral informational space and work as a virtual one. In the globalized world economy and evolving information technologies, virtual organizations enjoy a growing popularity. This particular case refers to partner networks (participants, stakeholders, etc.), which are not legal entities and where every member is highly competent within particular innovative technologies, which they jointly use in development, production, and distribution of globally competitive products: a virtual design bureau, factory, etc. A virtual organization does not confine its work to organizational structuring. It constantly attracts new organizations, which are competent enough to solve a major task.

The functioning of a digital organization should rely on methods used in life cycle management of a unique product, which helps the manufacturer to achieve rapid growth. These methods imply a concerted work of five mechanisms:

1. Outlining a product's concept and technical/economical image.

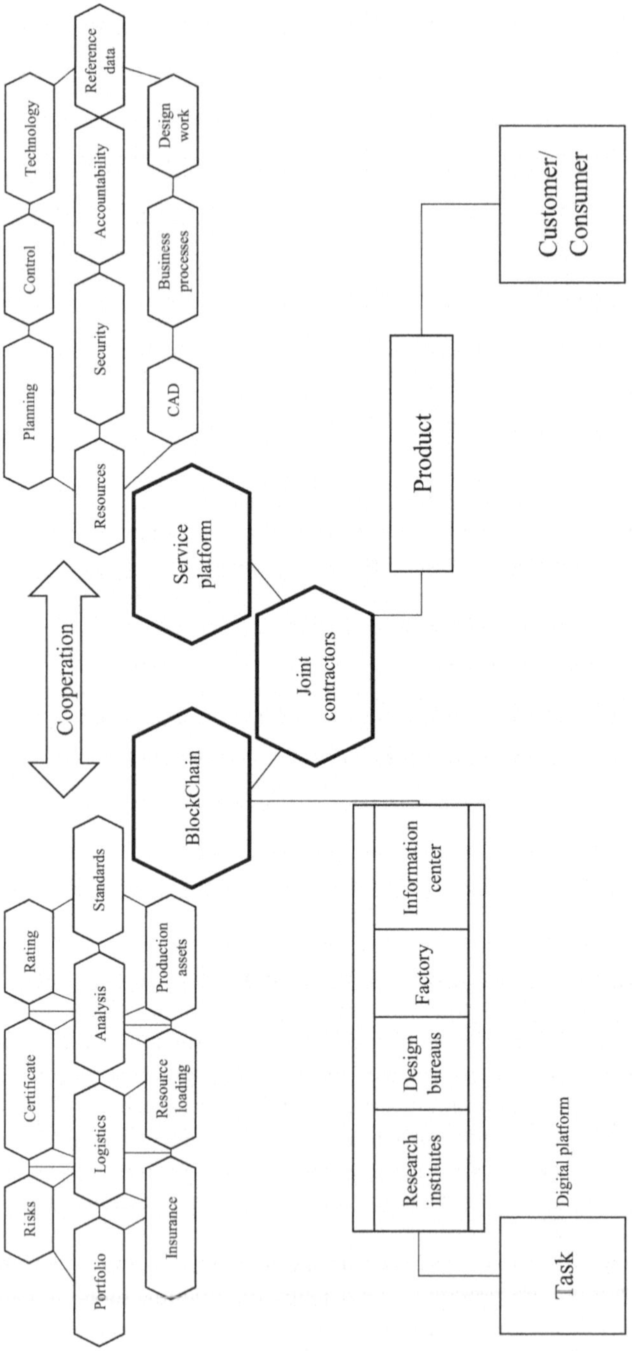

Fig. 1.17 Prospects of virtual industrial platform development

2. Complex assessment of resources needed for product development.
3. Management of the technological and industrial base.
4. Development, preproduction, and production cost optimization.
5. Estimates of economic effectiveness of modeling and production processes.

Effective interaction between these mechanisms will be achieved with the help of an informational backup of the automation of the life cycle management process with the use of digital, big data processing, and analysis technologies, which are also generated through integration into the global informational space.

Formation of mechanisms and methods of life cycle management for product that can contribute to rapid development of the company producing it can define solutions to the following topical tasks:

- Imaging of the "product of future" and the company's promising portfolios based on an analysis of consumer demand and quantitative assessment of science-intensive product's and the organization's competitiveness, to ensure rapid development.
- Boosting projects' investment attractiveness through helping high-tech companies' make promising goods, to keep them economically stable.
- Effective resource planning for development of science-intensive and hi-tech products.
- Analyzing manufacturing and economic activity of an organization and its readiness to switch over to the rapid development path. Building an optimal rapid development vector for a high-tech organization with account of having achieved a synergistic effect while managing the transfer of unique technologies, with the effect ensuring the company's economic stability and its product's competitiveness.
- Implementation of mechanisms that are used to attract stakeholder and venture capital and setting up cooperation with major foreign industry leaders to execute joint advanced product development and manufacturing projects.

1.3 Life Cycle Cost Calculation Methods

As follows from the research, to create a product that stimulates rapid development, it is necessary to solve tasks focusing on a company's economic stability, resource planning, gaining competitive advantages, and developing one-of-a-kind competences through the use of the innovative potential gained so far. It takes effective economic mechanisms and methodology to solve these issues effectively. These mechanisms have evaluation criteria, by which the effectiveness of a life cycle management strategy can be estimated.

Effective life cycle management is paramount for science-intensive products when it comes to making decisions predicating a high-tech company's success. In this context, effective management implies rational distribution of financial resources required for the development of a new product. Successful execution of

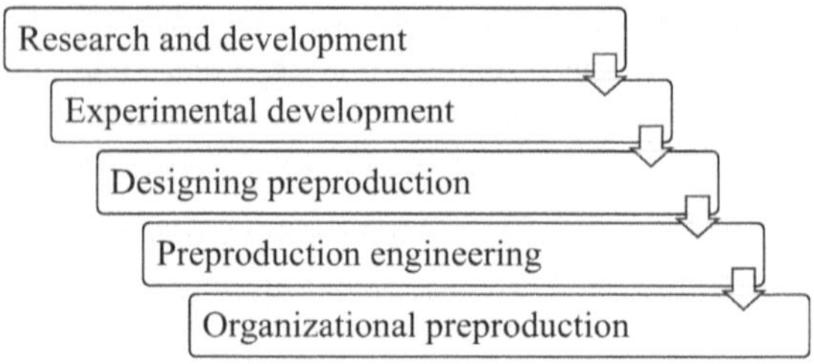

Fig. 1.18 Product life cycle stages

product development processes should result in the organization achieving the rapid development status, i.e., formation by it of a new rapid development market or backing up a decent share of an existing market with the company's current innovative potential and a set of key competences.

An organization, which has become a rapid development one, can ensure effective resource management and continue to produce advanced science-intensive products throughout life cycle. The product should meet the needs of particular customers, and it should be developed with account of the society's growing intellectual potential and competences.

The process of creation of advanced product comprises several stages, which are usually arranged in a consecutive order. They have a limited length, and each stage has a start point and an end point. Time points may require special attention. There are two main life cycle management models:

1. Predictive model
2. Adaptive model

The predictive model has clearly defined project limits (costs, resource potential, deadline, technical specifications) and output product or service parameters. Typical of the adaptive model is the iterative implementation procedure and detailed reviews of requirements made at each iteration.

Most high-tech organizations use the predictive life cycle management model. A predictive performance cycle is fully described by a set plan. Usually, every next point of the plan is different from the preceding one. Each stage has respective steps and operations. Each type of product has a fixed sequence of creation stages. The sequence may look as shown in Fig. 1.18.

Each life cycle phase uses a defined economic mechanism and methodic tools. The effectiveness of the stages' execution can be evaluated with the help of specific integrated criteria. Therefore, a change in these integrated criteria may reflect the quality of current managerial decisions, which concern the implementation of the current life cycle phase. Particular managerial decisions also influence qualitative characteristics of a product's life cycle. Variations of these characteristics (through varied managerial decisions) can provide an optimal plan for implementing a current life cycle phase.

The main goal of life cycle management in the predictive model is reaching the result set at the first stage, unless the product implementation plan was changed or modified. The downsides of the model relate to poor quality of the product's first version and lack of transparency; meanwhile, the rapidly changing outer environment requires a correction of the plan to preserve the product's competitive advantages.

This is possible *with the use of the adaptive life cycle management model*, as it ensures flexible handling of all stages and realization of effective cost calculation and cost optimization methods.

Meanwhile, the key point of building this model is outlining the methodology of life cycle cost calculation; the model should include several economic mechanisms—a set of methods and techniques that should influence and regulate economic processes.

As part of the methodology, *the number one mechanism* is a *mechanism that forms the concept of a product and its technical and economical image*. It makes it possible to predict an organization's product range, which could bring it to the point of rapid development. Integral to this mechanism should be *methodical prediction tools*, which provide an opportunity to predict the public's and organizations' future needs with account of the improving intellectual potential and competences. This kind of prognosis should rely on intellectual big data mining of the global information space with account of the current technological mode and predicted characteristics of the following technological mode.

A technological mode is a production cycle, which embraces obtaining primary resources, all resource processing phases and release of a set of output products, which live up to a respective public consumption pattern. Every technological mode witnesses creation of new technologies, machines, and mechanisms, and it contributes to the formation of new markets and further growth of existing ones, which form new consumer preferences, business models, professions, and labor organization forms. A company aiming for a rapid development mode should take the lead in generating new or expanding existing markets. To do so, the organization should have enough resources and shift its focus over to development of one-of-a-kind competences and technologies. These will provide a basis for creation of ultimately new types of products that will measure up to the challenge of the next technological mode and public preferences. All these products are made by companies that possess unique technological and organizational competences, which live up to a higher technological mode, and an innovative potential for creating new-generation products. These organizations grow fast and gain the rapid development status thanks to technological breakthroughs, which give them a competitive advantage in a newly created market (Fig. 1.19).

With characteristic features of a developing and prospective technological mode at hand and with the use of respective methodical tools, an organization can foresee a technical and economical image of a new product and schedule the development of appropriate competences. These processes should rely on major achievements of the fundamental science in the sphere of advanced technologies (biosystems, 3D technologies of neurosystems, etc.) and set vectors for the company's new key competences.

Methodical prediction tools should define the main technologies, which are going to be invested in the product, and define technical and economical parameters of the

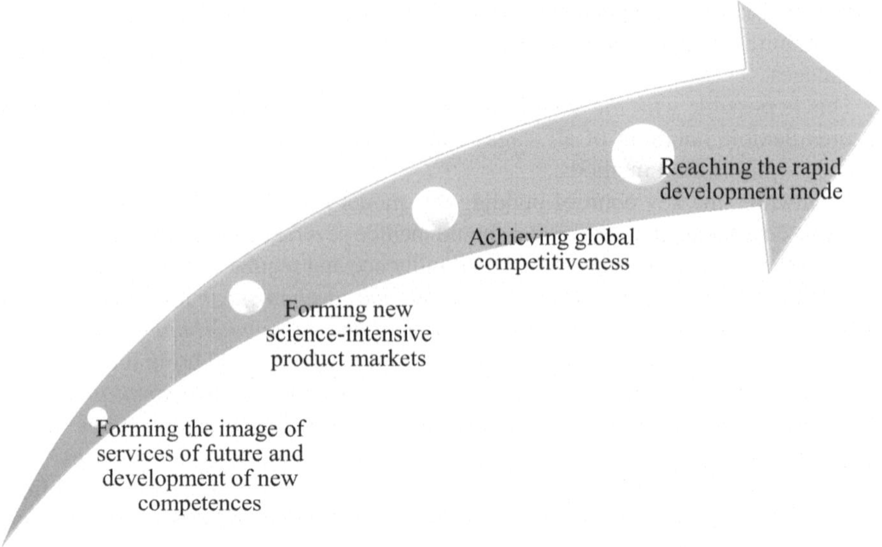

Fig. 1.19 The way up to the rapid development mode

product's image. It is at this stage that sound estimates of the product's future market value should be obtained.

Management of a product's value should be carried out at all life cycle stages, especially during the formation of its technical and economical image. It is because in machine engineering, cost of production, along with product quality and innovative potential, is key to successful competition on the global market. Therefore, it should be determined at the stage of product image formation. Having detailed information about costs (for maximal transparency) at hand at all life cycle stages is vital for effective production cost management. According to experts (particularly, the Defence Advanced Research Projects Agency (DARPA) of the USA), 80% of production cost is built into the whole sum at the first 20% of a product's life cycle. Popular resource supply evaluation methods (rate of return method, comparison to analogous products, etc.) cannot provide enough data to properly analyze costs sustained at early product development stages. Resource planning methods and software (Enterprise Resource Planning (ERP) systems) do not provide effective production cost analysis and evaluation tools at early product design stages, when analysis and evaluation data still can influence the rest of the development process.

The second economic mechanism, which is another important part of the integrated life cycle cost evaluation methodology, is a *complex resource evaluation mechanism* used to assess resources required for product development. Apart from material resources, key virtual ones, such as information and one-of-a-kind competences, must be considered. A respective economical mechanism should comprise a toolkit, which could be used to evaluate the level of available competences and technologies, determine ones that need to be found or developed, and analyze the availability of financial, manpower, and material resources that are necessary for creation of rapid development products.

Different stages of science-intensive products' life cycles require complex use of the resource potential and necessitate management of all types of resources with account of the specifics of organizations' activities. Enterprise Resource Planning systems, which ensure complex automation of the planning process, are used widely. In a digital economy, the most optimal resource use patterns can be generated through the use of systems based on neuron networking and machine tuition methods, which can generate an automated expert conclusion through big data processing and use of knowledge bases.

The *third economic mechanism*, which is needed for product life cycle management, is the *industrial and technological base management mechanism*, which requires development of tools to control its modernization aimed at achieving rapid development. In a digital economy, this modernization should pursue the following goals:

- Building a modern infrastructure for carrying out applied research to speed up the production of goods, which can prove competitive in the global and domestic high-tech markets.
- Mastery and implementation of new technologies to boost the effectiveness of industrial processes within organizations.
- Building a modern infrastructure for innovative activity within corporations, holdings, and separate organizations.
- Continuous work toward energy-efficient and eco-friendly production, incorporation of new quality control systems that meet international standards.
- Building a high technology commercialization and intellectual property right regulation system.
- Building an intellectual infrastructure to raise the effectiveness of business processes and the cost- and labor efficiency of work (virtual organizations, virtual design bureaus, and broad use of automated control systems).

Therefore, an industrial and technological base can evolve, for example, through a broader use of virtual economy technologies that boost management of business processes when creating science-intensive products. This process is based on brand new digital production technologies, which are the basis of the Industry 4.0 concept. The Industry 4.0 is The Fourth Industrial Revolution, following the steam-powered mechanization, mass production triggered by the advent of the assembly line, and automation based on electronics and information technology. This revolution focuses on transition from the "intelligent factory model," which is built in a modular fashion and boasts high transformability, effective resource planning, as well as an ability to integrate customers and business partners into business processes and value creation processes. Also, it fully focuses on customers' individual preferences, real-time use of data, as well as environmental and other characteristics of the production process.

A technological basis for the functioning of this "intelligent plant" is formed by cyber-physical systems (CPS), in which software components are linked with hardware and electronics via an infrastructure of data and the Internet of Things. Within these systems, physical objects (sensors and systems) can send/receive data via the Internet without any participation of humans. Automated decision-making is enabled by an advanced "object communication system," in which objects identify each other, describe a current state, exchange and process data. This excludes the human factor from the system of industrial objects and ensures autonomous, reliable, quick, systematic, and controllable interaction between these objects. Incorporated into industries, the Internet of Things provides several advantages, such as:

- Flexible production process thanks to departure from tough "pipeline" decision and an opportunity to accept and carry out individual orders, freely introduce new solutions, and outsource.
- Adjustable production process thanks to top-to-bottom control and its functioning on a single technological platform.
- Effective production due to lower human-factor-related costs resulting from errors, downtimes, manpower costs, etc.

The fourth integral economical mechanism that is part of the methodology is *production development cost optimization mechanism, preproduction, and production*. This mechanism should feature the following economic instruments:

- An instrument that analyses the influence of structural modifications on products' prime cost at any stage of its life cycle. This tool can help to detect "bottlenecks," components, which greatly influence the prime cost (this is crucial at early stages of the development process). This situation necessitates the use of prime cost reduction methods. Fact-based analysis of manufacturing costs could help to reproduce different production scenarios.
- An instrument for analyzing the cost of items to calculate the cost of components and raw materials obtained on the market. It is necessary to carry out an automated analysis of the costs and benefits, as well as a possibility and usefulness of manufacturing needed items by the organization itself (the necessity of purchasing new equipment/tools/accessories for production of one-off and short-run components).

The fifth economic mechanism within the integral methodology is presented as a *prime cost and labor intensity management mechanism*, as the management should go hand in hand with product design and setting up of production and technical facilities. At each product design stage, input data and information, which is used to evaluate the product's prime cost and labor intensity, should be particularized and modified. The more specific a project is in all aspects (technical characteristics, risks, etc.), the more precisely its prime cost and labor intensity can be evaluated. The designing process and economic analysis should go hand in hand, and data obtained during the analysis should determine the designer's work all the way down to his/her deciding upon admission, mounting, and other parameters of the item's structure.

Because prime costs and predicted value are must have data at the earliest stages of a product's life cycle, it is advisable to obtain an economic mechanism and informational decision-making system, which helps to:

- Give a precise estimate of products' prime cost with account of materials used in them (providing that the target technical characteristics have been achieved) and with account of set admissible working accuracy, level of assembly, and other parameters.
- Carry out a simulation modeling of prime cost and labor intensity with varied working accuracy, materials used in the product, etc., with account of having to achieve target technical characteristics and defining a maximally optimal structure.
- Define a complete structure of products' prime cost and component modifications based on specifications and operating–routing sequences.
- Define and specify direct and operating costs resulting from the use of various technological operations and processes, specific materials, and meeting parameters (for example, processing accuracy).

Finally, the *sixth mechanism* of the life cycle value calculation methodology for science-intensive products is the *evaluation of economic effectiveness of product structuring and manufacturing processes*. Methodic instrument should be used to evaluate how effectively the resources (staff competences, budget resources, production, technical facilities, etc.) have been used. The use can be deemed effective if it has resulted in the appearance of a new competitive product, which can put an organization in the rapid development mode.

All these mechanisms lying within the life cycle value calculation paradigm for science-intensive products should be adaptive; it is advisable to use an adaptive life cycle pattern for promising science-intensive products, as it provides for a flexible resource potential, deadlines, and other parameters (with technical specification requirements unchanged) at different life cycle stages. The developing advanced virtual economy tools enable to effectively adjust the product life cycle management system to rapidly changing internal and internal conditions.

Practical use of economic product life cycle calculation methodology suggests the use of data analytics platforms. They are built as automated expert systems, which should rest on a knowledgebase with data on costs resulting from the creation of science-intensive products.

Thus, the life cycle value calculation methodology for science-intensive products embraces a complex of economic mechanisms (Fig. 1.20).

The synergistic effect of simultaneous work of these economic mechanisms is observed at all stages of science-intensive products' life cycle. Advanced calculation systems and intelligent software complexes enable a simulation of the product creation process. Reference data would include information about an organization's resource potential. By loading a specific technical and economic image of a product in the software, it is possible to assess the required resource volume (including information and key competences), as the simulation will provide a selection of optimal design and technical methods. At the same time, a production and technology management mechanism could assess the required amount of modernization for preproduction, and the product development cost optimization mechanism will help

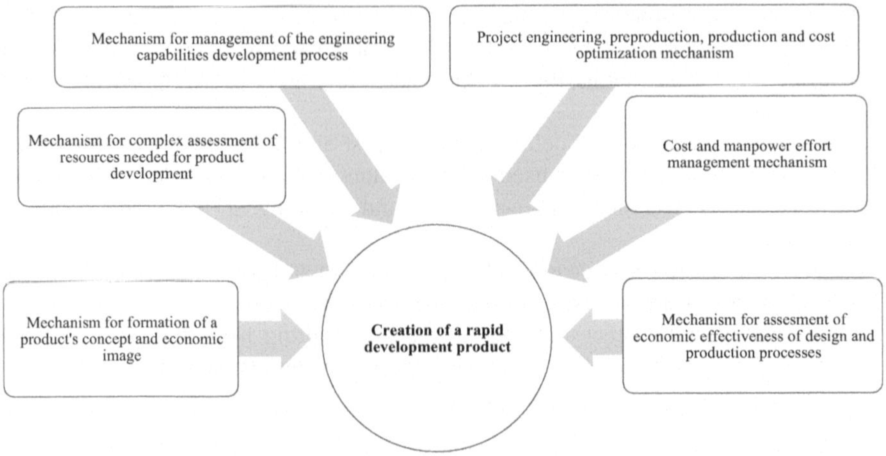

Fig. 1.20 Economic life cycle value calculation mechanisms

to minimize costs arising out of technological processes, providing that the product meets its technical characteristics.

Finally, the evaluation mechanism, which is used to assess the economic effectiveness of design and production processes, should help to monitor and analyze all processes accompanying the products' creation and manufacturing.

Should the economic effectiveness lack or exceed the required level, the mechanism will signal a necessity to take additional managerial decisions throughout the life cycle stages. This mechanism will define new life cycle conditions, to which other mechanisms should adapt.

There are several main principles, on which the adaptive life cycle value management system for science-intensive products relies.

The current phase of virtual economy brings 40 new forms of automated and intelligent production and business process management technologies applied within an organization. This creates a basis for much higher labor capabilities and deeper economic changes.

The adaptive product life cycle management system, in case whereby performance indicators deviate, launches a qualitative analysis of causes of the deviation, every time the management has to be carried out. Every actual or predicted under- or over-performance related to a particular activity triggers a system disturbance, which signals a necessity to boost the activity's effectiveness. It poses a necessity of taking a number of steps to bring back stable performance. The balancing process is often slow, and it is known as "relaxation." The length and intensity of disturbances is largely determined by the effectiveness of corporate product life cycle management systems.

In the modern digital economic environment, the best result in the field of adaptive product life cycle management can be achieved with the help of intelligent decision-making systems that use intelligent methods, neuron networks, machine learning, and big data processing technologies.

The most vivid example of an intelligent adaptive management system, which opens new horizons in project life cycle management and new venues for life cycle management and value optimization, is an intelligent project life cycle management system, which uses global data, including objective information obtained with the help of space technologies (Earth remote sensing data).

These systems can be a foundation for the entire virtual economy, and an opportunity to carry out online process management with minimal use of manpower is going to be their greatest advantage. This requires formation of a theoretical and practical basis for the use of Earth Remote Sensing (ERS) data for solving specific economic tasks. This theoretical and practical basis can give rise to new space services within different economic spheres through broad use of intelligent data analysis methods that ensure effective life cycle management and optimization.

Methods and algorithms, on which these services rely, will require enormous computing power from computers and supercomputers, because there is a huge amount of data to be analyzed. An example of an intelligent system designed for automated management of construction projects is shown in Fig. 1.21.

Intelligent mathematical methods, which form a basis for this type of system, may rest on an analytical tool. These methods help to minimize one of the biggest problems that impede the analysis and monitoring of projects (especially low-budget ones)—the human factor, which interferes in the pricing process, selection of responsible parties, performance monitoring, and quality control. These processes can be managed with the help of an automated system that uses intelligent techniques and does not use manpower. This can minimize the corruption element.

Intelligent space services can help to evaluate the adequacy of existing project evaluation methods, because reporting information may fail to reflect all violations that occur as major projects are executed. Therefore, even a small amount of unbiased information, along with machine learning tools and neuron algorithms, can test the traditional project evaluation and monitoring system.

The adequacy of an intelligent system depends on satellite-based evaluation and measurement tools. These tools should evaluate objects' geometrical dimensions with a precision that would be sufficient for a decent economic analysis. Current satellite image resolution standards provide low geometric precision, with image resolution not exceeding 30–50 cm. This results in inadmissibly high errors, because even a minor deviation in geometric measurements generates a serious discrepancy in economic data. A revolutionary leap forward in intelligent space systems requires that we get space data with errors, which, according to design documentation, stay within admissible limits. Therefore, new satellite tools need to be developed, and they should provide image data on objects' geometrical size with enough accuracy to be able to carry out an economic analysis.

The advent of these tools will endow the intelligent system with new competitive advantages, and it will contribute to the formation of a new global market of voice and data recorders, monitoring tools, etc. An organization, which manages to build such a system, will go into the rapid development mode.

Using such an intelligent system in the sphere of economy will greatly influence project life cycle costs, as the system will be managing projects. As a result,

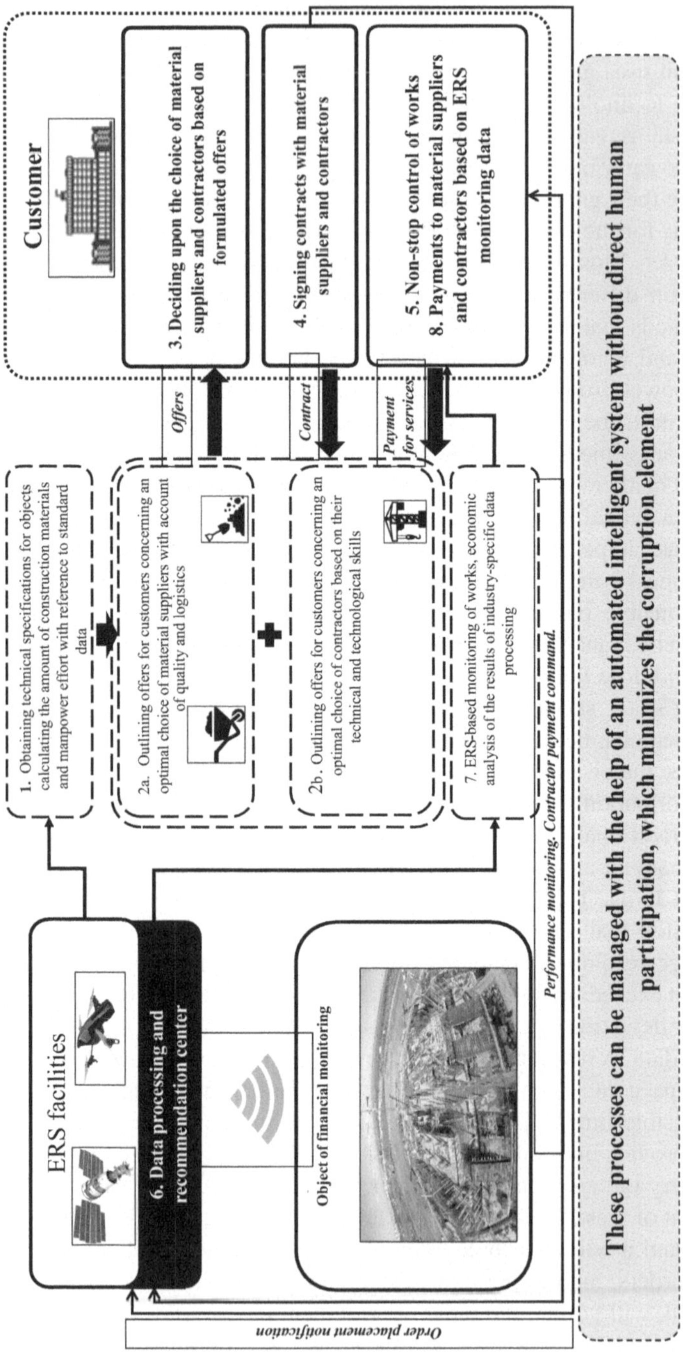

Fig. 1.21 Use of an intelligent space system in construction

processes, which are part of a project, will demonstrate a much higher economic effectiveness. The improvement will be due to the system's new abilities (both mathematical and analytical), and it is expected to seriously outperform man. They are already starting to do so, which is observed in supercomputers' ability to solve some tasks through intelligent processing of data from the global informational space, fundamental science, data obtained from analysis of the results of intelligent work and automation; this data will be the ground for effective design, technology, and management decisions.

The use of intelligent systems within the current techno-economic paradigm, their further modernization and development, will produce a significant economic effect within the seventh technological paradigm. Today, scientific research focusing on formation and development of the seventh techno-economic paradigm already suggests that the new paradigm's main difference from the previous ones will lie in incorporation into the production and economic processes of the human mind and biologic factors. Simply put, the human mind will become a productive power in the same way as science did.

1.4 Informational Support of Automated Life Cycle Management Processes with the Use of Digital Technologies

Given the growing competitiveness of organizations, their gaining of competitive advantages and rapid development products, it can be supposed that the most powerful sources of information, which enable management of these processes, are the big database of the global informational space, knowledge, prospects for further development of the global science, and the by-products of intelligent processing of this data with the use of modern mathematical methods (neuronal networks, machine learning, discrete mathematics).

Thus, the informational space can provide us data on specific design and technological solutions, advanced techniques, and technologies. This will allow organizations to execute design, preproduction, and production of new types of goods with minimal costs. Also, they will be able to continuously monitor the outer environment, competitors' activities, macroeconomic upheavals, and new scientific and technical achievements.

The digitization of industrial enterprises can be a foundation for a new organizational and economic environment, in which various aspects of economic activity relating to production of new types of goods will take on a more advanced shape. These processes will contribute to transformation of existing corporate management patterns and produce a need for doing an effective high-accuracy economic analysis of a much larger volume of information that it is required for traditional work and managerial decision-making.

In this situation, a new managerial decision-making system plays an important role, as it is based on advanced mathematical methods that rely on machine learning and big data processing. The system has several levels:

- Formation and implementation of advanced electronic telecommunication and data transfer systems (an electronic module).
- Formation of a new organizational sphere focusing on reforming and optimizing corporate management structures (an organizational module).
- Formation of new methods of management (a managerial module).
- Development and implementation of advanced data processing, analysis, and distribution systems, which use universal data processing toolkits and data transfer formats (an informational module).

This system is not intended for unification of automated information receipt, processing, synchronization, and storage tools, which are created in different ways with the use of advanced information and communication technologies. Instead, it should enable a balanced integration between existing management facilities and databases and advanced automated data analysis and processing technologies, which should take them to a whole new level. Creation of such a system in future can boost effectiveness and synchronize the key business processes of new product creation and, in turn, make an organization more competitive on an international scale.

Rational management of organizations' adaptation to the virtual environment by generating synergistic effects in the economic, scientific/technical, and industrial spheres, has gained ground for the following reasons:

- Poor integration of new product creation processes into the global information space and therefore limited opportunities for creating a product's technical and economic image that would otherwise put a company on a rapid development mode.
- Structural changes in the division of labor, emergence of cyber-physical system, and digital design methods; a substantial increase in digital information's significance being a stimulating factor of production of science-intensive goods.
- The fact that high-tech organizations are to a greater or lesser extent the most important stakeholders in the global information market (which has appeared due to rapid evolution of computerized technologies and electronic communication, big data analysis processing, and machine learning methods); this has greatly strengthened their interconnections and interdependence and therefore created vast opportunities for exchange of technologies and competences.
- Increased global economic instability and, consequently, risks of lower economic effectiveness.
- Difficulty coordinating and managing organization with different structures, size, and character of activities.

In this context, it is important to note that in today's environment, a company's organizational characteristics tend to be one of the main factors of competitiveness. Therefore, there is a need of ultimate transformation of life cycle management mechanisms for science-intensive products, which should rely on the use of digital management platforms and wide use of neuron networks to eliminate the negative influence of the human factor. This transformation should enable practical implementation of a complex approach to the formation of a versatile bundle of information/computer services and technologies, which make up a foundation for effective product life cycle management.

An intelligent multiservice information/computer infrastructure is one that is formed through incorporation of unified information-and-control networks (high-end computers, cutting-edge software, digital data transfer systems, etc.), to integrate processes that are part of different product life cycle phases into a single management space, which has a network-type architecture. A possible example of this multiservice intelligent information/computer infrastructure is shown in Fig. 1.22.

This intelligent multiservice information/computer infrastructure offers much greater competitive advantages than traditional management tactics:

- Real-time creation of a single digital model of interaction between business processes taking place when a new product is being created.
- Much quicker collection and distribution of information around structural divisions and contracting parties.
- A significant advantage over competitors in making and carrying out decisions and planning of work.
- Effective management of and quick concentration of resources in paramount spheres of work, etc.

It is certain that implementation of these novelties will help build a much more effective management system than the current one, thanks to its high mobility and automated decision-making features, which are based on real-time big data processing and machine learning.

The presence of an intelligent managerial infrastructure, which relies on telecommunication networks like the Internet and Internet of Things acting as an economic system, is crucial to a company's rapid development and its product's competitiveness.

The work of a company, whose managerial system combines all features of an intelligent multiservice information/computer infrastructure, can be demonstrated with the help of a common managerial approach, which is known as "Low signal management." The relevance of decisions that are related to the organization's work and evolution largely depends on its managerial staff's awareness of existing and impending internal and external threats, expansion of export and other competitive advantages. Low signal management implies an ability to detect negative tendencies in due time, thus providing more room for reaction and an opportunity to make timely managerial decisions to adapt to changes and maintain a company's competitiveness.

At the first phase of low signal management, when a potential danger is only beginning to emerge and when there is still little information, the response will be common and aimed at preserving strategic flexibility—the company's ability to adapt to temporary and permanent internal and external changes without losing room for export expansion, other competitive advantages, and general

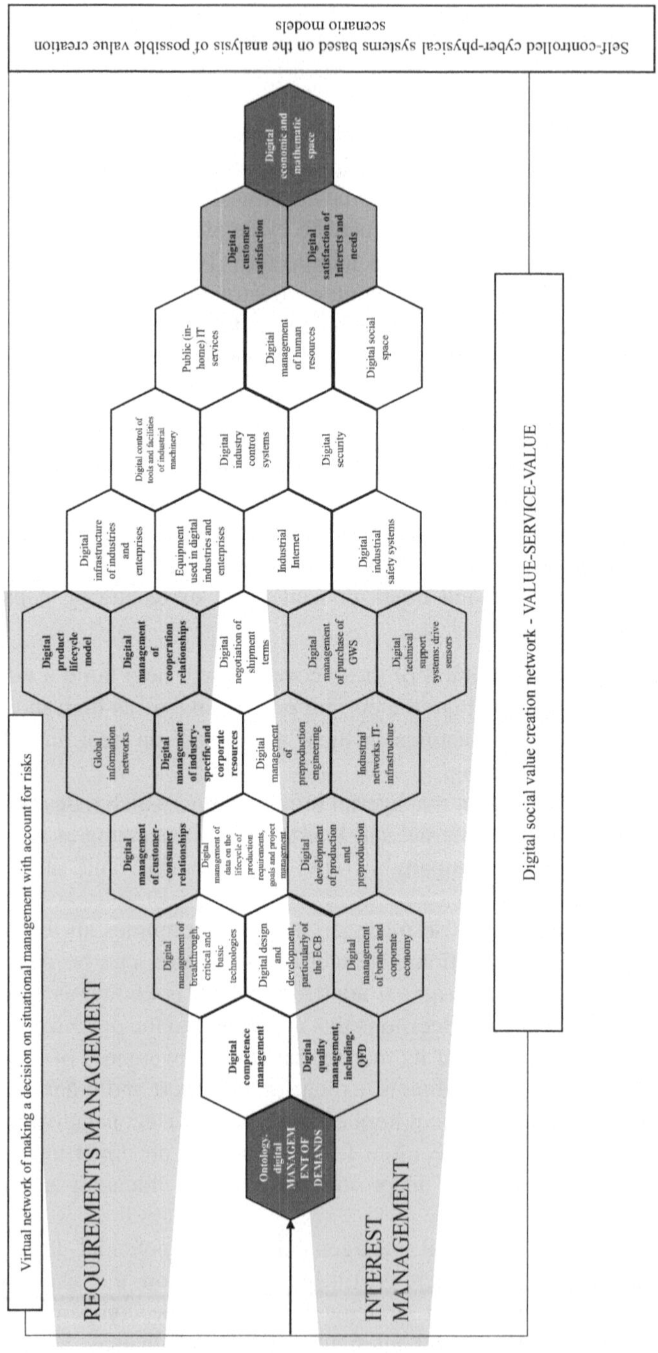

Fig. 1.22 A possible structure of a multiservice intelligent information/computer product life cycle management system

competitiveness. Recommendations concerning a rapid response can be received with the use of intelligent tools: machine learning and neuronal network technologies.

The following levels, with specific information arriving (as big data is accumulated), the managerial staff will activate response measures with the goal of either elimination of threats (for example, switching to another supplier), or using new opportunities (adaptation to a new political situation). In the context of strategic management, this approach is termed "stepping up response measures" and "responding to low signals."

Another important factor of an organization's competitiveness is an opportunity of quicker introduction of cutting-edge technologies, which provide for modernization and rapid development of an organization. For this reason, building an organizational basis for cooperation between business units with an emphasis on cutting edge intelligent management technologies is key to today's global economy. The effectiveness of a model for intelligent multiservice information/computing facilities plays a crucial role in creating an international ground for exchange of technologies, competences, and expansion of export.

This enables a practical implementation of the above-described adaptive approach to planning different types of financial and economic operations within organizations and product life cycle management through the development of automated intelligent mechanisms, such as:

- A mechanism for building distributed managerial databases.
- A mechanism for updating data with the use of the global information space.
- An optimizing managerial decision-making mechanism, also applying to different levels within the science-intensive product life cycle management hierarchy.

At the same time, corporate automated intellectual management mechanisms should be able to:

- Correct themselves on the basis of an intelligent multiservice information/computer facility model, which ensures a minimal length of the decision-making chainwork from the moment of perception of original information through the execution of the managerial decision.
- Correspond to a set science-intensive product life cycle management task aimed at putting an organization on a rapid development mode.
- Live up to tendencies and stages of the formation of new unique competences that are part of a product's added value.
- Provide for a final launch of an organization upon the rapid development mode through building a single multilayer business–process interaction system at different science-intensive product life cycle stages.

Information and telecommunication technologies, as well as information processing and transfer environments within organizations evolve through constant creation and introduction of innovative solutions. Naturally, we regard these information and telecommunication technologies as a tool for boosting competitiveness and organizations' effectiveness in terms of development of new products. This

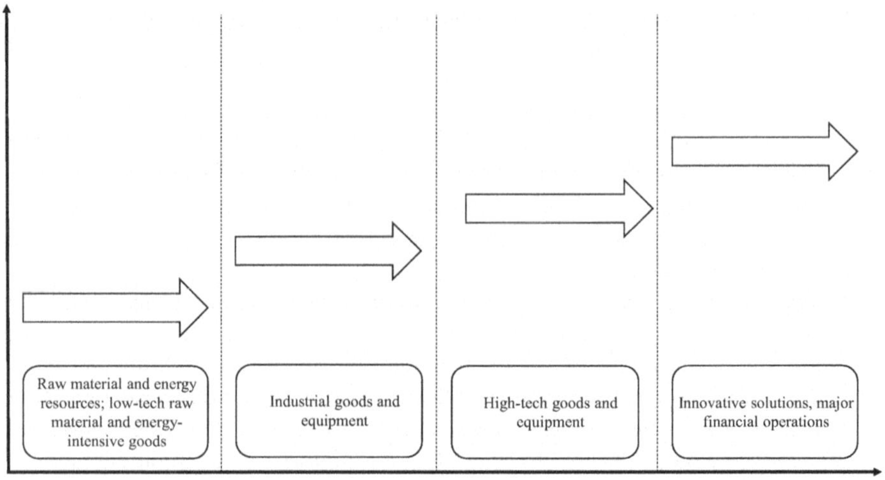

Fig. 1.23 Innovative transformation of an organizations' work within the global economy

creates a cyclic system, which is composed of creation and incorporation of information and telecommunication technologies, helps to find, develop, and introduce cutting-edge technologies, outline new competences and thus creates a technical and economic image of a rapid development product. Consequently, it heightens competitiveness, which ensures higher sales and provides additional profits, which, if managed rationally, can be partially invested in a company's development, creation, and introduction of new cutting-edge technologies.

Introduction of new management mechanisms will provide an advanced solution to the task of getting from the global information space the knowledge about patents, know-how, fundamental discoveries, state-of-the-art projects, production of, and providing goods and services with advanced characteristics.

Processing and use of this information will provide for production of highly competitive high-added-value goods and services and, in turn, contribute to a speedy increase in profits and give impetus to further innovation in general. This process is structured as shown in Fig. 1.23.

Informational support of this transformation and automation of product life cycle management processes with the use of digital technologies should help an organization achieve global competitiveness. Rational distribution of elements within the flexible structure of multiservice information/computer facilities, which are to be used in science-intensive product life cycle management, lays the foundation for a higher international competitiveness of organizations, also through gradual development of the synergistic effect during the formation of an intelligent digital managerial environment, as shown in Fig. 1.24.

Therefore, all work to provide informational support and digitize science-intensive product life cycle processes should focus on creation and development of a unified intelligent multiservice information/computer infrastructure. All informational support and automation tools must be viewed as functional or datalogical

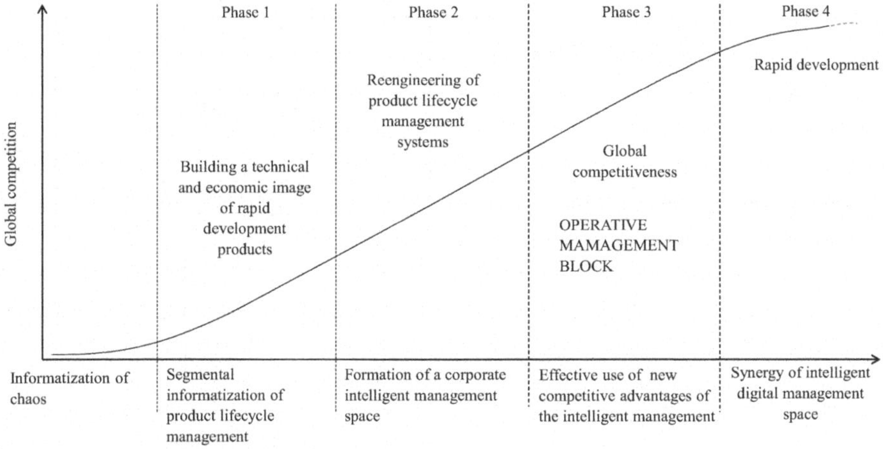

Fig. 1.24 The process of achieving global competitiveness through the synergy of the intelligent digital managerial environment

components of this intelligent multiservice information/computer infrastructure. Its creation and development should be aimed at making more effective managerial decisions at all product life cycle phases, which is now possible thanks to obtaining up-to-date information from the global space with the help of advanced software.

The main structural elements of an intelligent multiservice information/computing infrastructure include data- and knowledge bases on newly created goods and services, experience in organizing management process, phase of scientific and technical development, its direction, etc. Data- and knowledge bases are regarded as a foundation for a long-term strategy of preserving and boosting a company's innovative potential and competences within it. The data- and knowledge bases are the basis for informational support of automated life cycle management with the use of digital technologies. They make the core of expert systems—software tools, which in a way, through the use of neuronal algorithms, stand in for human experts in this field and serve as effective managerial decision making support systems. The structure of an expert system (ES) is shown in Fig. 1.25.

An expert system enables us to calculate specific values that are used in specific economical and mathematical models, which rely on the knowledge consisting of information from databases (big databases) and knowledge bases. The expertise, which makes a basis for a knowledge base, should be formed with the help of analysis of the global information space and through questioning of operating experts. Results of such analysis are formulated by knowledge engineers in the shape of specific knowledge bases.

Knowledge bases are a key element of intelligent information/computer systems, because this system module defines logical rules of inference when solving effective management tasks. Unlike databases, knowledge bases contain rules of inference based on input data, not factual information, which is contained in databases.

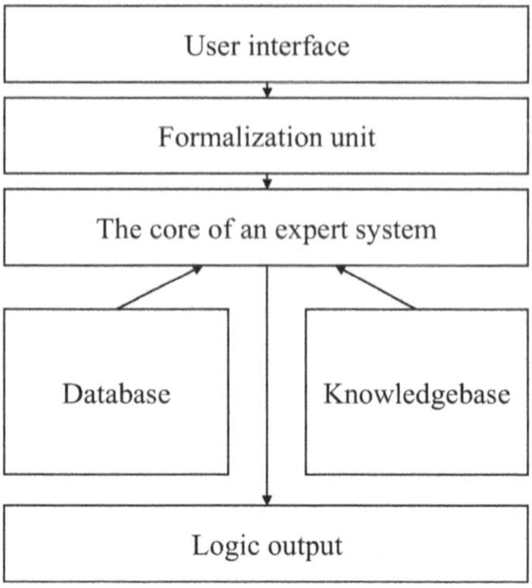

Fig. 1.25 Expert system use algorithm

Knowledge bases are used to launch a pipeline of logical conclusions. A logical conclusion is based on the results of questioning an intelligent system. Thus, the answers form a database of facts, which are to be processed by the system. Processing of this input data provides conclusions made by the intelligent system (optimal managerial decisions)—the output data.

The algorithm of work of this intelligent expert system, which provides informational support for automated product life cycle management with the use of digital technologies, can be described in the following steps:

Step 1: Defining input parameters (original conditions).
Step 2: Primary processing of the parameters and generation of facts.
Step 3: Processing of facts with the use of a knowledge base, drawing conclusions from the obtained facts.
Step 4: Obtaining the expert system's conclusion that is based on the facts revealed at Step 3.

To solve complex economic and management tasks, these steps can be carried out in a nonlinear fashion; repetitions can be made in order to clarify input data. Particularly, obtaining facts necessitates clarification of these facts to eliminate controversies, or to conduct a deeper analysis of the input data. The algorithm of solving science-intensive product life cycle management tasks with the use of an intelligent system is shown in Fig. 1.26.

When a logical conclusion is used within an intelligent expert system, it requires the use of different mathematical data analysis methods and input data processing models.

Modern tendencies, such as instability, globalization, black market, scientific, and technological revolution, testify to the important role of the formation of an intelligent multiservice information/computer infrastructure, which is based on an expert system. Effective management of major companies and their obtaining of

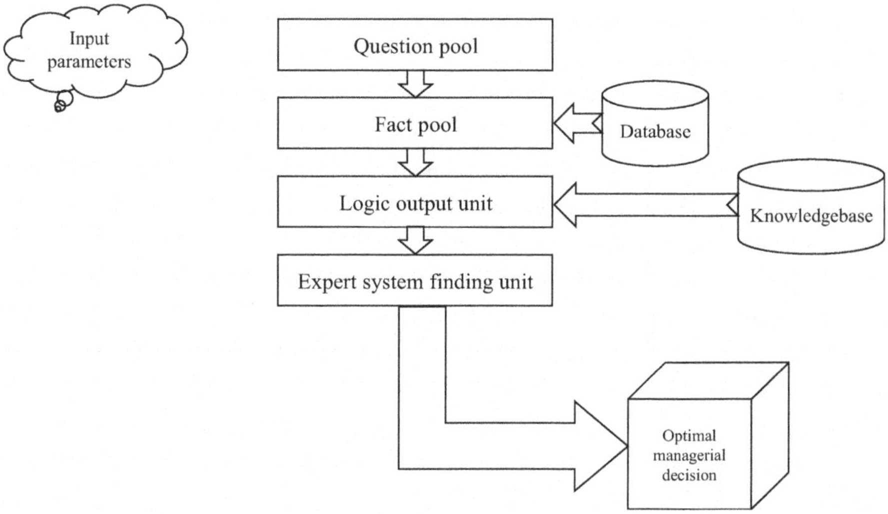

Fig. 1.26 Expert system use algorithm

competitive advantages throughout the design, production, and sale cycle, are only possible with the use of large automated organizational and economic systems, which operate in all the directions within an integrated structure, which rests on creation of an intelligent multiservice information/computer infrastructure. The latter speeds up management cycles and provides for a higher global competition with rapidly changing economic conditions and toughening global competition. This infrastructure makes it possible to create a promising product of future with the help of digital production technologies that are intended for reaching a specific level of competitiveness in the market with account of increasingly personified needs, growing intellectual potential, and evolving competences. Thus, enables an effective management of the production process at the earliest life cycle stages.

References

1. Schumpeter, J. A. (1995). *Capitalism, socialism and democracy*. Economics.
2. Sloan, A. (1964). *My years with general motors*. New York: Doubleday.
3. Drucker, P. (1969). *The age of discontinuity*. Piscataway, NJ: Transaction Publishers.
4. Solow, R. (1957). Technical change and the aggregate production function. *The Review of Economics and Statistics, 39*(3), 312–320.
5. Porter, M. E. (1980). *Competitive strategy: Techniques for analyzing industries and competitors*. New York: The Free Press.
6. Yerokhina E. A. (1999). *Theory of economic growth: The systemic and synergistic approach*. M.
7. Feigenbaum, A. (1951). *Quality control principles, practice, and administration*. New York: McGraw-Hill.

8. Deming, W. E. (1982). *Quality productivity and competitive position*. Cambridge, MA: Massachusetts Institute of Technology.
9. Drucker, P. (1993, August). *Post capitalist society*. Oxford: Butterworth& Heinemann.
10. Drucker, P. (1998). The next information revolution. *Forbes ASAP*, 24.
11. Senge, P. M. (1994). *The fifth discipline fieldbook: Strategies and tools for building a learning organization paperback*. New York: Doubleday.
12. Schein, E. H. (1996). Culture: The missing concept in organizational studies. *Administrative Science Quarterly., 41*(2), 229–240.
13. Ouchi, W. G. (1981). *Theory Z: How American business can meet the Japanese challenge*. Boston, MA: Addison-Wesley.
14. Wernerfelt, B. (1984). A resource-based view of the firm. *Strategic Management Journal, 5*(2), 171–180.
15. Collis, D., & Montgomery, C. A. (1995). Competing on resources: Strategy in the 1990s. *Harvard Business Review, 73*, 118–128.
16. Hammer, M., & Champy, J. (1993). *Reengineering the corporation: Manifesto for business revolution*. New York: Harper Business.
17. Teece, D. J., Pisano, G., Schuen, A. Teece, D. J., Pisano, G., & Schuen, A. (2003). *Dynamic capabilities and strategic management*. The Vestnik of the Saint-Petersburg State University. Ser. Management (Vol. 4, pp. 133–171).
18. Cravens, D. (1994). *Strategic marketing* (Vol. III). Burr Ridge, IL: Richard D. Irwin.
19. Tchursin, A. A. (2012). *Theoretic basics of competitiveness management*. Spektr: Theory and Practice. M.
20. Ephymov A. V. (2014). *Developing mechanisms of stable growth of industrial enterprises and business groups based on corporate knowledge management: Extended abstract of dissertation*, Candidate of Sciences in Economics.

Chapter 2
Rapid Development of an Organization

2.1 Rapid Development Management Law

An analysis of the evolution of the economic theory suggests that many of its hypotheses are going out of date. Today, the fundamental economic theory is behind practice. The theory base of the rapidly evolving digital economy is still poorly developed. Main economic laws and metrics, which describe the digital economy, were introduced and formulated in the nineteenth and the first half of the twentieth century, and they describe the traditional industrial sector quite well. From the second half of the twentieth century, the sector of services and nonmaterial production received a substantial impetus and eventually became a major one along with the industrial sector. The properties of production and consumption within the nonmaterial sphere greatly differ from each other; therefore, it is important to develop a theoretical base in order to properly describe the new type of economy.

New discoveries and transformations in techniques and technologies, highly effective solutions, information, and communication resources that are becoming globally available, as are neuronal networks, blockchain, the Internet of Things, VR- and 3D technologies, etc., modify the general concept of the global economy, which is being affected by globalization and imperfect competition resulting from the tyranny of particular states.

In this context, the politically biased modern economic science becomes a negative factor, as it enables speculative and intentional corruption of the general picture. The US sanctions policy is a bright example of such processes taking place in the global economy. Economic development is being presented as a byproduct of protectionist measures, not of competitive advantages obtained thanks to products' cutting-edge technical and economical characteristics. This creates new ideas of what an economic development management system is and necessitates a search for cause-and-effect links that would stimulate the growth of separate organizations, as well as the national and global economy.

A. Tyulin, A. Chursin, *The New Economy of the Product Life Cycle*,
https://doi.org/10.1007/978-3-030-37814-1_2

A key aspect of outlining a new economic theory is the choice of relevant integral parameters and setting up of new economic metrics.

Development of competitive management theory and practice is a foundation for the development of the corporate and branch economic theory and economy in general. The rapid development theory, which focuses on the processes of creation of highly competitive goods, is an outgrowth of the theory of competitiveness. According to the rapid development theory, practical implementation of product creation processes should result in an organization reaching the rapid development status, i.e., forming a new science-intensive product market or occupying a large share of an existing market with the help of its innovative potential and a set of key competences.

The concept of rapid development management is based on extremely important economic laws: the law of diminishing marginal utility, the law of rising economic necessities, and the law of interacting development of competences and consumer markets.

The law of diminishing marginal utility states as an item is consumed and the consumer's satisfaction (satisfaction of needs) grows, the item's marginal utility diminishes every time the consumer gets another such item. The law of diminishing marginal utility shows that a manufacturer always needs to *increase competitive advantages through rapid development, which implies designing and launching production of cutting-edge innovative technologies* that ensure creation of new products and upgrading of existing ones. Not only should new types of products keep the demand high, they also should stimulate new demand and generate new markets.

The law of rising economic necessities states an objective necessity of an increase and improvement of people's needs as production and culture evolve. It is also important to switch people's needs to their own education. Services and goods, which stimulate education, will always be competitive. The law proves that *creation and manufacturing of products that are based on new discoveries and technical/technological transformations will ensure formation of new needs and values for the society and contribute to the emergence of new markets*, in which the manufacturer is going to be a pioneer.

The economic *law of interacting development of competences and consumer markets* says that their mutual evolution goes along a spiraling trajectory. This ensures an innovative economic growth. This implies that emergence of unique competences increases high tech organizations' resource potential, which triggers a leaping growth of unique advanced technologies. Their use results in *the appearance of ultimately new products, which shape people's needs for new goods and stimulate new markets and economic growth through rapid development. In turn, rapid development creates an economic basis for the formation of new unique competences.*

This triggers a cyclical progress of competences and innovations, which form a basis for designing and manufacturing highly competitive high-added-value products, which contribute to the formation of new markets and are exceptionally valuable for consumers. However, based on innovative technologies, these products have a short life cycle and lose competitiveness once competitors reproduce an existing unique advanced solution or create a similar one equaling the original specimen in features and value.

A brief characteristic of these economic laws suggests that they do consider some basics of rapid development, so they create a foundation for the emerging law of rapid development management. It is natural to expect rapid development products, which define needs for new benefits and stimulate the emergence of new markets and economic growth, to dominate the market. The question is: which type of market situation does the described rapid development environment reflect? This work will focus on a situation, when an enterprise makes rapid development goods in an essential effort to satisfy growing public needs. In this situation, it is possible to temporarily maintain competitiveness through continuous upgrading of the product line. However, the main goal the rapid development management process is related to the manufacturer's striving to foresee public needs and create a product that can satisfy them.

Given the interest in an opportunity to manage the rapid development process of an organization and to create ultimate competitive advantages through technical and technological discoveries and transformations, not through protectionist measures, it is evident that this work focuses on a market with imperfect competition. Indeed, imperfect competition often manifests itself in monopolies, i.e., cases whereby markets are dominated by one single supplier; however, there are several types of situation, which are typical of imperfect competition:

- *Oligopoly* is a kind of monopoly with a small group of organizations that exclusively control production of and trade in a certain type of goods.
- *Pure monopoly* is quite the opposite to pure competition, as only one supplier operates in the market and possesses total control over the amount of goods it produces.
- *Monopsony* is a kind of market structure, in which a product has only one dominating buyer.

A situation with a rapidly developing organization gains possession of a large share of the market is called *oligopoly*. In cases whereby a company forms a new market and a highly competitive product pervades it, it is a pure *monopoly*. If a new type of product is created for the benefit of just one category of buyers (for example, building new types of rockets and space products for the state), the rapid development will result in a *monopsony*.

Here is an example of a market formed by a rapid development product. Supposedly, it is a market with imperfect competition. To understand how balance can be achieved and how incomes are generated within this market, it is necessary to understand discriminative and non-discriminative monopsony. The demand for a type of goods produced by an organization within this hypothetical market structure can be expressed as $\mathrm{MRP}_{\mathrm{VF}} = \mathrm{MP}_{\mathrm{VF}} \cdot P_Q$ (where MRP is marginal productivity revenue; MP is marginal product; P is the price of a product unit; VF is the production factor; Q is the quantity of product units).

Assume, $\mathrm{MP}_{\mathrm{VF}} = 1/\sqrt{\mathrm{VF}}$. Then, $\mathrm{MRP}_{\mathrm{VF}} = P_Q/\sqrt{\mathrm{VF}}$.

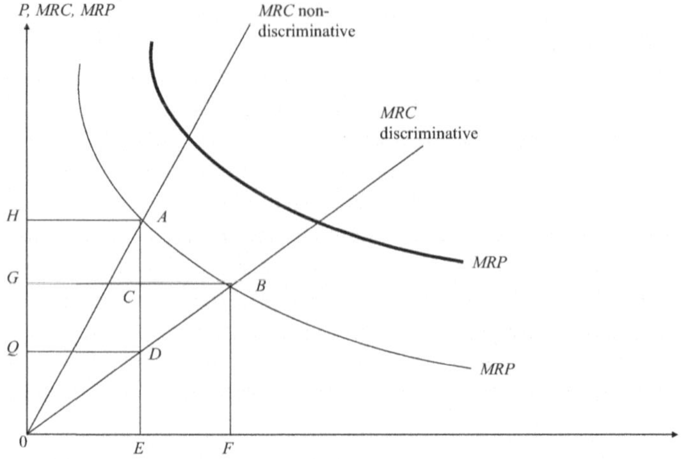

Fig. 2.1 Balance in a discriminative and nondiscriminative monopsony (*the thin line reflects protectionist monopoly dynamics, and the thick one—rapid development market dynamics*)

Values that characterize the balance in the new product's market with imperfect competition are: $\text{MRC}_{\text{VF}} = \text{MRP}_{\text{VF}}$, where MRC is the marginal revenue cost, which depends on whether it is a discriminative or non-discriminative monopsony. If non-discriminative, balance can be achieved providing that: $P_{\text{VF}} + \text{VF} \cdot d(P_{\text{VF}})/d(\text{VF}) = \text{MP}_{\text{VF}} \cdot P_Q$, where $\text{MRC}_{\text{VF}} = P_{\text{VF}} + \text{VF} \cdot d(P_{\text{VF}})/d(\text{VF})$.

The balance in a discriminative new product's market is expressed as $P_{\text{VF}} = \text{MP}_{\text{VF}} \cdot P_Q$, where $\text{MRC}_{\text{VF}} = P_{\text{VF}}$, $\text{MRP}_{\text{VF}} = \text{MP}_{\text{VF}} \cdot \text{MR} = \text{MP}_{\text{VF}} \cdot P_Q$, so $\text{MRC}_{\text{VF}} = \text{MRP}_{\text{VF}}$ (MR is the marginal revenue).

Events that take place in the market when an additional product is introduced are best illustrated in the form of a diagram. Figure 2.1 shows curves MRP_{VF} and MRC_{VF} reflecting discriminative and non-discriminative monopsony. The intersection point of curves MRP_{VF} and MRC_{VF} is actually the balance point. From Fig. 2.1 it is possible to see the dynamics of income in a discriminative and non-discriminative monopsony. Consequently, it is possible to judge the usefulness of boosting competitiveness through introduction of additional product units.

The value of "economic" types of income within the newly introduced product market with a non-discriminative monopsony equals the square of quadrangle QUAD plus the square of the figure beneath curve MRP and above line *HA*. The value of "economic" types of income within the newly introduced discriminative monopsony equals the square of quadrangle *GHAC* plus the square of figure *ABC* and the figure beneath line MRP and above line *HA*.

This situation may occur, for example, if the government takes protectionist steps and to keep the market open for only one organization. However, from the standpoint of the economic theory, a more interesting situation occurs when an organization dominates a market and actually monopolizes it through rapid development. A rapid development product, which is supposed to satisfy new public needs, brings a significant profit at the beginning of sales and continues to do so until competitors appear in the market. The thick curve in Fig. 2.1 describes this situation.

Management of competitiveness and rapid development in a market with imperfect competition is a complex process. To analyze competitiveness, it is necessary to describe an organization's productive function, which should link economic resources used in production to the volume of production achieved through their use. The effectiveness of production will stand for the use of economic resources (factors), which the organization possesses, such as labor $L(t)$, capital $K(t)$, innovative potential (revolutionary technologies and one-of-a-kind competences) $A(t)$, allowing for an optimal volume of production $Q(t) = f(L(t), K(t), A(t))$ (optimality parameters will depend on the goals of optimization), where the time moment $t \in T$ (T is the period of time, during which the analysis of an organization's competitiveness and rapid development status is carried out). This analysis implies the use of the term "marginal production" MP. Marginal production demonstrates a change in productivity of a production variable providing that at least one of other production factors remains constant.

This work considers three most important factors used in the production process: manpower, physical capital, and innovative potential. In rapid development management, innovative potential plays the most important role. Marginal production, which is backed by innovative potential—cutting edge technologies and competences used in production (MP_A)—can be achieved in cases whereby production is driven by changes in technologies (A = var) with labor and physical capital values unchanged (L = const, K = const).

The following mathematical expressions define marginal production level achieved with the help of innovative potential: $MP_A = \partial Q/\partial A$ or $MP_A = \Delta Q/\Delta A$. The diminishing productivity of the variable resource is explained by the law of diminishing marginal productivity. In the context of marginal production with reference to innovative potential, the law of diminishing productivity is actually the law of diminishing performance of a technology: $A \uparrow MP_A \downarrow$ (L = const, K = const). In other words, the longer a rapid development technology is used by an organization, the less effective an output they produce, which is observable in lower competitiveness.

This means that *the potential for rapid development dynamically changes over time; therefore, to maintain rapid development, an organization should manage it.*

On the whole, management of an organization's competitiveness should be carried out through controlled introduction of additional resources (those can be competences and innovative potential), as well as through controlled choice of the highest priority resource application areas. The management process can be deemed effective when an organization has sold a large amount of higher quality goods. The price of the goods should be defined by a minimal yet effective use of resources.

At the same time, to assess the effectiveness of rapid development management, other estimation criteria need to be applied. These estimates can be based on a statement that innovative work and creating one-of-a-kind products (or new markets) with resources available should be in sync (in terms of speed and intensity) with the evolution of the production system and business processes, which are accompanied by a company's economic stability. In practice, this happens as follows: the growing competitiveness of the goods produced by an organization will entail a higher sales profit. It will provide new assets for modernization of the company's production and technology system, as well as for creation of new types of products. More effective functioning of the scientific, design, production and technology, and other organizational systems will result in a higher general competitiveness and, consequently, stable economy.

This suggests that there is a cause-and-effect link between the emergence of new public needs and an opportunity of satisfying them through creating a product relying on discoveries and technical/technological transformations that are backed by a substantial resource base and a bunch of unique competences.

With reference to the main economic laws and approaches to management of competitiveness in different types of competition, it is possible to formulate the law of rapid development of an organization:[1]

> The law of rapid development of a manufacturer states that the product being designed and created should have a consumer usefulness (consumer value), which can boost the public needs that stimulate new markets and ensure the manufacturer's stable development.

There is a mathematical model of the law of rapid development. It describes the following most important parameters:

1. $W(t)$ is a function describing the management of a company's rapid development. This function can be deemed positive, and its higher values entail better management of the rapid development process. The zero value means complete absence of such management.
2. $Q(t)$ is a function describing the integral competitiveness value. Supposedly, it gives an integral description of all competitiveness management activities taking place within an organization.
3. $D(t)$ is a function describing availability of resources for an organization. The function is supposed to be positive.
4. $F(t)$ is a function describing an integral value of a company's economic stability. This value is alternating.
5. $C(t)$ is a function describing an organization's current innovative potential and available competences. This function is positive.

[1]In economy and economic process management, the concept of law is conditional. Any strict definition of a law is limited by certain conditions even in natural phenomena. For example, the law of Ohm does not work at extremely low temperatures, when materials obtain superconductivity. In economy, laws, such as "internal, substantial, stable, recurring cause-and-effect links within a system of production relationships" (the New Economic Dictionary. M.: Institute of New Economy, 2008) are even less likely to be absolute and independent from coordinators of work. The author uses the concept of law in the context of management of competitiveness to emphasize a newly discovered regularity, which manifests itself the harder, the greater number of reverse causalities (consequence to cause) overlap an economic structure; this already lies within the managers, the process coordinator's toolkit. However, disregarding the regularity is likely to produce the same effect as a shift from indicative planning toward a greater number of missed opportunities.

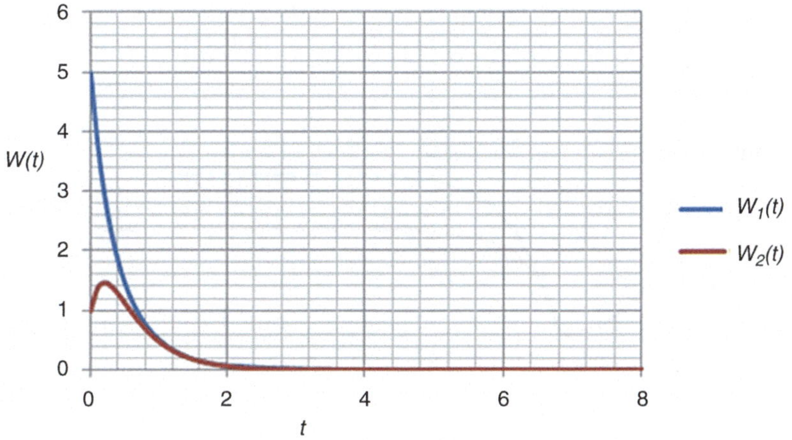

Fig. 2.2 Loss of potential for rapid development due to lack of management

A general rapid development model used by an organization should describe the influence of controlling activities on product's competitiveness and the manufacturer's economic stability.

Management of rapid development can be described by the following expression:

$$\dot{W}(t) = A(t)W(t) + B(t, W(t)),$$

where $A(t)$ is a matrix describing mutual influence of the components of vector W. At every timepoint (t), elements of the matrix obtain new values.

$\dot{W}(t)$ is a vector, the components of which are an integral indicator of competitiveness, availability of resources for an organization, an integral indicator of an organization's economic stability, innovative potential, and competences.

$B(t,W(t))$ are the controllers of organizations' activities that are aimed at reaching rapid development.

The model gives a mathematical description of how economic influence of all these parameters on the end result—an organization's [potential for rapid development.

This potential is going to diminish unless additional steps are made to support it (i.e., in case whereby $B(t, W(t)) = 0$). Assume at the initial stage an organization does have some potential for rapid development.

An approximate solution to this case (for varying initial values of potential for rapid development) is demonstrated in Fig. 2.2.

In the former case, a steady decrease in the potential for rapid development is observed ($W^1(t)$). In the latter case, it is ($W^2(t)$) $A(t)$—an accelerative increase in the potential for rapid development at the initial stage followed by a steady decline. The accelerative growth observed at the beginning may be due to competitors' inactivity during this period or to high demand for rapid development products, which contribute to an increase in the organization's resource potential. This situation

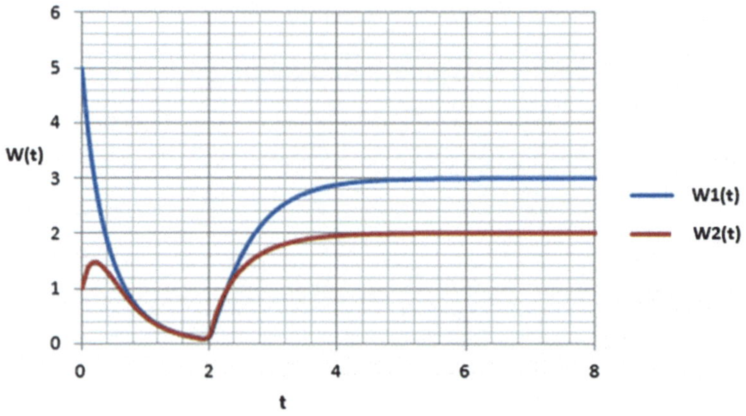

Fig. 2.3 Gaining a stable ability for rapid development thanks to effective management

can be presented within the model through the choice of a certain type of matrix $A(t)$. At timepoint $t = 2$, the potential for rapid development falls down to the critical point ($W(t)$ close to zero), when emergency control measures are required to control the potential for rapid development.

Sometimes, steps to maintain the potential for rapid development are taken on a constant basis and with unchanging intensity. In this case, functions $B(t, W(t))$ will look in the following way:

$$B(t, Q(t)) = \text{const.}$$

The use of regular measures to maintain the potential will lead to stable economic progress of an organization and keep its potential for rapid development at the current level. The results of these measures are shown in Fig. 2.3, where control over rapid development, starting at timepoint $t = 2$, stabilizes the potential for rapid development.

This mathematical model enables high quality research of the dynamics of economic indicators in keeping with the law of management of an organization's rapid development. With the help of this model, it is possible to demonstrate a number of ultimate results from the general law of management of an organization's rapid development.

Particularly, the following results can be formulated:

1. Rapid development management should be sufficiently strong and stable to ensure an optimal level of rapid development.
2. Lack of management leads to a dramatically low intensity of a company's economic development.
3. An organization's potential for rapid development can be maintained more effectively through stable control of the ability than through urgent (transient) control.

4. Insufficient or zero management of rapid development creates a critical point, hitting which either results in budget and economic indicators falling dramatically or necessitates emergency rapid development measures to regain the lost budget and economic status.

2.2 Axiomatic Fundamentals of Management of Rapid Development

The rapid development management process can be described with the use of an axiomatic method. This approach helps to formulate some postulates that make it possible to outline the main principles of the management process. Its distinctive feature is the use of mathematical models describing rapid development in real-time mode and the influence of this management on rapid development indicators.

Mathematical models used herein are based on the theory of dynamic systems. Although the dynamics of individual rapid development indicators is described in a dynamic system, management may fit in these models in a nonlinear fashion. The mathematical models described below are nothing less than a qualitative illustration of axioms (hypotheses) of rapid development management, which are intended for demonstration of these new concepts. On the other hand, mathematical models have provided a more straightforward approach to management of the rapid development process.

2.2.1 The Main Axioms of Rapid Development

Rapid development is viewed as an evaluation criterion for the possibility of forming a new science-intensive market or occupying a substantial market share. A numeric vector, which is termed an indicator of rapid development, can be viewed as a yardstick for rapid development:

$$U = \begin{pmatrix} u_1 \\ u_2 \\ \vdots \\ u_N \end{pmatrix},$$

where u_i, $i = 1, \ldots, N$, are individual indicators of rapid development, which have positive values:

$$u_i \geq 0, \quad i = 1, \ldots, N.$$

Like any other type of vector evaluation, this one poses the problem of partial order with a multitude of indicators of rapid development. Usually, partial ordering implies the use of utility functions. This work does not describe any issues resulting from the use of utility functions; it only describes a very simple variant of partial order:

$$U^1 \leq U^2 \Leftrightarrow u_i^1 \leq u_i^2, \quad i = 1, \ldots, N.$$

In the current model, a higher value of the individual indicator of rapid development reflects a favorable tendency in the rapid development process. Here comes the first axiom of rapid development.

Axiom 1 The yardstick for rapid development is the *vector of individual rapid development indicators*. Individual indicators of rapid development are nonnegative values.

The subject of rapid development management focuses on changes in rapid development in time, which are caused by various factors. Many economic processes are discrete, but individual rapid development indicators should be viewed as uninterrupted, because lack of controlling influences should cause individual rapid development indicators to diminish. This is the consequence of a "diffusion," which is typical of all quality indicators. Thus, the axiom of uninterrupted diffusion can be formulated.

Axiom 2 Individual rapid development indicators are *functions of time*:

$$u_i = u_i(t), \quad t \geq 0.$$

Besides, lack of factors influencing rapid development results in *diminishing individual indicators* of rapid development *in terms of time*:

$$u_i(t') > u_i(t''), \quad t' < t''.$$

Needless to say, a situation when rapid development indicators diminish due to natural diffusion cannot be satisfying; this leads to a fundamental thesis.

Product of Axiom 2. Maintaining the rapid development process at a set level *requires nonstop rapid development management* to produce higher individual indicators.

In itself, the yardstick for rapid development is an indicator, which reflects a given object's (an organization, branch, country, etc.) advantage over competitors. Therefore, to calculate this indicator, not only this object's state should be considered, but competing ones' as well. On the other hand, there is often a need to calculate rapid development indicators with no reference to competitors. For this reason, it is necessary to distinguish between external and internal (subjective and objective) potential for rapid development.

Axiom 3 The potential for rapid development can be *external*, which is calculated based on an object's *relationship* with competitors, and *internal*, which is calculated through *comparison* of its objective indicators.

Axiom 3′ In case with *perfect competition, external potential for rapid development* is important; in case with *imperfect competition,* the focus should shift toward the *internal potential for rapid development.*

2.2.2 Dynamic Mathematical Model

According to Axiom 2, rapid development indicators are temporal functions; therefore, it is possible to build a mathematical model describing the dynamics of these indicators based upon differential equations. Assume $U(t)$ is a vector rapid development indicator detected at time t. Then, when there are no factors influencing competitiveness, this indicator reflects the following linear homogeneous differential equation:

$$\frac{dU(t)}{dt} = B(t)U(t),$$

where $B(t)$ is the square $N \times N$ matrix.

According to Axiom 2, which states that absence of external factors causes individual rapid development factors to diminish, individual matrix values $B(t)$ match the statements:

$$\mathrm{Re}\lambda_i(t) < 0, \quad i = 1, \ldots, N; t \geq 0.$$

According to known theorems of the theory of common differential equations, with the initial statement:

$$U(0) = \begin{pmatrix} u_1(0) \\ u_2(0) \\ \vdots \\ u_N(0) \end{pmatrix}, \quad u_i(0) \geq 0, \quad i = 1, \ldots, N$$

It is

$$\lim_{t \to \infty} u_i(t) = 0, \quad i = 1, \ldots, N.$$

To maintain the potential for rapid development at a certain level, it is necessary to always take steps to manage the rapid development process. This mathematical model expresses it in the following way. Assume it is possible to take the *M* number of measures to influence the *N* number of individual rapid development indicators, which are described by the abovementioned system of differential equations. Therefore, rapid development management is a time-dependent *M*-dimensional vector:

$$G(t) = \begin{pmatrix} g_1(t) \\ g_2(t) \\ \vdots \\ g_M(t) \end{pmatrix}.$$

Within the equation, this vector is linear:

$$\frac{dU(t)}{dt} = B(t)U(t) + C(t)G(t),$$

where $C(t)$ is a $N \times M$ square matrix that is defined by the potential for controlling action. The role of this matrix consists in ensuring influence of the controlling action on the speed of changes in individual indicators of rapid development. The dependence of the potential for management on the matrix's time is reflected by the fact that the influence of the controlling action on individual indicators' behavior can change over time.

Also, this model suggests that apart from natural diffusion and controlling actions, rapid development indicators are influenced by competitors' activities. Supposedly, the influence of external factors on individual rapid development indicators within this model is adaptive. Assume external factors are described by the K-dimensional vector:

$$F(t) = \begin{pmatrix} f_1(t) \\ f_2(t) \\ \vdots \\ f_K(t) \end{pmatrix}.$$

Just like it is with controlling actions, the external factors' vector should use the transitory $N \times K$ matrix $D(t)$, which can be termed the "external factor damping matrix." Its economic role consists in its elements reflecting our object's potential, which is defined by external factors' reaction. Therefore, the final expression of the rapid development management model is:

$$\frac{dU(t)}{dt} = B(t)U(t) + C(t)G(t) + D(t)F(t). \tag{2.1}$$

Any management task suggests that management has a purpose. In mathematical models, this purpose is expressed in a so-called objective functional. In this case, the target function should be set in solutions $J(U)$ to Eq. (2.1).

Important features of management are limitations to management and the "price" of the choice of management. Mathematically, it is:

$$G(t) \in W, \quad t \in [0, T],$$

i.e., the choice of a management is limited to multitude W. In a number of situations, the multitude of possible variants of management W may depend both on the timepoint and rapid development indicators and previously selected management variants. Assume that the chosen variant of management has its "price" $I(G)$.

Now, we can also formulate a mathematical task for rapid development management:

$$(J(U) - I(G)) \to \max_{G(t) \in W}. \tag{2.2}$$

This mathematical model does not focus on the task of optimal management alone, because the dynamic Eq. (2.1) also comprises external factors, which are formed by competitors and, generally, work against the competitiveness indicator.

2.2.3 *The Hysteresis Type Rapid Development Model*

As noted above, the main focus of rapid development is maintaining the rapid development indicator at a certain level. Given the diffusion of rapid development indicators predicated by matrix $B(t)$, maintaining rapid development at a due level requires nonstop management.

According to extreme task Eq. (2.2), optimal management is a complex mathematical problem. Many difficulties occur due factors that are purely mathematical and are not always decisive in solving economic problems. Economic methods of rapid development management often show a limited potential during implementation. For this reason, this section will study a rapid development management method in keeping with this model; the method is based on hysteresis management. This approach has a clear economic expression and helps to solve the problem of maintaining the potential for rapid development at a set level.

The idea of managing dynamic processes through hysteresis lies in the fact that at each fixed timepoint along the vector an integral scalar function is calculated from individual indicators $U(t_0)$ of rapid development:

$$IU : R^n \to R,$$

The function defines the extent of rapid development at a period of time. Assume that management of rapid development is aimed at building a potential for rapid development, i.e., a state of vector U, which meets the following condition:

$$IU(U(t)) \geq U^*_{\min},$$

where $U^*_{\min}$ is a minimal fixed measure of rapid development. Assume $U^*_{\max}$ is an extent of rapid development, with which $U^*_{\max} > U^*_{\min}$. Then hysteresis-based management is organized in the following way:

$$H(U(t)) = \begin{cases} 0, & IU(U(T)) \geq U^*_{\max} \\ G(t), & IU(U(t)) < U^*_{\min} \end{cases}.$$

Here $G(t)$ denotes a management vector, which aims for a higher degree of rapid development. Thus, hysteresis-based management activates measures aimed at increasing the degree of rapid development when the rapid development indicator drops below $U^*_{\min}$, and it continues until the integral rapid development indicator exceeds $U^*_{\max}$. The resulting "fork" between $U^*_{\min}$ and $U^*_{\max}$ creates the renowned "hysteresis loop", which must be used to reduce the sliding management mode.

It can be demonstrated that if matrix $C(t)$ allows rapid development management activities to raise the degree of rapid development up to the set value, hysteresis-based management will provide for an acceptable management mode, which solves the task in the most efficient way.

Now, a rapid development axiom can be formulated.

Axiom 4 Steps to increase the potential for rapid development *should be carried out* when the extent of rapid development *drops below the set* minimal value and *up until* the indicator *exceeds the set maximal* value.

Hysteresis-based rapid development management has a clear economic purpose. Major economic activities, which are aimed at increasing the degree of rapid development, should continue for a period long enough to achieve the required effect. The hysteresis principle helps to tailor this standard to possible rapid development management patterns.

2.2.4 Management of Rapid Development in Imperfect Competition

The models described above apply to perfect competition. These models imply management of competitiveness through steps that are expressed through addend $C(t)G(t)$. In case of imperfect competition, this management strategy may fail to solve set rapid development tasks, or it will prove to be economically unjustified. Imperfect competition requires innovative management methods, as it is stated in Axiom 5.

Axiom 5 In case of *imperfect competition,* management of rapid development should be based on an *innovative approach.*

Assume that matrixes in Eq. (2.1) are production matrixes: *B*(*t*), *C*(*t*), and *D*(*t*). Using the product matrix terminology, it is possible to formulate an axiom describing the management process that is based on an innovative approach.

Axiom 6 *Innovation-based* management should *alter production matrixes.*

Axiom 6 states the following differential equation, which reflects an opportunity to manage rapid development through modifying production matrixes:

$$\frac{dU(t)}{dt} = B^{G(t)}(t)U(t) + C^{G(t)}(t)G(t) + D^{G(t)}(t)F(t) \tag{2.3}$$

Management by way of modifying matrix *B*(*t*) should result in diminished real parts of the matrix's eigenvalues, which is going to reflect a lower diffusion of rapid development indicators. A slight decrease in the diffusion indicators can make rapid development management much more effective.

A change in matrix *B*(*t*) should significantly increase competitiveness, because this matrix reflects the technological potential of management of individual competitiveness indicators. According to the linear theory of differential equations, positive summands in the right-hand part of Eq. (2.3) exert an exponential influence on the values of respective vector *U* components. Thus, even a slight change in matrix *C*(*t*) coefficients has an exponential influence on rapid development indicators.

Finally, matrix *D*(*t*) is responsible for countering negative factors that could influence the rapid development process. Not infrequently, the lowering of the potential for rapid development occurs due to these negative factors, and it is important for an organization to develop an ability to reduce these factors' influence.

As noted above, innovation-based management can result in rapid development providing there is a sufficient supply of resources through the accumulated experience. Thus, it is possible to formulate the following rapid development axiom based on the innovative approach.

Axiom 7 *Innovation-based* management of rapid development, i.e., with modified production matrixes, results in *nonlinear* influence of investment volumes on rapid development indicators.

2.2.5 General Management of Rapid Development in Imperfect Competition

The above paragraphs describe the rapid development management process in imperfect competition applicable to a single object. Meanwhile, there is a need to maintain this process within a whole system. In competitive systems, objects strive for a continuous growth of competitiveness and the potential for rapid development. However, it is not uncommon for a perfect competitive environment to be replaced with an imperfect one, which slows down the growth of objects' ability to maintain rapid development and reduces the effectiveness of the whole system.

In this situation, management of rapid development is not a problem of separate competing object, but of specific state regulation of a competitive system. The goal of this regulation is creating a perfect competitive environment, which should eventually raise the competitiveness of separate objects, their potential for rapid development, and ensure a higher economic effectiveness of the whole system.

Within the dynamic model's framework, imperfect competition has the following mathematical signs:

Diffusion matrix $B(t)$—Decreasing matrix modulus eigenvalues real parts—results in the "conservation" of rapid development indicators.

Matrix $C(t)$—Decreasing modulus of matrix elements, and it causes a drop in effectiveness of rapid development management.

Matrix $D(t)$—Increasing modulus of matrix elements; it leads to a more pronounced negative influence on rapid development indicators due to external factors.

The signs of imperfect competition described above reduce indicators of rapid development in all objects, and rapid development management tools lose effectiveness.

Thus, it is possible to formulate an axiom of general rapid development management in imperfect competitive markets.

Axiom 8 *Imperfect competitive markets* require *management of competitive environments* related to increasing matrix modulus of eigenvalues real parts $B(t)$, an increase in matrix elements $C(t)$ and a decrease in matrix elements $D(t)$. This *management mode will result in a larger competitive environment, higher competitiveness, and a potential for rapid development for all participants.*

2.2.6 A Relationship Between Rapid Development and Competence Management by Organizations

There are some common patterns relating the evolution of one-of-a-kind key competences to the formation of new consumer markets and manufacturers' rapid development status. These patterns emerge due to a transformation of the modern economy into an economy of knowledge, when organizations' competences and innovative technologies become the main economic resource. This economic pattern contributes to creation of a large number of new goods and services, rapidly growing new markets and manufacturers' rapid development. In turn, the growth of markets and the necessity to follow the path of rapid development creates a demand for higher competences and use of innovative technologies. The resulting spiraling progress has objective conditions.

Axiom 9 Creation of unique competences increases high-tech organizations' resource potential, which contributes to a leaping growth of one-of-a-kind innovative technologies that are used for creation of brand new rapid development products.

The market-growth-market-reduction cycle in modern high-tech economy contributes to (accumulates) a sufficient stock of competences (intellectual wealth) that preconditions a new market and competence growth cycle. Qualitative development of advanced technologies results in creation and leaping growth of new consumer market, which, in turn, boosts investment into key competences and innovative technologies. This boosts a dual-track spiraling development of new technologies, acquisition of new key competences, and stable economic growth of manufacturers. Also, new consumer markets emerge and evolve and manufacturers attain the rapid development mode.

As it is known, new markets emerge not only due to increasing consumer demand for new types of goods; they are fueled by the supply of new consumer goods. During the industrial era, there would be years and decades between the inception of a new item and creating a stable market, while now this period lasts only a few months.

Axiom 10 Emergence of rapid development products increases the demand for new goods. Economic growth stimulates further demand for state-of-the-art technologies. This law demonstrates a kind of spiraling competences-resources-products-needs-competences cycle.

Competition in the sphere of advanced technologies and unique competences is driven by mechanisms that are different from those of the traditional industry. These mechanisms are much quicker to respond to changes in competitors and thus make organizations want to quickly and effectively modify their behavior. These changes also apply to consumers. While in the nineteenth and twentieth centuries many goods were advertised as durable or even eternal (for example, watches and clocks), today even high-value goods are short lived (for example, a flagship smartphone stops being flagship after about a year). Those durable and eternal goods are giving way to cutting-edge ones, which can be achieved only through constant rotation of goods. This approach generates concepts, such as "rapid development goods," and their constantly updated versions. In the twentieth century, it was common for us to wear our grandfathers' watches. This does not apply to today's smartphones and laptops. These changes in consumer behavior are closely related to changes in the high-tech industry.

The axioms set milestones for developing a theory of rapid development management. The use of a mathematical mechanism to develop and demonstrate rapid development management axioms has helped to clearly and concisely formulate statements, which can be used for economic research of this phenomenon and for building mathematical models of rapid development management.

2.3 The Role of Competences in Management of Organizations' Rapid Development

As a part of research of the concept of rapid development, the concept should be completed and extended. Preparation for rapid development can be characterized as an intentional collection, systematization, and analysis of specific information about an organization's internal potential, its competences and the external environment, with the purpose of finding and rationally using its competitive advantages and minimization of risk factors. Such commercial risks as rapidly growing competition in the high-tech market, loss of a market share, and changes in consumer needs are the strongest risk factors in an unstable economy.

Rapid development of a high-tech organization, being crucial for dynamic development, predicates the necessity of creating an ultimately new advanced system for managing business processes: product design and development, preproduction, production, and sales. This system should be based on extensive use of artificial intelligence and machine learning, virtual economy, and the Digital Earth concept.

What we understand as rapid development is control and management of scientific and research, design and experimental, innovative, manufacturing, organizational, and other processes taking place within an organization, which ensures that the organization builds a potential necessary for creating a brand new product boasting consumer properties that enable it to win a good share of an existing market or create a new science-intensive product market.

The law of rapid development described above states that any product that is being designed and manufactured should have the usability (value), which should stimulate public needs contributing to the formation of new markets and helping manufacturers to achieve stable economic development.

What makes a foundation for stable economic development of a manufacturer are effectively organized practical design of complex science-intensive products, which are usually performed by several organizations and various design and technology divisions within those. These companies' ability to follow individual technical specifications and create an end product with required technical characteristics, proper materials, and components, will define the general result of high-tech project implementation. This is about achieving target technical and economic parameters, which ensure maximal customer- and market satisfaction or form a new product sales market.

It is understood that it is not before a product has been manufactured that it can enter a market. This does not pose a necessity of creating new production assets of modernizing existing ones, although these are integral to a manufacturer's stable economic growth. The head of an organization should choose the direction while relying on available competences, budget resources, product delivery deadline, or entering the sales market, with consideration of possible risks and the likelihood of introduction of a similar and more technically advanced and usable product by competitors ahead of time.

Therefore, a manufacturer's economic development implies mutual industrial and technical cooperation with partners and allies, transfer of technologies, exchange of one-of-a-kind competences, and measures aimed at building up the organization's innovative potential.

When improving the organizational structure, it is necessary to ensure a complex integration of all divisions' competences, which can help to produce a powerful synergic effect. This effect will facilitate active incorporation and use of strategic innovations that provide for high added value, usability, effective competition in internal and external markets, tailoring of the corporate management strategy to rapidly changing customers' needs and market environment, and ensuring stable economic growth within the organization. Effective use and development of unique competences can bring a global advantage and give an opportunity to create a new market of innovative products and technologies and, based on these, give rise to new market segments.

The key factor determining an organization's potential for rapid development is its ability to achieve new levels of branch presentation. This demonstrates the organization's ability (competence) to develop and introduce advanced technologies for different branches. As part of competence profile analysis, it is necessary to analyze competences at three branch positioning levels: sector-specific positioning (focuses on competitiveness within a single area), broad inter-branch positioning (focusing on competitiveness in several areas), and interdisciplinary positioning (competitiveness in a large number of branches). An organization's current branch positioning level can be determined by its unique technological competences. The life cycle of a company's competences is shown in Fig. 2.4.

Evidently, the life cycle of a unique competence starts from its inception, and it originates due to a global challenge or prognosis of scientific and technological development, which is expressed in issues and goals having specific technical requirements. Should a product reach a global dominance thanks to its technical characteristics, the competent team can become a foundation for a global competent center capable of developing highly competitive products. These products can generate new markets. Practice shows that after a while competitors do embrace a given competence, which will bring this competence from unique down to a center of global competitiveness. Next, a large group of experts and organizations will embrace the competence and turn it into a broad-spectrum competence, which, depending on the situation, either carries on or gives way to another one.

Thus, the basis for creation of new unique competences is formed by current problems and long-term objectives, which are solved with the help of corporate resources by science and technology experts. They have respective technological competences, and that can be used as part of R&D efforts, or a company can attract competences from outside. In turn, the fundamental basis for setting new goals consists of global challenges. They define management of the rapid development process, which is their main mission, as getting an organization ready to counter these challenges.

One of the biggest problems relating to strategic management of rapid development of a high-tech organization is the serious time gap between getting fundamental

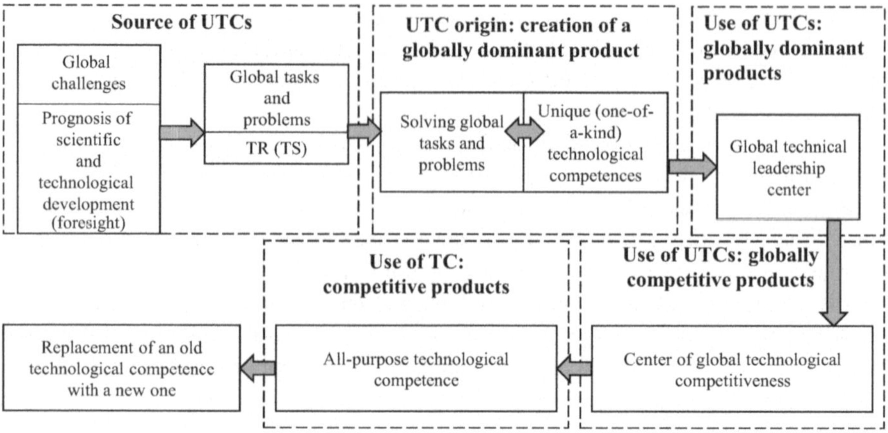

Fig. 2.4 The life cycle of unique technological competences

scientific research results (building a unique technological competence), creating a science-intensive prototype and mass production. Therefore, today's rapid development requires investment not only in production, but, first of all, in creation of new unique competences, which create a basis for developing advanced products, as well as in innovative strategies of marketing the new unique products. These processes are typical of both micro- and macroeconomic environments; therefore, they reflect a shift of the competition toward unique key competences of numerous sectors of economy.

This effect is observed not only in production, but also in education, because the demand for competences and high level of education generates a supply of new education services. The system produces another effect: competences evolve and produce new technologies, and this creates a breeding ground form new consumer markets. There is a model of interaction between the development of competences and consumer markets generated by rapid development. The link between the level of unique competences and emergence of new needs and markets triggers a spiraling cycle, which fuels rapid development of businesses and innovative economic growth. The scheme of this process is shown in Fig. 2.5.

This scheme demonstrates that high-tech production mechanisms create new advanced technologies resulting from the growth of unique competences within science-intensive organizations' R&D divisions. Laws of rapid development suggest that new reaches in the fundamental science produce applied projects, which form the foundation for new innovative technologies. The appearing innovative technologies are materialized in new goods and services. Commercialization of innovative technologies forces high-tech organizations to create or stimulate the creation of new markets, so that they can sell their goods (Fig. 2.6). This requires using various advertising and marketing mechanisms.

Once new consumer markets of innovative products are created, respective industries begin to boom. Attracting investments in organizations stimulates both

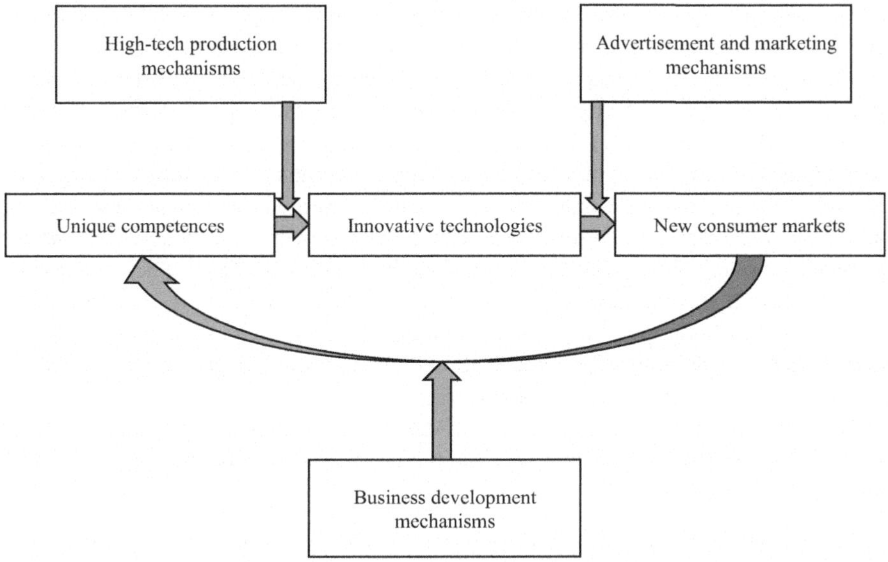

Fig. 2.5 Relationship between unique competences and appearance of new consumer markets

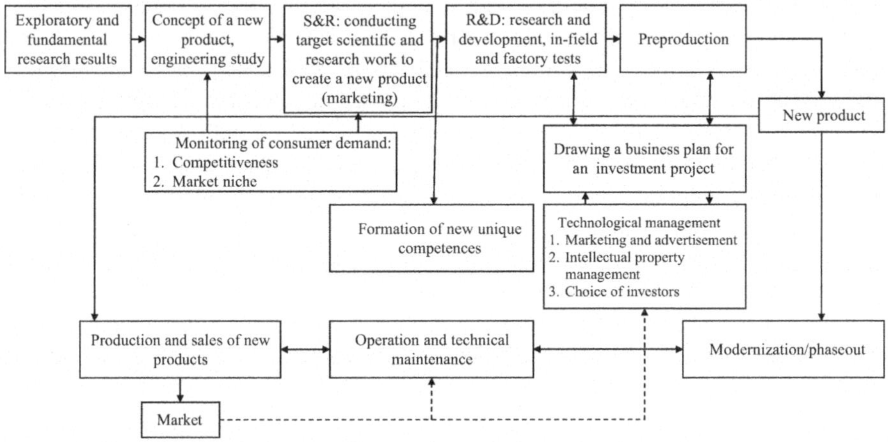

Fig. 2.6 Commercialization of technologies and promotional support of innovations

production processes and further investments in organizations' R&D efforts. Consequently, there is a demand for key competences for new branches. Definitely, an increase in this demand will raise the supply of key competences. For instance, the appearance of new mobile application markets has added new directions to lots of education programs for training specialists. Now, this segment is attracting huge investments, which produces lot of new key competences. These, in turn, contribute to the production of new types of goods and, consequently, new consumer markets and rapid development of manufacturers.

The competence-innovation-market model demonstrates that high-level competences broaden consumer markets through production of new sorts of goods, which satisfy newly emerging needs. This helps businesses achieve rapid development. Meanwhile, market mechanisms hugely increase investment in breakthrough technologies, which are used for production of such goods. As stated by laws of innovative economy, these investments fuel the development of competences within this field. The explosive growth continues for but a short while as consumer markets boom, because there are marginal laws, which start working once consumer markets become saturated. The saturation shows up at a point of time, when consumer markets slow down dramatically, and once innovative technologies become ordinary. In this situation, to maintain rapid development, an organization should incorporate new technologies, which should precondition further growth of consumer markets and therefore competences.

There is a mathematical model, which describes the process of managing competences that precondition a company's rapid development. The most common key business competences are:

- Technological competence A
- Production competence B
- Scientific-research competence C

With the help of a dynamic system, it is possible to build a model simulating management of competences, which are needed to rapid development. Because the model is dynamic, the time indicator needs to be introduced. Most economical models use discrete time, because many processes develop on a long-term basis. However, adequate results can be conveniently obtained through the use of continuous time, because this method provides more compact mathematical models. In any event, there is always an opportunity to generate a discrete-time model with the help of a continuous-time model.

This model focuses on continuous-time processes (t), yet with a limited interval of time, which will suffice for this model.

Values $A(t)$, $B(t)$, and $C(t)$ describe a dynamic competence management model. The system of differential equations describing a dynamic system is expressed as follows:

$$\begin{cases} \dot{A}(t) = F(A, B, C); \\ \dot{B}(t) = F(A, B, C); \\ \dot{C}(t) = F(A, B, C). \end{cases}$$

To build a formal basis for economic and mathematical models, main numerical characteristics should be chosen to express respective economic indicators. We are going to use nondimensional values, because they reflect economic indicators effectively.

Competence indicators may be economically meaningless, while comparison of these values does have an economic meaning. For example, it is possible to state that the competence level observed at time t_1 is higher (or lower) than the competence level observed at time t_2.

The dynamics of values that describe competences can be expressed as a system of linear differential equations:

$$\begin{cases} \dfrac{dA}{dt} = \alpha_0 A(t) + \alpha_1 C(t); \\ \dfrac{dB}{dt} = \beta_0 B(t) + \beta_1 A(t); \\ \dfrac{dC}{dt} = \gamma_0 C(t) + \gamma_1 B(t). \end{cases}$$

In these equations, each dynamic variable has a coefficient to reflect a visible gradual decline of all indicators resulting from general economic, scientific, and technical progress, without any competence management activities.

In this model, the coefficient of technological competences increases, as it is influenced by the coefficient of scientific and research competences; likewise, the coefficient of industrial competences—by technological competences; the coefficient of scientific and research competences—by industrial competences. This competence management scenario enables a complex development of all competences, improvement of indicators describing them, and therefore maintaining an organization's rapid development status.

Outer influences, which describe competence management, as well as the influence of random factors can be described with the help of varying coefficients. These coefficients describe qualitative changes in competence indicators, which result from competence management activities. With reference to controlling influences ε, the dynamic system can be described in the following way:

$$\begin{cases} \dfrac{dA}{dt} = \alpha_0 A(t) + \alpha_1 C(t) + \varepsilon_A; \\ \dfrac{dB}{dt} = \beta_0 B(t) + \beta_1 A(t) + \varepsilon_B; \\ \dfrac{dC}{dt} = \gamma_0 C(t) + \gamma_1 B(t) + \varepsilon_C. \end{cases}$$

Therefore, with original values of the competence dynamics model available, it is possible to simulate the competence management process. The modeling can help to select optimal competence management scenarios, which predicate rapid development of an organization. To carry out strategic planning of rapid development, an organization can build projections to predict changes in these indicators. Should real indicators greatly deviate from projections, competence management has been ineffective and strategic decisions will need to be made.

Research of global challenges, new business, and public needs, is a management instrument, which can reduce risks when there is uncertainty. The research is always aimed at solving specific issues and therefore they are generally regarded as part of applied research work, which is intended for use in practice.

The fundamental difference of such research (as a line of activity) from market research procedures is that the process implies a multi-aspect study of an internal corporate potential. The research becomes a broad category, which is needed for outlining a strategy for rapid development management.

The main criterion of studying expected business and public needs is variability. It is extremely important when it comes to interpreting the main objective of a study, which is informational backup for improving the competitive environment within an organization and providing for its science-intensive product's competitive advantages in markets. Such research is based on detailed analysis of the global information space. Principal schemes of analytical work focusing on the growing public needs, which are based on data obtained from the global information space, if generalized, may be different (Figs. 2.7 and 2.8).

After that, all information about prospective needs is collected, analyzed, and distributed within the informational system (Fig. 2.9).

The process of studying prospective needs of the public and businesses, as applied to science-intensive products, has its specifics:

1. The process of studying foreign markets is hardly different from the process of studying national markets. However, in international marketing, analysis of market chances, and risks requires more careful collection of data on international market potentials and external factors, such as import limitations, international laws, etc.
2. Practice shows that most desk studies of foreign markets relying on second-hand information can prove to be enough for reaching the goals of studying foreign science-intensive product markets. Field study of foreign markets is one of the costliest market research methods; however, it provides a maximal amount of fresh and trustworthy data.
3. The length of most rapid development products' life cycles, tough competition and anti-import policies exercised by most financially reliable markets necessitate studying of not only potential sales markets, but also know-how markets, biggest scientific and technological reaches, fundamental scientific stockpiles in respective fields—a full-scale marketing research of the innovation market.

Now that global economic tendencies are the basis for planning a rapid development strategy, it is necessary to focus on the development of economic relationships setting up thanks to the influence of the progressing globalization. Today, rapid growth of the global market of goods and services, capital, and manpower, is accompanied by internationalization of high-tech production, the breaking down of administrative barriers that impede the transfer of technologies, and an increasingly free exchange of capital between civilized and developing countries.

Within the global economy forming right now, rapid mastery of scientific and technical reaches, creation of science-intensive goods and technologies that enable

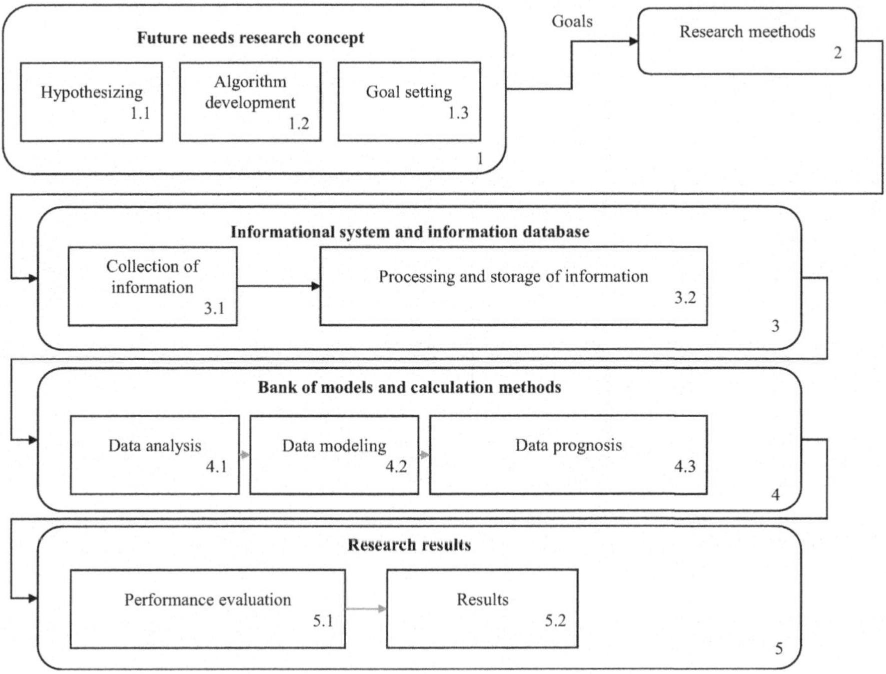

Fig. 2.7 Generalized future needs research method

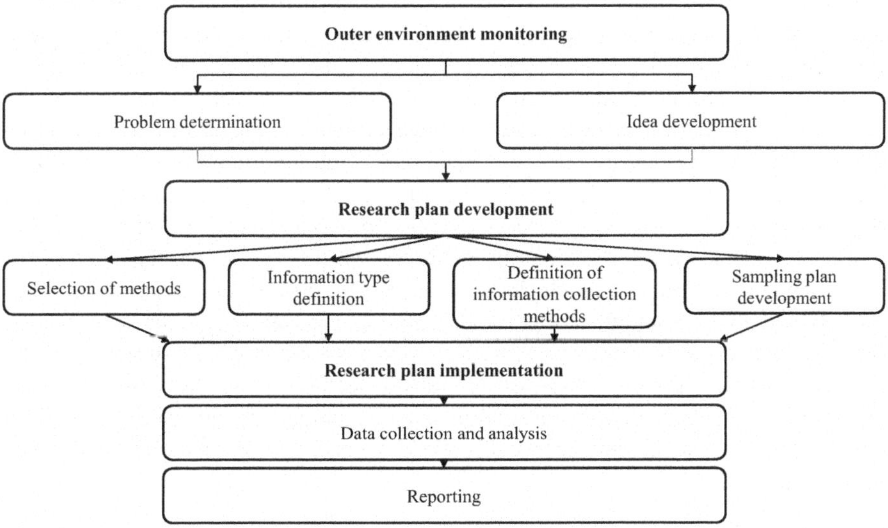

Fig. 2.8 Generalized future need performance research method

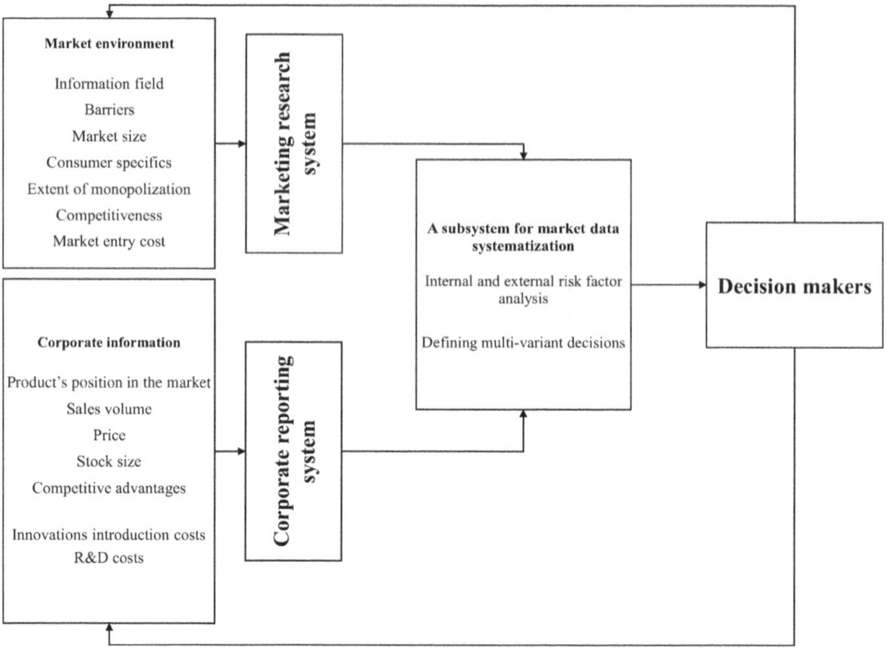

Fig. 2.9 System of information on prospective needs

their production, their entry into the global market and active joining of international integration processes, make up a strategic economic growth model for most countries. This helps them to retain the right degree of economic and technological independence. The necessity to outline a rational rapid development strategy for the national production sector, which should facilitate a transition to an innovative economy, is also proved by the fact that during the economic crisis, innovative economies were better off. Thanks to stockpiles made during less stressful periods, they are recovering quickly. Effective entry into the system of global market relationships based on innovations in technological processes and management of science-intensive industries, which have access to promising international sales markets and to integrating national and regional innovative systems, is the most important aspect of the competitiveness of Russia's economy. The direction of the national economy directly depends on how effectively branches use advanced technologies, methods that ensure competitive advantages for goods and services, and organizations producing them; this gives a general idea of the county's economic potential.

2.4 Strategic Rapid Development Approaches and Instruments for Science-Intensive Companies and Branches

Globalization eliminates outdated economic mechanisms and technological solutions, and only advanced business structures, thanks to previously gained competitive advantages, find new economic and technological niches, create breeding ground for new markets, and innovative goods. The strategy and tactics of increasing competitiveness, being a foundation for rapid development, needs to be supported by an arsenal of economic rapid development process assessment tools.

A rapid development strategy, which focuses on raising a company's competitiveness, is a controlling action performed to balance external and internal factors of the development of this controllable system. This ensures gradual growth and more effective development of an organization along with detection and prevention of negative consequences of economic activity in the foreseeable future. The ultimate goal of rapid development is formation of new markets or gaining a large share of existing markets.

Thus, strategic rapid development of high-tech branches should rely on tendencies observed in economic relationships, which result from globalization, and it should rely on nonstop introduction of advanced solutions at all production and production management stages. The direction of a national economy largely depends on the development of high-tech branches. Competitive advantages of goods and services and companies producing them and the economic potential of a country depend on how innovative branches evolve in terms of effective use of innovative technologies.

Here is the standard, which the rapid development strategy for the high-tech sector of Russian economy, which is looking to compete on a global market, should meet:

1. A rational rapid development strategy for high-tech branches should be based on effective introduction of breakthrough innovations at all phases of launching the production of a science-intensive product.
2. The rapid development strategy for high-tech branches, which is aimed at boosting the competitiveness of the branch and companies operating within it, must be closely tied with and embrace quick response to possible changes in the local or global regulatory environment and the current state policy in the sphere of innovation. It should not run counter to any officially accepted programs.
3. The strategy must be fully focused on more effective and quicker introduction of innovative solutions. Regular qualitative and quantitative assessment of this effectiveness, being part of the strategy, is mandatory at all management levels. Because the resource invested/end result ratio is the main criterion of effectiveness, raising effectiveness implies both reaching higher end results and reducing resource investment without detriment to target effect, or combining these activities. Various methods can be used as instruments, and various functional fields—as application

objects. The rapid development model for high-tech branches is structured as shown in Fig. 2.10.

There are four aspects of successful innovation:

- Effective corporate management with the use of modern information and innovation technologies.
- Effective development of science-intensive products with characteristics, which ensure competitive advantages in sales markets.
- Effective production technology for science-intensive goods, when all must-have goods and services are produced at the lowest possible costs.
- Effective marketing part, i.e., breakthrough innovations in distribution of science-intensive products.

A rapid development strategy of an organization or branch, which comprises a complex and integrated solution to all abovementioned problems, will also have a synergic effect that will contribute to stable innovative development of an economic entity.

4. A rational rapid development strategy, especially one intended for distribution of products in external markets, should be outlined with account of not only global economic tendencies (globalization, changes in the structure and direction of financial flows, industrial outsourcing, etc.), but also with account for a possibility of unexpected risks. Therefore, it should include a well thought-out element of strategic management of rapid development in case of a sudden crisis, unstable economy, inflation, etc.; in other words, it should have a full-scale subsystem of strategic alternatives.

The most effective mechanisms of rapid development management for a manufacturer, including those needed for its stable economic growth, can include proactive mechanisms, particularly state regulation mechanisms, which should exert a preventive influence on factors that can undermine economic development. The main purpose of these proactive mechanisms is the development of spheres and sectors that raise general economic stability and prevent negative influences of globalization (like the current global financial and economic crisis).

These tasks can be solved through the use of two types of proactive mechanisms—strategic and operative. Strategic mechanisms should focus on diversification and modernization of the Russian economy to make it less resource dependent and therefore more stable in its development.

Most operative mechanisms are economic rather than organizational. At the current phase of global economy, given its specifics explained by the well-developed financial infrastructure, financial mechanisms deserve a special mention. They can be termed as priority mechanisms, also because in this globalized world, it is unstable and highly volatile financial markets that pose the biggest danger for stability. Proactive operative mechanisms must focus on countering these threats. These mechanisms may include hedging unfavorable price tendencies with the help of binary options, insurance, and reinsurance of export, optimization of currency relations aimed at minimization of losses sustained by Russian exporters and borrowers due to currency fluctuations.

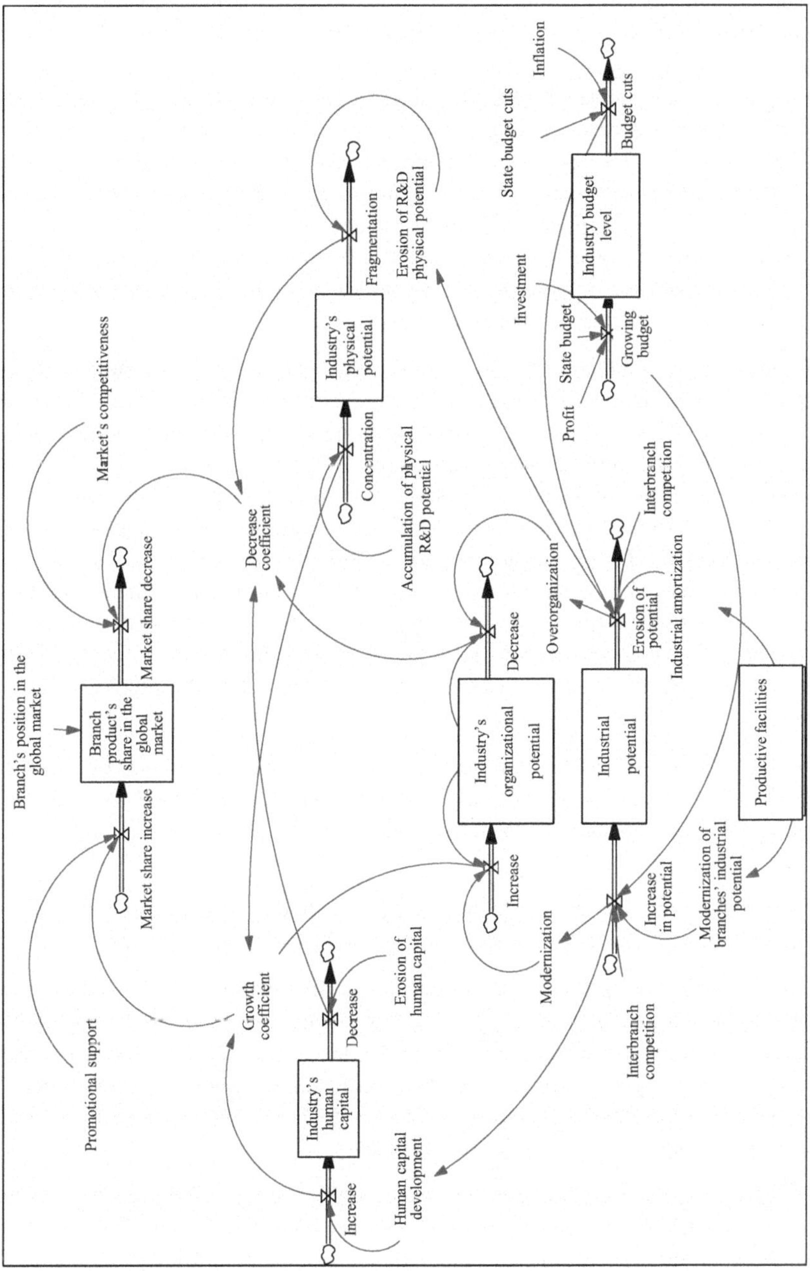

Fig. 2.10 A rapid development management model used for a high-tech branch

The use of advanced finance technologies aimed at reducing risks, should also be practiced by governmental structures—from the Ministry of Finance, as it issues bonds for foreign markets (a risk of exchange rate fluctuations) to financial and investment structures, such as the Investment Fund (risks resulting from investing in foreign assets) and national industrial organizations.

The concept of corporate rapid development management can be presented as a tree of goals, based on which it is possible to draft a corporate rapid development management system. The tree of goals is built based on tasks, which the corporate rapid development management system is going to solve. A draft of the tree of goals of corporate rapid development management is shown in Fig. 2.11.

This tree clarifies the priorities, which should be included in the work statement. Based on these priorities, it is possible to formulate the main principles in designing a corporate rapid development management system:

- Designing of an organizational system should be based on and start with the modeling of organizational processes. These models are referred to as normative, and they define basic principles of the system's setup and functioning, which should be followed in the best possible way.
- When designing an organizational system, it is necessary to rely on the particularities and specifics of the main and auxiliary production, as well as technological backup and all factors influencing an organization's work.
- An organizational system should be flexible and self-adjusted so that it could use every way to ensure a quick approach of actual indicators to planned ones.
- An organizational system should be structured with due regard to management levels and layers. The deeper down it goes, the more detailed a picture it should present.

These systems have specific missions and interconnections, which define their effectiveness. Figure 2.12 demonstrates the main goals that are solved when designing a rapid development management system.

Task 1 Setting the goals of building and improving a corporate rapid development management system. The process should start with defining the purpose of its creation and embrace the development strategy and main characteristics.

Task 2 Creating a description of the system. It should contain not only a description of a particular system and its parameters, but also a setup diagram demonstrating this system's relationship with subsystems within it, which are to fulfill the system's task. Therefore, the description will also contain information about the hierarchy of systems, which could ensure reaching of the set goal (or goals in case the system is multipurpose), and a rapid development strategy.

Task 3 Revealing external and internal accountable factors affecting the system. The task implies not only admissions and boundary conditions used for the system, but also suggestions concerning the outer environment; particularly, scenarios or explanations of outer influences, which the system may experience. If more than one scenario has to be considered, they should be listed.

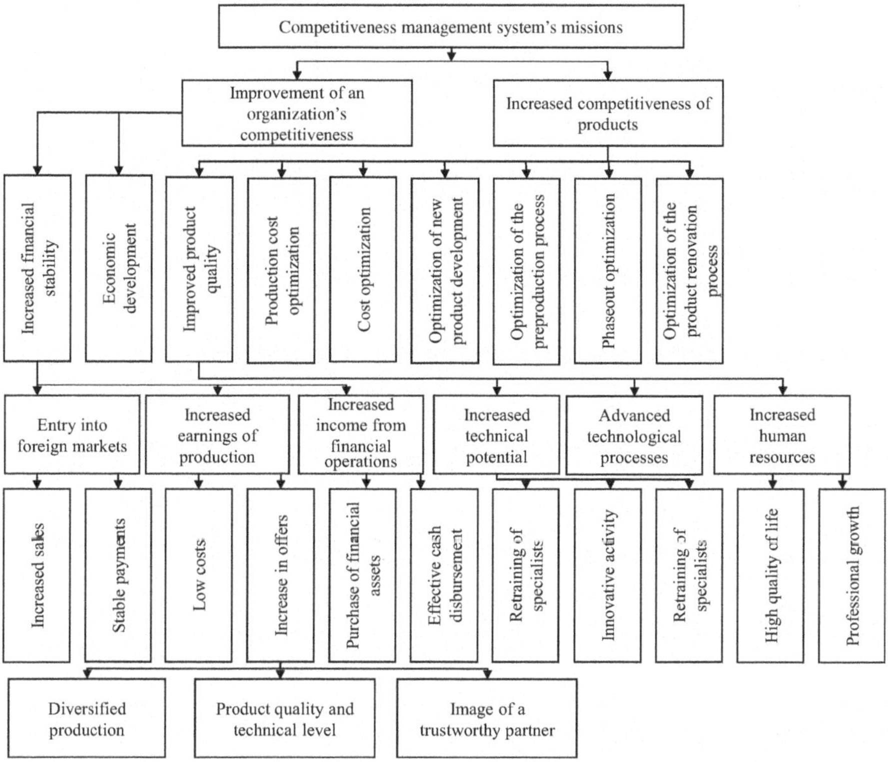

Fig. 2.11 The tree of goals for creating a corporate rapid development management system

Task 4 Operative description. Based on the set hierarchy and main systems' main characteristics, a mathematical model will be built. It is used as a basis for assessment of the effectiveness of decisions made during previous phases of work, which embraces various possible events, the hierarchy of events and different indicators of the system and characteristics of the outer environment, which are related to each model.

Task 5 Building a mathematical model. It is a two-stage task. Stage 1 is development of a system building mode and an operation model. Stage 2 is assessment of quantitative ratios set by some mathematical and/or logical equations, which link different factors contained in each of the operation submodels.

To clearly define the parametrical relationships between factors and data that is needed for obtaining quantitative results, the analyzer should select a scenario describing one of these systems and study the interaction between the system and the outer environment. Then he/she should closely analyze each event that is part of general operations reflected in the operation model that is built beforehand for step-by-step assessment of specific factors and relationships between them. It would be more effective for the analyzer to build so called "test hypotheses," which are

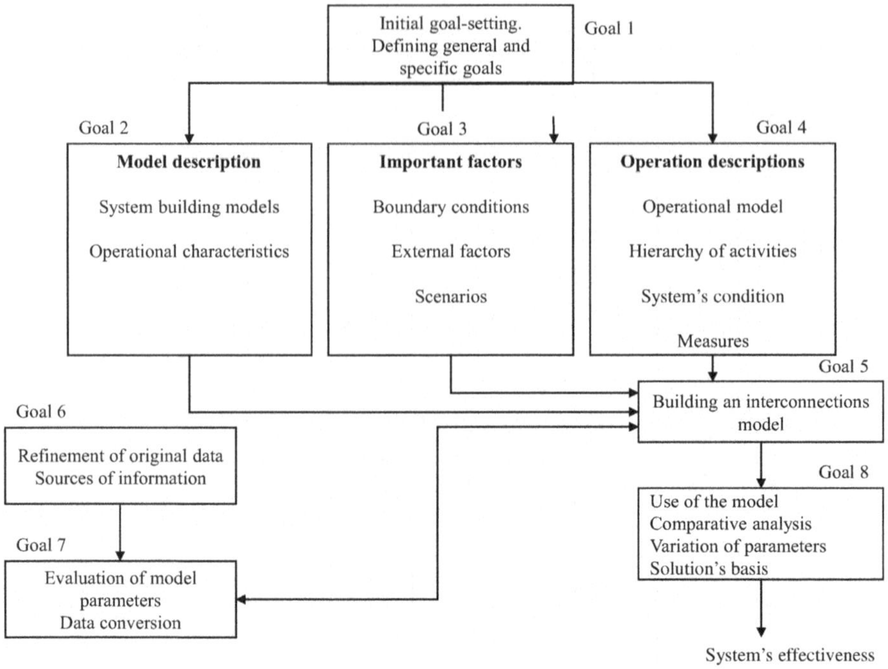

Fig. 2.12 Interconnections between different goals observed at the stage of designing for a rapid development system

defined by the interrelation between the events, which is expressed in a parametrical form. This interrelation relies on the analyzer's awareness of the type of the operation, which is the result of his/her carefulness and experience.

Task 6 Refinement of original data based on the results of modeling. This task defines events, which provide an opportunity to correct original data (for example, to increase funding, upgrade equipment or computing devices), based on the completed mathematical analysis of factors influencing the system.

Task 7 Evaluation of a model's parameters. When using existing data or information, or that obtained during work, with account of limited time and resources, qualitative evaluation of respective parameters is carried out. The evaluation process may imply use of known equations, which connect respective factors, statistics methods, or subjective estimates that are based on personal opinions. The evaluation method being used shows the greatness of a possible input data error. Once the task is solved, it is necessary to specify a method that can be used to extrapolate the data when evaluating the model's parameters.

Task 8 Using the model. The process implies unification (joining) of submodels that are parts of the operation model, and obtaining high-quality output data with set input data. Also, it implies the use of various economical and mathematical methods, including general simulation. The selected effective method is a function expressing the quantity and integrity of available data, integrity of expected results, and amount of resources provided for the analysis.

Solving these tasks makes it possible to build a corporate rapid development management system, which provides fairly adequate information about the organization and its competitiveness to persons, who are responsible for making decisions.

It is possible to use medium estimates of effectiveness as a reference model of an organization's work while referring to a group of organizations that have proved successful within a particular segment. Obviously, different types of organizations use different types of models, so the modeling block (Fig. 2.13) should contain a basic set, in which each organization has its own reference model.

A potential for self-training is a must-have property of each block. After each diagnostic examination of an organization, apart from the correction of the reference model, the system, which evaluates the significance of criteria of effectiveness, is modified.

The following level of rapid development management is the level of a branch, which unifies manufacturers.

To ensure successful completion of tasks in the sphere of rapid development, it needs to use a rational strategy of improving a science-intensive product's AND branch's competitiveness. This strategy should include development of mechanisms that ensure abidance by activity plans and strategy implementation programs, as well as evaluation of its effectiveness:

1. A mechanism of funding innovative projects aimed at increasing science-intensive branches' competitiveness.
2. A mechanism of selection of powerhouse branches for special-purpose financing. In each particular case, both official and nonofficial financial resources can be used. This section should provide a detailed review of mechanisms of funding and sources of money, as it is specified by the new priority funding model.
3. A mechanism of preparing the public for a transition to competitive and science-intensive production, which includes:
   - Assessment of manpower to be used in science-intensive branches.
   - Current level of education and science.
   - Training of the intellectual elite.
4. A mechanism of evaluation of branches' innovative potential, which includes:
   - Evaluation of competitiveness of leading science-intensive branches and organizations based on indicators formulated as required by the global standard.
   - Evaluation of scientific, technical, and innovative base.
   - Recommendations concerning creation of a unified product standardization and certification base.
5. A mechanism of evaluating the degree and quality of the effect produced by the institutional, public, informational, telecommunications, and transport

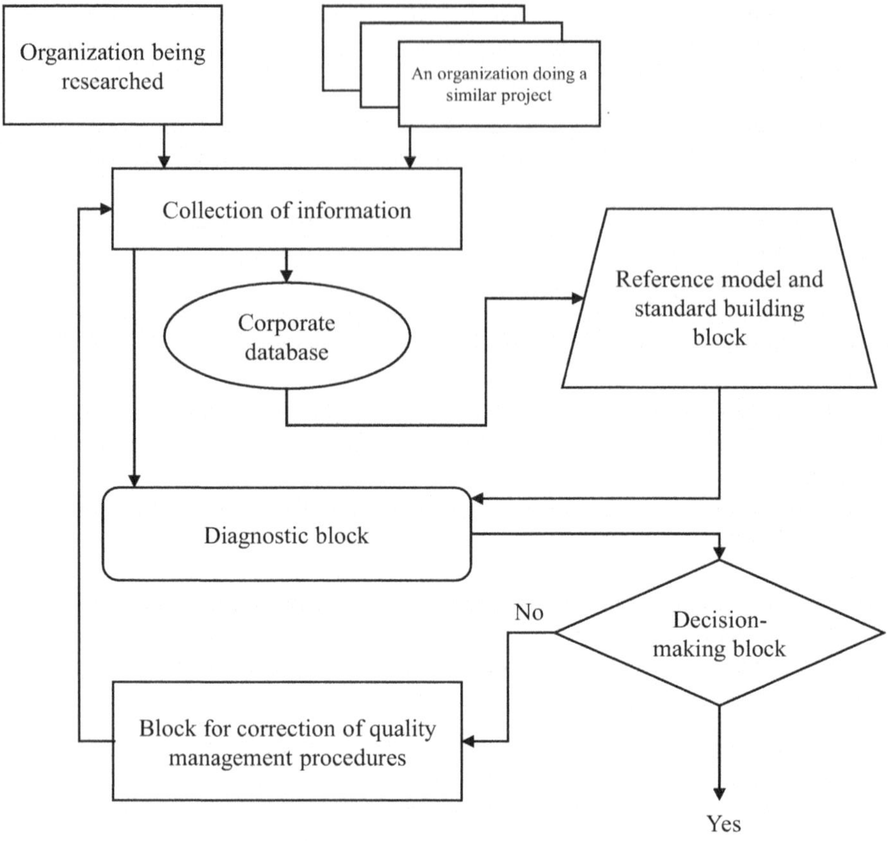

Fig. 2.13 Operative management process correction system

infrastructure as external components of the strategic management of science-intensive product's and branches' competitiveness.

6. A mechanism of evaluation of science-intensive branches infrastructure:
   - The level and prospects of the energy base
   - The level of information technologies
   - The level of telecommunication technologies
   - Transport infrastructure

The rapid development strategy should be implemented according to event schedules that are part of the high-technology branch development programs. These event schedules should reflect qualitative characteristics of the current stage of innovative development of a country and outline particular steps to implement the strategy's points.

Also, to outline a rapid development strategy for high-tech production, it is necessary to provide a methodological basis for such crucial elements as:

- Specifying rapid development indicators on a national and global scale, because nations are integral parts of the global economic system. The economic concept of these evaluation indicators, which are designed as part of the competitiveness evaluation strategy, should be similar to those applied abroad. It is necessary to build a system of specific indicators of rapid development for science-intensive branches with account of modern economic situation, global market dynamics, and frequency of introduction of innovative products and solutions.
- Formulation of ultimately new scientific approaches to effective evaluation of quantitative and qualitative indicators, which characterize managerial decisions that are made at all levels as the rapid development strategy is implemented.
- Developing qualitative indicators reflecting management and the growth of competitiveness, these include grading science-intensive industries, selection of the most important ones, and long-term pricing of science-intensive products.

The main points that must be included in a rapid development strategy for science-intensive industries are:

1. Formation of governmental rapid development management tools, which should include an integrated governmental innovation stimulation system for industrial enterprises. First, it is an effective priority investment mechanism for science-intensive branches, which incur high technology implementation costs, as well as for formation of innovation clusters, which create a basis for public–private partnership, integration between large, medium, and small science-intensive businesses.
2. Creation of an integrated automatic rapid development management system. The strategy, which is mainly aimed at development of science-intensive branches, should include diverse steps to catch up nations that occupy leading positions in implementing innovations. Second, it should include building of large business systems, which give broad opportunities for collection, accounting, prognosis, and processing of information coming in from the inner and outer environments. This information may describe the modern level of science, advanced technical solutions, know-how, innovative industrial, corporate and managerial solutions, as well as qualitative characteristics of top-quality product specimens. As major business systems process and utilize this information, they enable organizations to promptly create new markets and saturate them with their products, which boast great competitive advantages and make state regulation mechanisms more effective. Creating a mechanism for state regulation and management of science-intensive industries' competitiveness with the help of an automated management system is only possible with the use of existing competitiveness management mechanisms held by major organizations within industries.
3. Transformation of the available cluster of the most competitive science-intensive organizations, which operate in global markets on a long-term basis and are part of the global scientific and technological system.
4. Elimination of unfair competition and prevention of unlawful modification of the "economic space" parameters. This will require building of a high-quality innovative infrastructure, possibly organized as special territorial and corporate

formations (business parks, technology parks, development centers), as well as state insurance systems for innovative projects.

5. Compulsory formation of a strategic stockpiles for further development of the innovative potential for small and medium business operating in science-intensive industries and producing science-intensive goods and services.
6. Creation on a governmental level of an effective system for certification of science-intensive production technologies and science-intensive services that could ensure stable product quality and consumption dynamics. The state should strictly select and effectively support technologies, which create a basis for strategic economic interests and national security, and which are used in a multipurpose and multi-industry mode.
7. Creation of a favorable environment to attract financial resources for the implementation of advanced products by providing tax exemption and subsidized loans and inclusion in the production cost of expenses on innovations with a multiplying coefficient. When using a preferential resource supply policies for target scientific and technical industries and projects, design and mastery of complex technologies, the practice of leftover funding of science should be avoided, and this must be a preferential budget item. There should be a mechanism supporting organizations, which accumulate and direct financial resources to high-risk innovative projects from the early stage of R&D. Also, it is possible to create a preferential environment that would help to build an investment potential and enable organizations to purchase domestic technologies that are more effective than foreign ones.
8. Creating an effective mechanism for integration of the fundamental science into the innovative process, i.e., working toward elimination of the dangerous gap between the theoretic science and high-tech production, transformation of theoretic research data into real competitive products, also through regulation of private–public partnership within the R&D system.
9. Outlining a labor market regulation order to satisfy both current and future demand for qualified staff. This will require creation of a system of research centers, which would create in technically advanced regions conditions for improving expertise and ensure manpower influx.

Implementation of a rapid development strategy gives the following short- and long-term benefits:

- Quick transition from an export-oriented to a public- and innovation-focused economy.
- Higher nationwide investment and innovation activity.
- Effective distribution of resources and much better funding of social services.
- Upgrading the main industrial capabilities.
- Reducing structural/logical imbalance and raising the economy's investment potential.
- Diversified high-tech production based on qualitative and quantitative evaluation of products', organizations', and industries' competitiveness.
- High GNP share of high-GVA manufactured goods.

- Calculation and reduction production costs with the help of formalization, actualization, and prediction of prices for science-intensive products to make Russian science-intensive products more price-competitive in the global market.
- Providing conditions for import substitution of science-intensive products in the domestic market.
- Raising the proportion of high-added-value products in the market.

Chapter 3
Personification of Needs as a Landmark for Creating Future Goods

3.1 Development of Technologies and Public Needs with Technological Waves and Information Technologies Enhancing in All Spheres of Public Life

As economy grows dynamically, industrial, managerial, and other technologies, as well as public needs, have to change and stimulate each other. In other words, new one-of-a-kind technologies, which appear as part of innovative development (some discovered unexpectedly), stimulate new needs and thus create demand. However, the evolving public needs should stimulate nonstop technological progress and creation of new types of goods and services, which can satisfy them.

Meanwhile, technological progress and public needs are largely dependent on cyclical economic fluctuations, which influence the innovative process. Recurring economic crises, which affect some nations, motivate experts to analyze their causes and factors contributing to them. Many negative economic events, such as progressing recession, slow economic growth, poor labor efficiency, etc., have one fundamental and decisive cause—the technological progress and, particularly, its ever-changing effects.

Economic crises and depressions following them stimulate scientific and technical progress in nations, national industries, specific enterprises, and they lay ground for creation of brand new products that are based on advanced technologies and scientific discoveries. In other words, an economic crisis acts as a starting point and stimulates the cyclical change of technological waves (innovation waves).

A technological wave is understood as a complex of well-studied revolutionary technologies, innovations, and inventions, which explain a quantitative and qualitative leap forward in the development of public productive forces. The concept of technological cycle or wave was introduced by a Russian economist N. D. Kondratiev. As he studied the history of the capitalist economy, he pointed out lengthy (50–55 years) economic cycles, each one characterized by a certain level of productivity. The concept of a technological wave is identical to the concept of a

A. Tyulin, A. Chursin, *The New Economy of the Product Life Cycle*,
https://doi.org/10.1007/978-3-030-37814-1_3

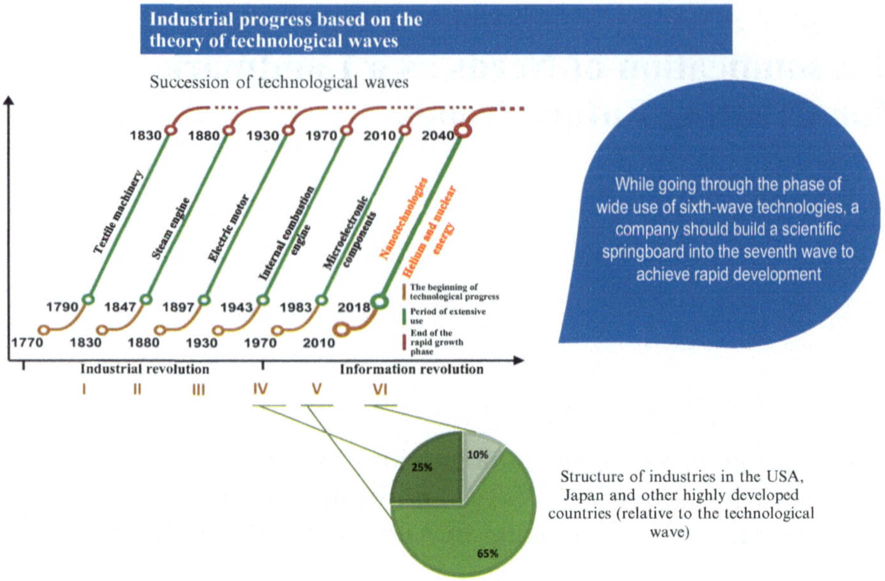

Fig. 3.1 Development of technological waves

technological system. A technological wave comprises groups of technological complexes, which are related to each other via single-type manufacturing chains and form self-reproducing integrated entities.

A technological wave has a core, key driver, and organizational/economic adjustment mechanism. So far, six types of technological waves or orders have been described, and some experts expect seventh-wave technologies to appear quite soon.

Figure 3.1 demonstrates technological waves and main inventions resulting from scientific and technical breakthroughs.

Processes that are observed in well-developed economies demonstrate the specific importance of the relationship between cyclic economic phenomena and innovative work, which is dictated by the growing public and corporate needs. The discovery of these stable and regular relationships between technical/technological innovations and economic fluctuations has created a basis for development of effective tools to counter and reduce the negative influence of these cyclic changes on national economies through consecutive reforming of the innovation sphere.

The key driver of these reforms should be creation of ultimately new products that are based on advanced functional principles, which can satisfy personified needs. The improving quality, lowering production cost, as well as appearance of new types of goods and services are part of the nonstop innovation process, which implies development and implementation of new technologies and equipment, and using advanced management methods.

The sixth technological wave is based on nanotechnologies and alternative energy (Fig. 3.1); also, there are biotechnology and "living" artificial system projects, robots, materials with predefined properties, advanced medicine, high humanitarian technologies, which teach man to adapt to the rapidly changing environment.

Digital information, virtual reality, digital twin, artificial intelligence, 3D, and other technologies are booming, as they help to automate corporate business processes, which include production. They reduce labor intensity and production costs while increasing labor efficiency and productivity. Meanwhile, the use of existing innovative scientific and technical potential, obtaining of information about brand new production technologies and methods, and rapidly changing public needs, have shortened the product development period. This allows to flexibly meet market requirements and helps industries to promptly switch to satisfying the increasingly advanced needs, which are the backbone of demand stimulated by the production highly competitive goods. The use of information technologies at the early product design and branding stage sets up a link between two important elements—production and economy. As a result, the economic effectiveness of creation and implementation of a product, the adequacy of its production cost and, later, market price and high-quality indicators can be observed at the earliest life cycle phases.

Given the ongoing global digital transformation, a modern organization should be focusing on running an economically stable, innovative, competitive, and diversified company capable of solving strategic goals of improvement and development and occupying its rightful place in the global market.

The pervasion by digital technologies of all industries and spheres of life is a specific feature of today's world. It is explained by progressing microelectronics, digital technologies, and telecommunication. The main spheres of public life—economy, management, and security—are going to take a new form and substance in the foreseeable future.

The current global information space started to form and evolve rapidly with the advent of the Internet in 1982, which came as a reaction to the scientific community's need for exchanging scientific information. It can be deemed the starting point of the digital space formation process. Since then, this space has been growing dynamically and obtaining new components: forums, social networks, the Internet of Things, etc. Each of these blocks is a structural element of the virtual world and, at the same time, a kind of bridge that links it to the real world. Evidently, these two realms are both interrelated and interdependent in the same fashion as a real person is related to and dependent on his/her virtual social network image.

Today, we can identify every little being and attribute it to both worlds; however, after a while this division will not be possible for many objects. A modern smartphone can store a huge load of data: telephone numbers, birthdays, pictures, passwords, etc. We have surrendered part of our memory function to an electronic device. Should a person lose one, he/she will feel lost and helpless. Even though man and the smartphone do not make a physical whole yet, they do make a functional whole. The merging of the real and virtual worlds has gotten up and running and it cannot be stopped.

Global digitization is modifying national economies. Until recently, organizations specializing in a variety of branches, such as energy, production (electronics),

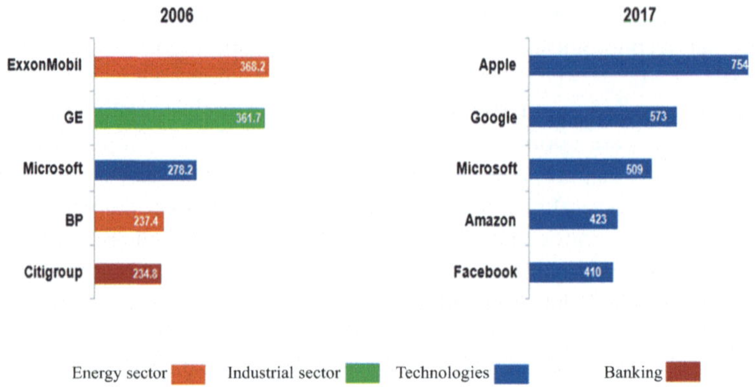

Fig. 3.2 Companies with the world's highest market capitalization (USD billion) in 2006 and 2017

banking, etc., boasted the largest market capitalization. Today, the majority of capitalized whales are companies focusing on the development of digital information technologies and related promising goods (Fig. 3.2).

Figure 3.3 shows the changes and revolutionary technologies, which have helped these organizations to achieve supremacy over their competitors and the rapid development status. Organizations tend to burst during periods when they create or purchase digital technologies or businesses, which shapes a digital economy on a corporate or global scale.

A digital economy is a spectrum of activities, in which digital data is a key production element; processing of big data blocks and applying analytical results can make different types of production, technologies, equipment, storage, sales, delivery of goods and services, much more effective than in a traditional economy. A specific feature of an economy developing in an era of digital transformation is maximal satisfaction of all its subjects' needs through the use of information, including personal. It is possible thanks to the development of information, communication, and financial technologies, as well as the infrastructure, which together provide for a complete integration between all participants: subjects and objects of creation, distribution, exchange and consumption of goods and services.

Informational interaction can produce a maximal effect providing that all subjects and objects obtain a sufficient digital element. For example, even nowadays, the digital component of automobiles (sensors and software), which significantly raise their marketability and safety, accounts for more than half of a car's price. The next stage of digital interaction is cars "communicating" with each other and road infrastructure elements, which fuel the self-driving car technology.

Eventually, most products' and services' value will largely depend on their digital aspect. These products are termed as "smart things," as they are organized into

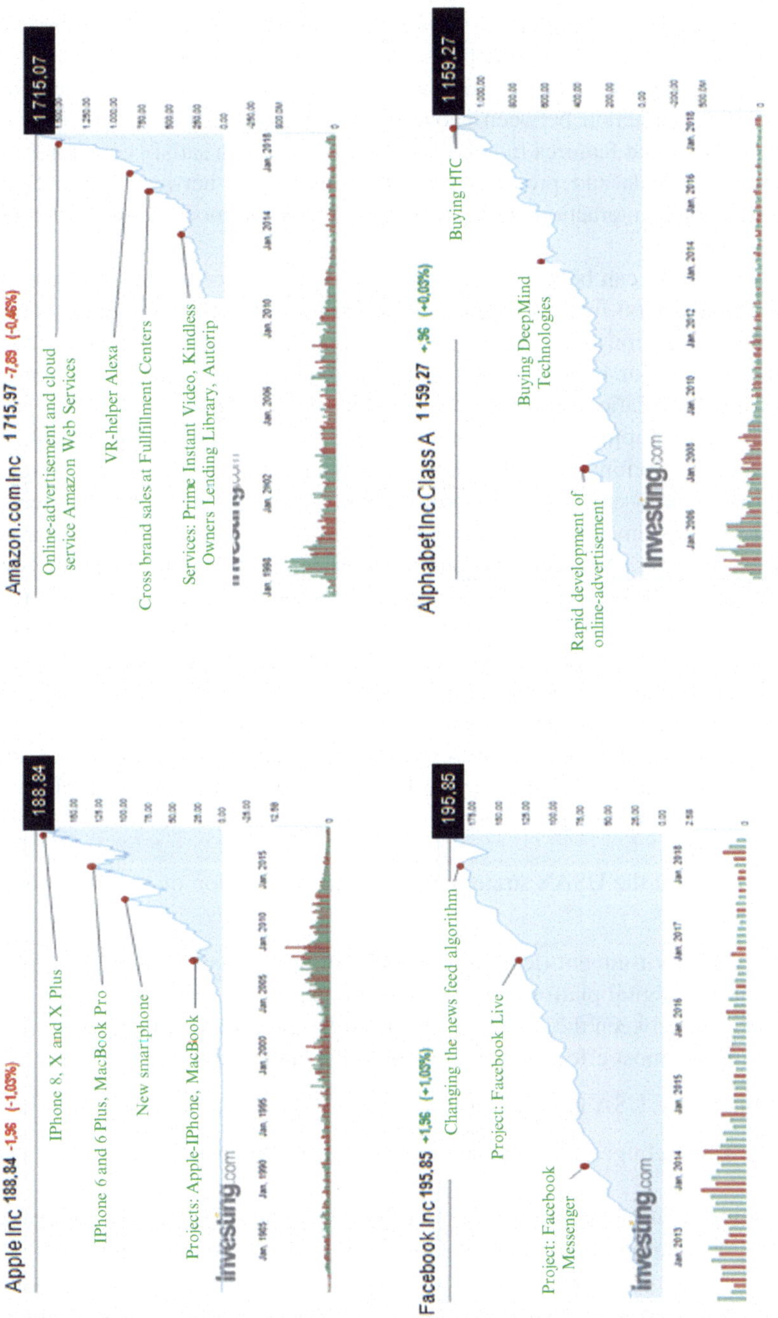

Fig. 3.3 Corporate capitalization dynamics

complex systems (for example, a smart home, smart city, smart organization). Digitization should either improve their main properties (for example, cars should become safer and less maintenance costly) or add new ones (self-driving, voice control, remote Internet- or mobile phone-controlled driving, etc.)

Evolution of digital communication technologies enables direct communication and facilitates interaction between economic agents. This requires formation of a digital ecosystem that features functions and services, which satisfy consumers' and manufacturers' needs and provides for direct interaction between these. Surely, mechanisms of this interaction, which form a kind business model, have been around for quite a while.

A classic market can be used as a simple example: sellers and buyers (manufacturers and consumers) find each other. Digitization pushes the interaction between manufacturers and consumers up to a new level. This reduces costs and provides additional toolkits for both suppliers and consumers, contributing to exchange of data and, finally, creation of innovative products and solutions.

In a digital economy, business models that focus on providing personalized services tend to dominate. Technologies like Big Data, targeted marketing, 3D printing, etc., allow enterprises to produce goods and services, which meet the each particular customer's, not only an ordinary customer's needs. The development of information and communication technologies brings a manufacturer and every end user together. Now it is possible to significantly shorten chains of intermediaries. A good example is the BlaBlaCar service, which uses a smartphone application that enables users to locate cab drivers moving in the needed direction. This pattern enables implementation of a mechanism, which is economically beneficial for all participants excluding transport operators, which act as intermediaries.

Many highly developed economies have realized that the changes are inevitable and began to deliberately shift toward digital transformation. The USA and China—unofficial information technology leaders—were the first states to move in this direction.

A closer look at the USA's strategy reveals that digitization of economy consists of four blocks:

1. Creating an environment (legal framework) for digital economy
2. Formation of digital platforms in the most prepared industries
3. Competition between the platforms followed by their gradual integration
4. Projecting the most effective solutions onto the entire economy

Most likely, the USA has chosen this strategy for the following reasons:

- The USA has a huge economic and technological supremacy over the rest of the world.
- In building an infrastructure for a digital economy, the USA can rely on high-tech transnational corporations, such as Google, FaceBook, Amazon, Intel, etc.
- The USA is home to the required critical number of privately owned companies, which can trigger a spontaneous growth of digital economy and use its potential for their own and the country's benefits.

The other digital technology leader—China—has chosen a strategy, which is in many ways opposite to the USA's: planned digital economy. A closer look at it reveals that China's strategy has two parallel and hardly connected vectors:

1. Digitization of production through the use of the industrial Internet
2. Use of the Internet's potential for further extension of sales markets

This strategy has four components:

1. Total digitization of the production and logistics sectors
2. Development of the legal framework
3. Digitization of management systems and creation of digital platforms
4. Assimilation of digital platforms and ecosystems into an integrated space

However, in different countries (the USA, Great Britain, etc.), many digital transformation programs have social focuses, such as the Digital Medicine and the Digital Health. These projects do not produce a significant economic effect, but they have several explanations.

First, any large-scale social program needs public support. This is the reason why digital economy is branded by social projects.

Second, large industries will be digitized by themselves for economic reasons. On the contrary, social projects need support from the government.

Third, most developed economies have a substantial technological background, which allows them to launch digital economy to a certain extent. Full-scale social projects will receive a massive feedback thanks to large communities of nonprofessional users. This will help to improve technologies from users' perspective and make them available for a large part of the public.

Fourth, introduction of digital technologies in industries (for example, introduction of the Internet of Things at factories) embraces but a narrow range of goals. Implementation of the Digital Health and the Smart City projects implies a higher diversity and complexity. All advanced technologies must go through this kind of "social stress test," particularly for their management systems.

Today's digital economy and high-tech trends explain the urgent necessity to analyze personified needs while creating advanced products through convergent use of knowledge derived from the global information space and revolutionary academic and scientific solutions.

It should be noted once again that information plays a crucial role in digital economy, as it acts as the main and high-value nonmaterial asset. Large businesses (Internet whales like Google or Yandex) have long been deriving data from user searches and building customer profiles. Being aware of potential needs allows companies to create new consumer niches, make competitive products and services and effectively implement their and their customers' services. Although not all organizations have as big marketing and analytical potentials as Yandex AND Google do, some companies, for example, communications providers can obtain information on the use of services by their customers.

This helps them to get coherent information on potential consumer groups, which can be presented on digital platforms (Fig. 3.4).

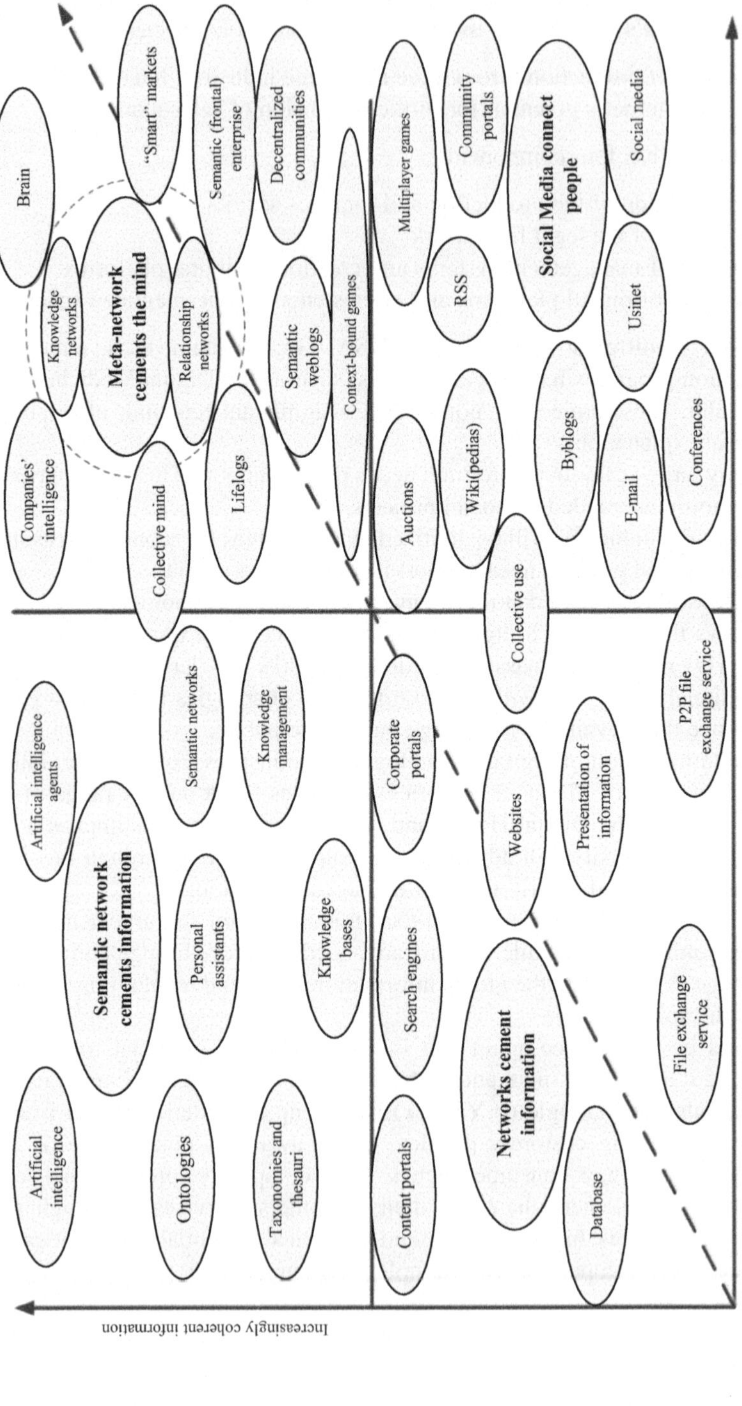

Fig. 3.4 Coherence of information

Information becomes a new strategic resource (informational resources), which can be obtained both from inside of an organization and from data coming in from outside, as well as from the global information space. This resource becomes the key one in the process of finding and personification of the needs of the market or its separate segments.

Therefore, informational resources should be managed for the purpose of obtaining the biggest possible data to analyze and personify needs, as well as to evaluate an organization's and industry's scientific and technical potential for satisfying newly emerging needs.

Information resources can be managed more effectively with the use of an integration and logistic approach. It implies interaction between different scientific and industrial companies and consolidation of their expertise and competences, which can be used for creation of a marketable product or service to satisfy current needs. A network-type system for management and unification of information resources within an integrated logistical and informational space is shown in Fig. 3.5.

After a while, management of information can create an ultimately new pattern of people's life and interaction between them, as well as of economic and political governance. This is not only about creation of new products and services, but also about creation of a social environment stimulating personal fulfillment; and, from the perspective of the global community, an environment that would eliminate aggression and self-destruction, preserve nature and resources for future generations.

This statement is proved by the fact that developed and rapidly developing countries are modernizing their governance infrastructures to create favorable conditions for informational development and to meet the global economy's needs, which are changing toward the use of cognitive managerial competences and optimized informational and governance relationships between national and transnational structures and international organizations.

Information and communication technologies and the information space are evolving rapidly thanks to large sums invested in the use of innovative technologies in all spheres of economy. This helps to build an information and computation management services, which emerge thanks to the use of strategic and ultimately new managerial competences. They develop due to the growing demand for information and knowledge, as well as due to the scientific and technical progress, which is accelerating within this segment.

It is also important to mention *digital convergence*, which is interpenetration and merging of digital computation and data transfer facilities based on the primary capture of different messages. Digital convergence reduces data processing and delivery costs with a greater number and quality of tools used by information and telecommunication complexes and networks. Digital convergence helps to generate new needs, stimulates a more complex integration between global networks and higher industrial cooperation. For instance, convergence between telephony, computerized technologies, cable TV, broadband streaming technologies, and publishing work enables a new kind of transnational integration within the global economy.

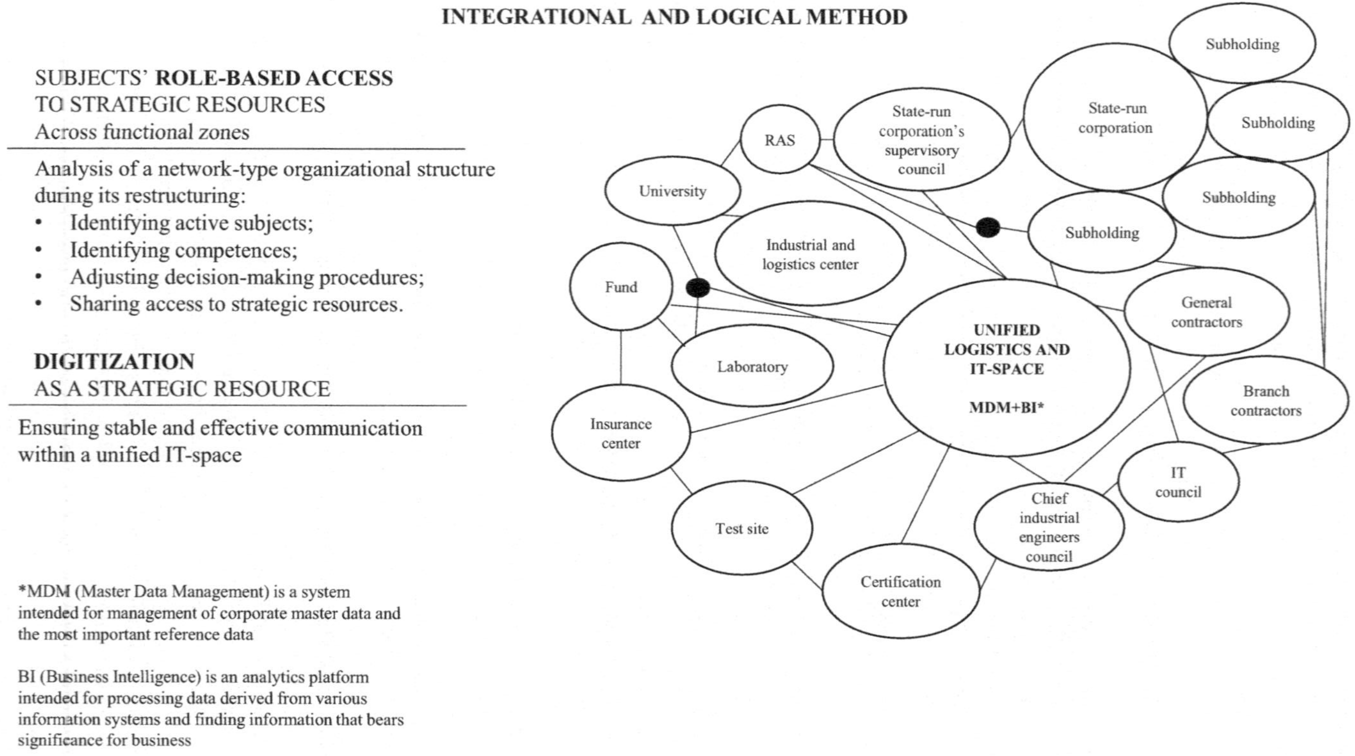

Fig. 3.5 A network-type information resource management system

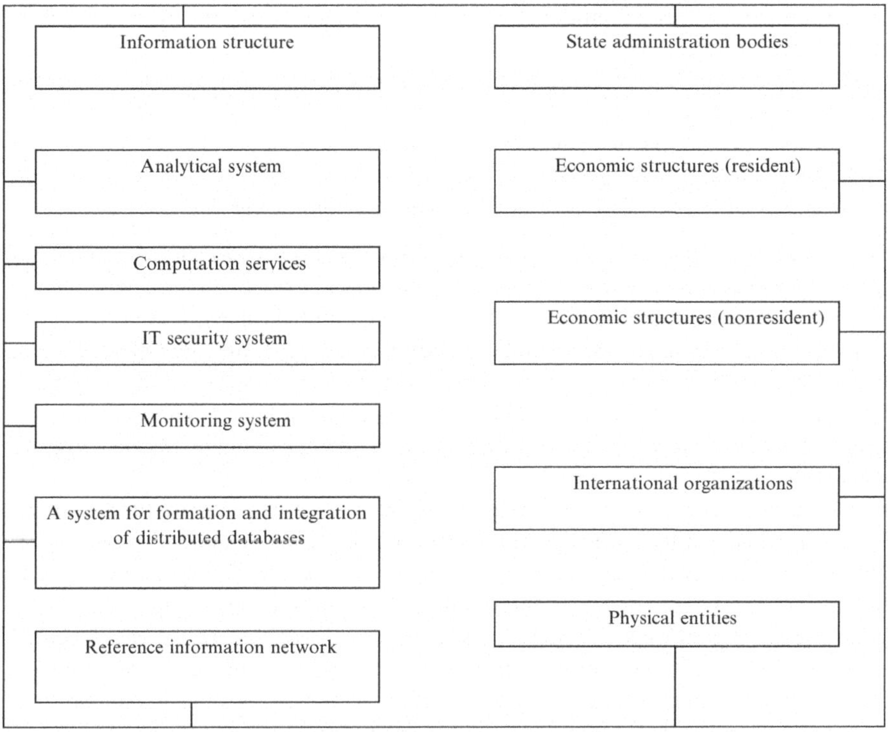

Fig. 3.6 An integrated digital environment with a network-type architecture

Meanwhile, modern convergence technologies provide for effective interaction between objects and subjects of management and for creating a convergent managerial infrastructure through integration of segmental information and communication networks and distributed cross-boundary electronic managerial environments (corporate, industrial, regional, clustered, etc.).

This type of infrastructure makes an integrated managerial environment with a network-type architecture (Fig. 3.6).

This integrated digital managerial environment provides significant competitive advantages compared to traditional management, such as:

- Real-time creation of an integrated international picture of controlled objects
- Much quicker collection and distribution of information across structural divisions and between partners, a quicker product design and development process
- Significant advantage over competitors in making and execution of decisions and scheduling work, which should finally put an organization on the path of rapid development
- Production of science-intensive, export-oriented, high-added-value, and labor-efficient goods

- Reduced import of IT products and a higher share of export-oriented science-intensive products
- Development of a legal framework for electronic government and electronic state administration services
- Quick concentration of resources in the most important sectors that meet prospective public needs
- Prognosis of public need development with the help of big data analysis

According to research carried out by McKinsey & Company, a consulting agency, nowadays it is possible to predict future needs, which will inevitably emerge given the rapid change of technological waves and global digitization.

First of all, these needs are related to building information and analytical services, which aggregate diverse information and provide consumers with ready-made solutions. Therefore, this is not just about providing data. It is also about collecting, systematizing and structuring it in such a way as to make it valuable for the consumer at this point.

So far, this approach can be illustrated in an application used by the Commonwealth Bank of Australia, which provides mortgage loans. The service operates as follows: first, a user makes a smartphone picture of a house. The system recognizes and locates the house, provides information about similar objects that are located close by, pricing and tax information, and the area's position in the terms acceptability list. The bank not only had to go beyond providing financial instruments; it went beyond the banking industry itself through mere analysis and personification of its potential clients' needs and deriving data from different sources.

When presenting information it is necessary to consider personal needs and preferences, i.e., the information should be personified based on the stored data. Credit services, whose scores depend on social network profiles, do exist nowadays. These profiles are already used for building users' credit records. Collection of data from networks like Foursquare, Instagram, Pinterest, etc., can help to aggregate useful and personified information.

However, the evolution of the economic system and its transformation into a digital ecosystem do not reduce material production; they significantly change product characteristics, which is the result of satisfying developing and newly emerging public and corporate needs.

The ongoing transformation of the economic system is giving products the properties of "smart things," which can integrate into economic systems (smart homes, smart cities, etc.). It is expected that the big data, which is formed by visualizations and measurements, as well as information from the global information space, being merged into a single integrated environment, is going to make (and is already making) a foundation for the Intellect of Things. Big data processing, which uses artificial intelligence and self-training neuronal networks, ensures penetration of digital economy and space data into all traditional economic spheres, increases labor efficiency and reduces costs. Therefore, changes occur at all layers of the global economy.

The most traditional sector—agricultural—uses self-driving farm machines, which operate on navigation systems, such as GPS and GLONASS. Also, they use automated intelligent watering, fertilization, pest control, and other systems to ensure real-time control of these operations based on ERS data. There are similar examples applicable to other traditional sectors.

Today, the world is witnessing global changes resulting from the space activity, which is, starting from the fourth technological wave, one of the most science-intensive spheres of economy. This sphere is characterized by a nonstop innovative process coupled with deep fundamental research. Personification of needs that are satisfied by the results of space work necessitates, on the one hand, realization of global projects and, on the other hand, personified targeting of the end user (a person, organization, or state).

Particularly, the Artificial Moon, which is being launched in China, can light up vast areas at night time by redirecting sunlight with the help of orbital complexes. The realization of such a project can selectively solve urban illumination issues, light up agricultural areas at night, etc. Another global space project is expected to provide access to the Internet from any part of the Earth. It will make the global information space constantly available for the public and elements, which unify within the net on an Internet-of-Things basis.

Space technologies are entering our life, and information, which we get from space, integrates into individuals', organizations', and nations' everyday life and provides increasingly adequate economic solutions. In the foreseeable future, data will continue to gain adequacy, and its cost will fall in keeping with current economic laws.

Digitization's influence on the process of product design and manufacturing is observed in a tendency toward a shorter production cycle. For instance, 3D-printing technologies shorten the preproduction period and, consequently, product-to-market time for highly competitive goods. The reduction of the design and production time gives manufacturers another opportunity to reach the rapid development status and take the lead in a newly created market or segment, where the product can satisfy unique needs.

These tendencies demonstrate that changing technological waves and enhanced digitization of all spheres of life will inevitably fuel technological progress and public needs; this will change the scientific, technical, and technological priorities in different countries.

In order to make these priorities effective and widely chosen in light of new economic tendencies, a priority-formation mechanism should be implemented officially. This mechanism is shown in Fig. 3.7.

The implementation of the mechanism presented in Fig. 3.7 can be helpful in listing the most promising technologies, which meet the current personified market needs of groups of consumers, thus stimulating organizations to design advanced products that can create new market niches. A manufacturer operating within these niches will be a "temporary monopolist" until competitors come up with similar, higher quality or more functional items.

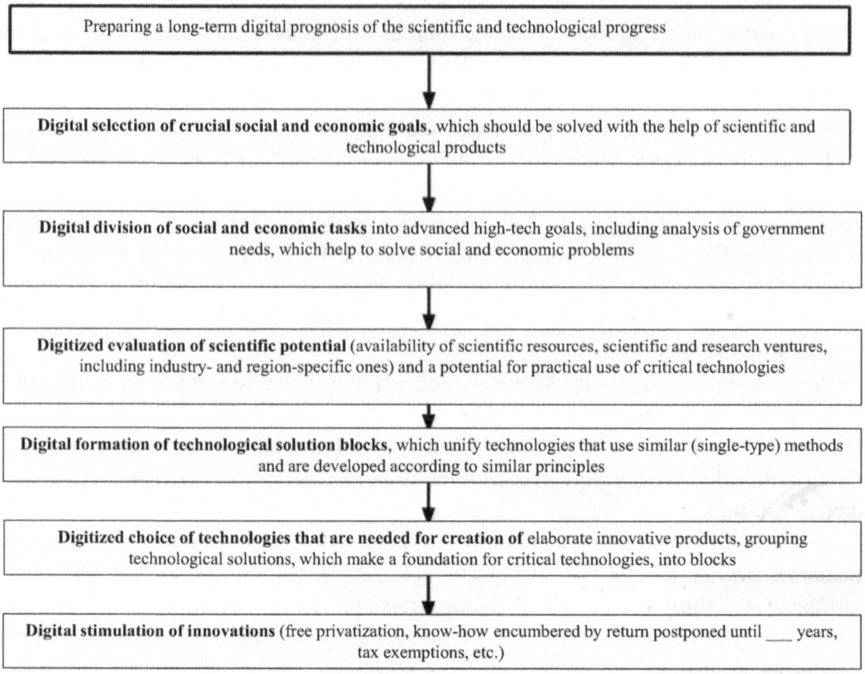

Fig. 3.7 Mechanism of building priorities in science, technique, and technology

Meanwhile, temporary monopolization that results from high competitiveness and use of information technologies can occur as described in Fig. 3.8.

As follows from the above, digital economy opens broad venues for development of industry-level, national and even transnational economy management systems. Modern technologies enable the creation of a high-tech digital economy management platform that will minimize the human factor and therefore corruption and errors, automate statistical, tax and other reporting and make decisions depending on real situation analysis. Economic process management will be structured upon intelligent digital cloud platforms featuring open inter-machine communication interfaces and providing room for a broader communication between business units, as they are going to create platform-based proprietary applications (supplied with mandatory security and legal certificates).

Development of intelligent communication systems, which enable real-time remote management of industrial units, will incorporate effective and versatile resource planning and management models and systems. It is not going to be a directive planning system. Instead, it is going to be a dynamic real-time feedback-based system (it should adapt to end user's preferences).

To describe and predict economic transformation that results from digitization, it is necessary to develop theoretical economic models describing the digitization-related economic growth and redeployment of labor and financial resources to the

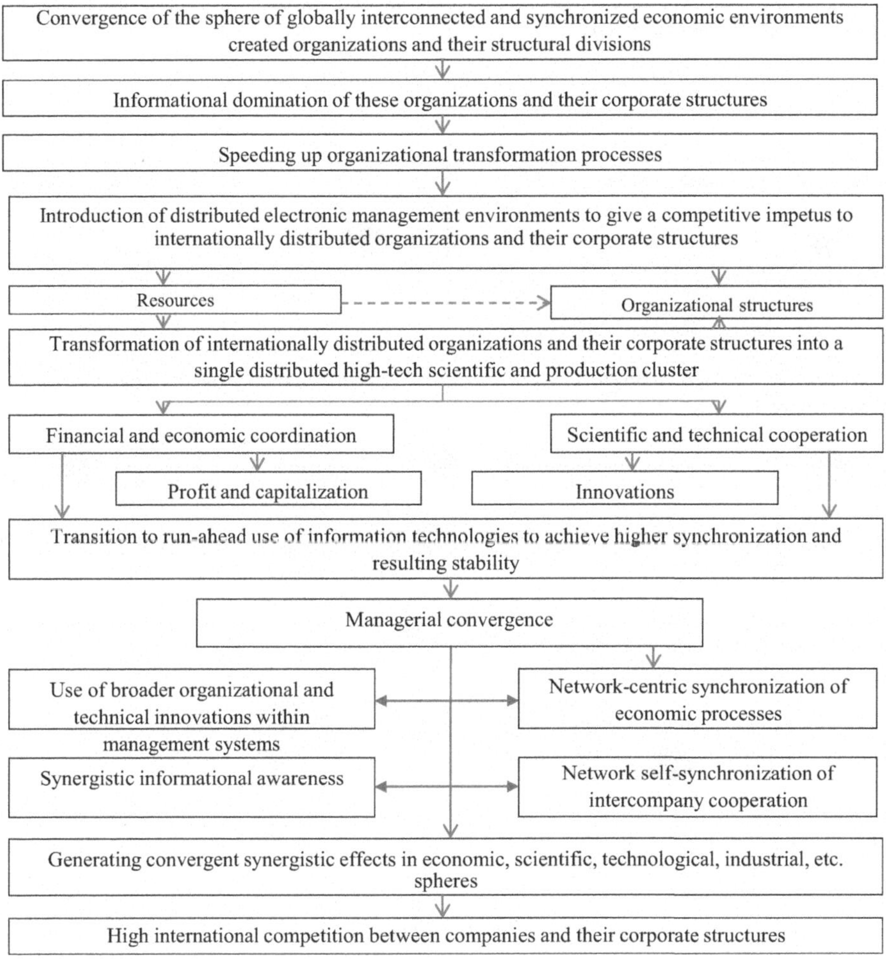

Fig. 3.8 Reaching high competitiveness through the use of digital technologies

new sector. Another key aspect of the theory of economic process management is the choice of relevant integral parameters and formation of new metrics reflecting current tendencies of digital economy (for example, information-to-product transformation, goal-to-value transition, etc.).

Meanwhile, the accumulation of competences and progressing communication technologies (they are growing rapidly year by year) will create a basis for a new source of energy, a new driver of breakthrough, and rapid transition to the next technological wave. Founded on virtual management technologies and integrated information space will be new business models, which are going to take companies to leading market positions, stimulate technological modernization of key sectors, and ensure balanced and stable evolution of the research and development sector, which has an optimal institutional architecture and guarantees extensive production

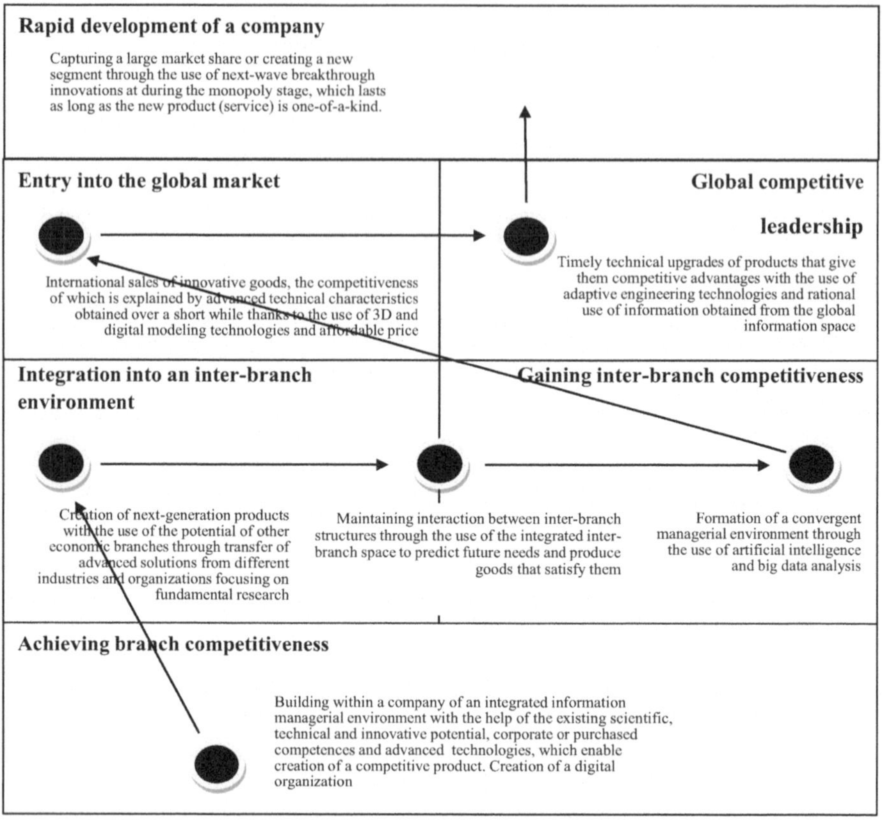

Fig. 3.9 Achieving rapid development

of knowledge. Also, it is going to raise the effectiveness and performance of the infrastructure, which commercializes scientific research results and provide for an open national innovation system and economy.

This should help large organizations reach a new level of competitiveness, which will be a springboard for rapid development and reaching a global competitive leadership as described in Fig. 3.9.

As follows from the above, reaching leadership in the innovative segment at an integrated global level, which can be a springboard for rapid development, is explained by bringing new scientific and technological solution to a commercial level and development of advanced digital technologies.

Meanwhile, these technologies should provide a basis for developing new methods of designing and production of one-of-a-kind competitive goods, which are of great value for consumers, can satisfy new market needs, create new market segments and new demand. To solve this task, it is necessary to develop adaptive product design methods with the help of long-term demand analysis and big data processing technologies. It takes a kind of economic mechanism to help companies achieve breakthrough rapid development through implementation of the *economy–production–economy* chain.

3.2 Big Data Analysis-Based Methods of Assessment of the Effectiveness of Next-Generation Product Creation Processes Aimed at Satisfying Future Needs

At present, one of the driving forces of scientific and technological development is high-tech production, which sets "growth points" for all branches and economies.

Given the current macroeconomic instability and rapidly growing global science-intensive and high-tech markets, manufacturers have to face numerous challenges, which affect the competitiveness of industries and the economy in general. A key activity area for organizations, which preserves the existing market niche and helps to create new markets, is stimulation of their innovative work through rapid development of innovative potential and use of advanced technologies in planning, management, and control of all corporate business processes. This requires companies to master new production process planning techniques that use big data processing and analysis systems, industrial IT technologies, artificial intelligence, expert analytical systems, etc. This necessitates continuous investment in innovative projects that focus on the production of science-intensive goods and raising their investment appeal.

However, the market is receiving innovative solutions, for example, software tools and robots, which can adapt to personified needs given the ongoing nonstop transformation. These products give foreign high-tech organizations a competitive leadership.

It is also important to analyze the structure and content of an innovative process (or project), which is aimed at developing a product that should satisfy future needs. Being a single system, this process can be divided into subsystems and separate segments, which helps to reveal factors that either stimulate or impede successful commercialization of a novelty. Products created as part of the innovative process should help the manufacturer achieve stable rapid development on a global scale. Therefore, it is necessary to evaluate and research the innovative potential of projects focusing on producing the goods of future according to an algorithm shown in Fig. 3.10.

Along with the innovative potential, it is necessary to assess these products' influence on how soon an organization can achieve a rapid development status based on its innovative and resource potential. The analysis of this influence is shown in Fig. 3.11.

The analysis of advanced products' influence on the speed of a company's achieving rapid development as described in Fig. 3.11 provides sufficient information about the planned innovative process, which pursues the goal of developing and

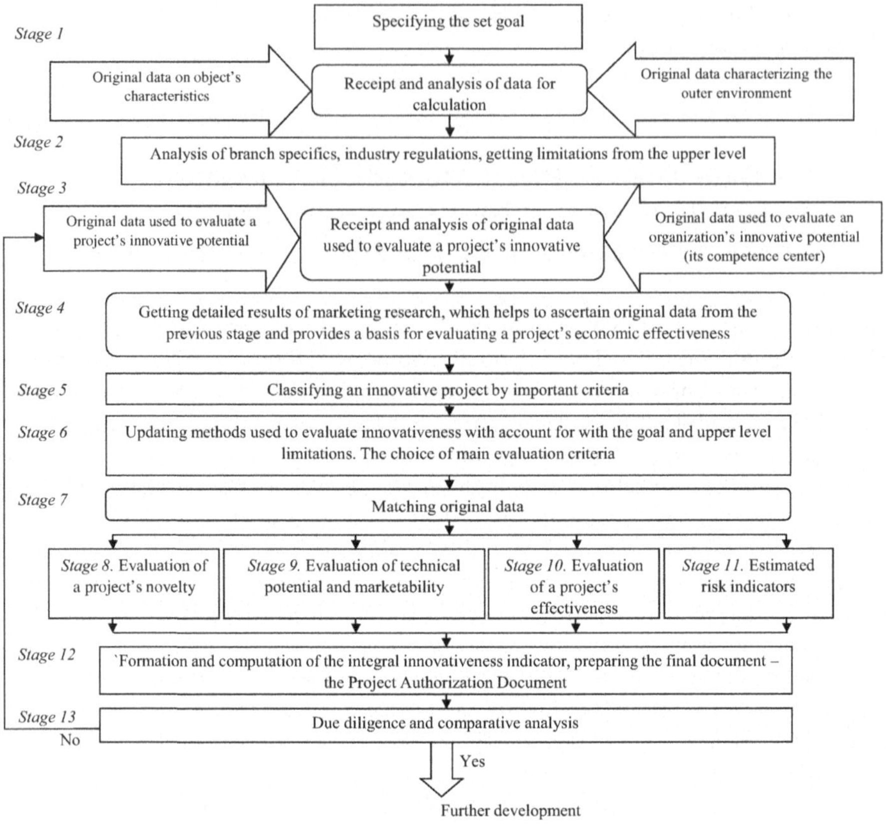

Fig. 3.10 An algorithm for evaluation and research of the innovativeness of projects' that focus on creation of advanced products

manufacturing advanced goods in so far as they influence the possibility that the organization will take the path toward rapid development.

The innovative process, which results from project execution, will be dynamic. It is not just a set of empirical elements; it is divided into diverse subsystems, and it changes over time. The innovative process has internal business logic and a set order, as the new product goes all the way from its inception through creation, distribution, and consumption. The innovative process never stops interacting with different subsystems that are part of the outer environment—natural, technical, economic, social, cultural, etc.

As we analyze the structure of the innovative process, we suppose that innovation is a kind of process, which takes place within a limited period of time. This period is divided into consecutive stages, which differ in the types of activity, which enable creation and use of the new product. This pattern represents the innovation's life cycle.

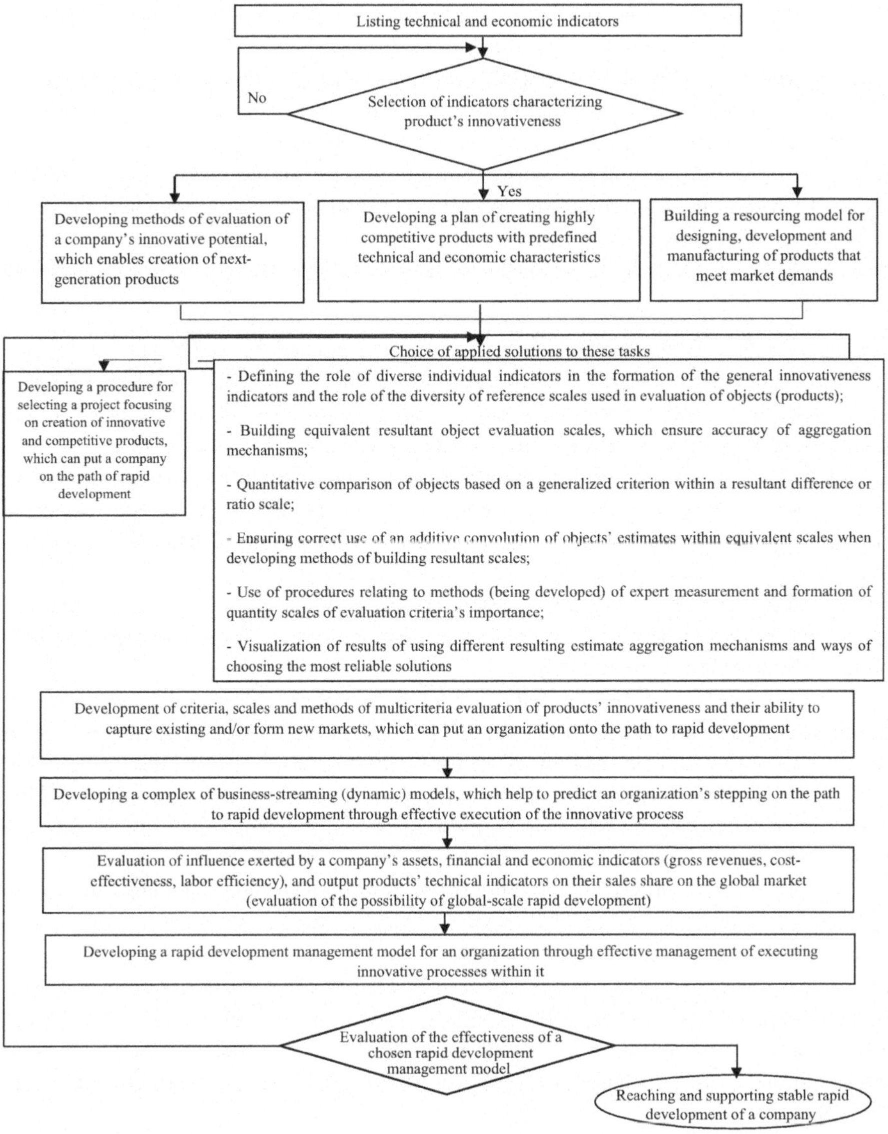

Fig. 3.11 Advanced products' influence on an organizations' ability to quickly achieve rapid development

Analysis of the innovative process's structure and meaning is an important thing. Dividing it into subsystems and separate elements helps to identify factors, which either stimulate or impede successful commercialization of an innovative product. Studying, creating and using this product take a bit of time. These activities should be organized in a set order of several stages.

Stage 1: the birth of a concept, which results from fundamental or applied studies, an "instant insight" or an accidental event.

Stage 2: the invention stage, at which an innovation materializes into a physical object that reflects the innovation's material nature (a new product, technology, service, etc.) This stage confirms a novelty's feasibility.

Stage 3: implementation of a new product, method or human tool.

Stage 4: diffusion of innovations consists in broad implementation and growing usability of a product in various fields.

Stage 5: the new product's domination in a specific industry, consumer market and activity.

Stage 6: decreased use of the new product as it gives way to a more effective one.

This sequence can be characterized as linear, and it is the result of a logical division of the entire process into separate functional or structural elements. It looks like a simplified scheme of a real innovative process.

Analysis of the innovation distribution process is difficult because studying the dynamics of only one localized innovation within its life cycle, or the statistical structure of the innovative space, is not enough. The inception, creation, development, and distribution of a specific innovation depends not only on its own dynamics and relationships with other innovations, but also on the prehistory of alternative, competitive, supplementing, and modifying innovations and on the entire permanently evolving innovative space.

The linear model of the innovative process has serious downsides. However, it dominated in all economies at the stage of developing innovative strategies and it would be confined to promotion of an innovative solution at all innovative cycle phases: fundamental research, applied research, commercial development. This approach also marked the boundaries of the information base, which sets estimation patterns for innovative activity. It describes but a limited segment of innovative activity and reflects only the first stage of the innovative cycle—generation of new knowledge. The linear model has been repeatedly criticized because it disregards many factors and elements of innovative activity. It does not reflect the influence of the marked and economic conjuncture, the relationship between production and science, innovations, risks and opportunities for their use in the current technological and social environment, entrepreneurs, motivations, and opportunities. There is an advanced tool, which helps to research these processes—intelligent big data analysis.

According to the McKinsey Institute "Big data: The next frontier for innovation, competition and productivity" report, the term "big data" refers to data sets, whose volume exceeds typical databases' potential for introduction, storage, management, and analysis of information. In fact, the concept of big data implies management of huge volumes of diverse information, which is constantly updated and stored in a variety of sources with the aim for more effective work, creation of new products, and higher competitiveness.

Unlike traditional business analysis, which describes the results achieved by a business over a certain period of time, big data technologies can make analysis

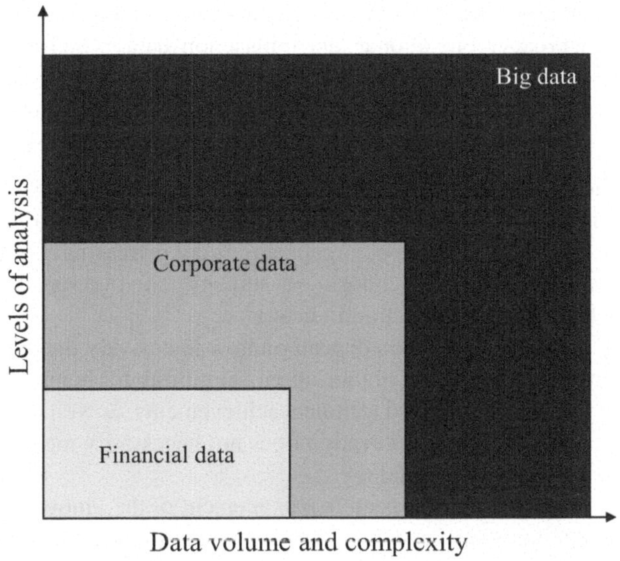

Fig. 3.12 Big data analysis process

predictive and capable of giving businesses recommendations for future. Although big data and business analysis have similar goals, they differ from each other in several aspects:

- Big data implies processing of a much larger amount of information than business analysis.
- Big data implies processing of more quickly obtained and changing information, which requires deeper research and interactivity.
- Big data implies processing of unstructured data and therefore development of algorithms for the search of tendencies contained within these data blocks.

Operating big data is not like the common business analysis process, where simple addition of known values yields a result. When operating big data, the result emerges as the data is refined through consecutive modeling: first comes a hypothesis, then a statistical, visual or semantic model is built, based on which the hypothesis's adequacy is checked. Next, the following hypothesis is offered. This process requires that the researcher should either interpret visual values and compile knowledge-based interactive inquiries, or develop adaptive algorithms based on artificial intelligence and machine learning, which can provide the needed result.

According to McKinsey, an increase in data will trigger a new wave of "innovations, competition and productiveness" in the business sphere, and "the use of big data will help to make actual data-based decisions instead of intuitive ones. For this reason, it can create a revolution in the sphere of management."

Figure 3.12 shows a ratio between traditional data, which is used in business analysis, and big data.

Financial data: highly traceable and clear standard financial indicators.
Corporate data: broader operational and transaction data, which can be used to support analytical conclusions and for prognosis purposes.
Big data: new types of internal and external data, most of which is unstructured, but part of it can create a new idea of effectiveness, potential and risks.

According to research carried out at the MIT Sloan School of Management, organizations that make big data-based economic decisions show a 5–6% increase in productivity. At the baseline level, organizations are increasingly eager to rely on data analysis when it comes to choosing a strategy, solving specific operational issues and reaching higher production indicators.

Companies' competitive statuses depend on how intensively they use knowledge and information in their work. An organization's eligibility for being termed a leader depends on the use of scientific and technical achievements, as well as their ability to predict future needs. However, innovation does not necessarily mean possession of or intrusion in an advanced technology.

An innovative strategy and respective management of the innovative process are very important. An innovative activity can be viewed as practical learning. From this perspective, standard ideas of simulative work as of competitors' less important role are no longer convincing. A nation's well-being greatly depends on the use of knowledge, which originates outside of its national boundaries. Supposedly, if a group of 20 nations participate in creation of advanced technologies, 19 out of 20 inventions will be created outside of this group after a long while, and each of these nations will have only 5% of the invention. Even a top-rate scientific potential will not produce the needed effect unless the company has a competitive potential in other spheres, particularly in production. Transformation of scientific and technical advantages into commercial ones requires development of extra assets and conditions.

When an economic entity develops and releases a new product, it has to face a dilemma: it can either rely on its own innovations, scientific and research work, or use those provided by other companies. About 90% of information, knowledge, and cutting-edge scientific achievements are now available to the global scientific and research community. However, it takes a bit of training for a user to perceive this knowledge. Surely, 10% of technological achievements made by the world's top organizations are proprietary or even top secret. However, most B-league organizations are doing quite well, as they rely on the "midcourse" 3 to 5 year old technologies, which make up the biggest share of those used in practice. These second-league organizations also develop their own innovations, and the original "midcourse" potential, which is based on their own achievements and makes them competitive. Not infrequently, corporate achievements and innovations do help B-league organizations attain ultimate leadership (as Apple and Intel did in their contest with Microsoft and others).

Thus, all economic objects (organizations), as they work and progress, take advantage of innovative process to a certain extent. First-league organizations have to work on their own projects, because they are too far ahead of others to borrow from them; second-league organizations borrow existing innovations from a broad spectrum of ones—patents, know-how, etc. With the openness of the global informational space, a company's ability to predict potential customers' needs and respond to them ahead of time is a key driver of competitiveness.

Indicators that characterize the success of these processes include a company's technological status and the speed of its progress, as it reflects the company's ability to adapt to ever-evolving needs and demands.

Global leaders define what the cutting-edge level is within their segments, and, as they compete, they raise it all the time and foresee the direction of the demand within a segment (Canon and Nikon; Apple, Samsung and Huawei; Airbus and Boeing). Market requires that technological progress and adaptation to the changing competitive environment should go at similar rates. Competitors, which can boast a larger number of innovations, can satisfy a greater range of needs and, consequently, win. Therefore, a slowdown in rapid development is inadmissible. In an oligopoly, competing organizations show similar rapid development rates.

Second-league organizations must show a more intensive innovative activity providing that their strategic goal is winning a larger market share through rapid development. Soaring sales and growing market shares are preceded by an increase in the innovative potential and development rate: large investments in innovative projects (a strategy chosen by Airbus in the late 1990s and 2000s in relation to the new A380 and the SuperJet project are examples) contribute to a rapid growth of these companies' product quality and potentials and help them par with or outperform leaders. It should be noted that timely and successful new-generation passenger planes A380 and A330, which have been tailored to the prospects of the airline market, have not only upgraded the company's technological base, but also launched a new inter-branch development strategy, logistics, and even rebranded it (changing the EADS to Airbus).

These projects have succeeded because the organizations executing them never cease to improve their innovative potential, competences, economic stability management methods, resource planning and accumulation of knowledge and results of their innovative activities, such as patents, know-how, etc. Their effective following of these directions has helped them to develop a scientific and technical platform for production of rapid development goods.

Rapid development, being a special kind of functioning model for an organization, explains the necessity of building an ultimately new innovative system for management of its business processes: product design, technological and material preproduction, production, output product sales, human and capital resources, economic, financial, investment, material and technical potential. In the sixth technological wave, this system relies on IT technologies:

- Artificial intelligence and machine learning
- Digital replications of objects and corporate processes
- The industrial Internet and the Internet of Things
- Big data analysis
- Key technologies, which make a basis for digital economy

Products developed by high-tech companies cannot be marketed before they are actually made; therefore, it is necessary to predict its technical and brand image, prepare ultimately new or modify existing production facilities beforehand. The required direction should be chosen based on a company's competences, financial resources, shipment and/or sales dates with account for possible risks, including a risk of losing the deadline to competitors coming up with a similar yet more technologically advanced product. There are different methods, which can be used to manage these risks. The renowned "low signal management" approach relies on the analysis of big data derived from the global informational space. Low signal management implies an ability to detect negative situations ahead of time and thus provide room for reaction and taking managerial steps to adapt to such changes and avoid loss of the competitive edge resulting from competitors' efforts.

At the first level of low signal management, when a potential danger just begins to show up and there is little information yet, the response will be general, and it will focus on strategic flexibility, which is a company's ability to adapt to occasional or regular changes in the outer and inner environments with the potential for a higher export, other competitive advantages, and competitiveness intact.

At the following levels, as specific information arrives (i.e., big data builds up), the managerial staff should respond in a more specific way to either eliminate the danger (for example, find another supplier) or use new opportunities (for example, adapt to a new political situation). In the context of strategic management, this approach is referred to as "gradually intensifying response steps" or "reacting to low signals."

To develop algorithms of reacting to low signals, it is possible to use a variety of data analysis methods, which use tools derived from statistics and computer science (for example, machine learning). For instance, in A/B testing the control selection of indicators (low signals) is compared with other selections one at a time. This helps to find an optimal combination of target indicators to ensure, for example, the most effective response to competitors' activities.

Big data helps to execute a huge number of iterations and thus obtain a statistically adequate indicator. The *predictive modeling method* helps to create a mathematical model, which foregoes a specified scenario that is likely to occur in the competitive market. The *time series analysis method* is a set of methods, which are derived from statistics and digital signal processing and used to analyze iterating data sequences. One of the most likely uses is monitoring a market conjuncture and competitors' activities in the market.

To manage rapid development processes relating to production of advanced goods and services, companies use tools for management of planned, financial, production-and-market, organizational, administrative, innovative, and other business processes. The synergistic effect of complex management of business processes will enable active introduction and implementation of strategic innovations providing for high added value, effective competition within domestic and international markets, adaptation of management strategies to rapidly changing customer needs and market conjuncture, corporate economic growth, and high productivity.

Effective use and development of unique competences can provide for global supremacy, formation of a new market of cutting-edge products and technologies, and creation of new market segments.

The complex of the main expected and promising goals and objectives posed by corporate rapid development and negotiated with account for all types of resources and deadlines should be the core of the corporate development management process. Rapid development is crucial for successful and effective functioning of high-tech organizations and science-intensive industries.

The most significant indicators characterizing the basis of a company's rapid development and reflecting the stability of its technical progress are ones that characterize its competitiveness and its ability to tailor its work to the ultimately new task of creating next-generation products. If these indicators are high, they can help to make a highly competitive product within a predicted period of time. Consequently, the company will enjoy new opportunities for opening new markets and increasing sales in an existing market.

There are methods used to evaluate the effectiveness of corporate rapid development programs, which can result in the creation of products that will suit future needs.

The project method regards a rapid development program as a set of innovative investment projects. In this case, the program's effectiveness is the sum of these projects' effectiveness indexes, and the more revolutionary, important, and advanced these projects are, the higher its effectiveness is.

The variability of different rapid development programs for a single organization consists in the presence of different sets of projects that are subject to execution; in effort-driven projects (for example, in case of winning a tender or state order)—in differences between types of resourcing and project implementation measures.

In this case, the effectiveness of rapid development is understood as cumulative effectiveness of investment and innovation projects executed by businesses.

$$\boldsymbol{E}_{\text{rapid dev.}} = \sum_{i=1}^{n} \boldsymbol{E}_{\text{pr}.i}\boldsymbol{\xi}_i, \tag{3.1}$$

where ξ_i is a project's significance in reaching goals set by an organization; n is the number of rapid development projects.

This approach does not reveal the innovation component in project, as it reflects their equal novelty (innovativeness). In fact, different innovative projects result in production and sales of products that have a different degree of novelty. A more effective and therefore more economically important project may be less innovative. To take the innovation component into account, an innovativeness (novelty) coefficient (ν_i) of project (j) can be introduced into the formula (3.1) (See Table 3.1):

Table 3.1 A scale for defining the novelty coefficient v_i

Novelty of concept	Novelty of product	Range of values
World-class novelties (discoveries and patents) used	Ultimately new product in the market	1
International patents used	Outperforming all competitors	0.8
Concepts used nationwide	A product that is new in a domestic market	0.6
Corporate novelty	Modernization of old assets	0.4
Innovation in technologies used in a commercial product	Commercial product	0.2

$$E_{\text{rapid dev.}} = \sum_{i=1}^{n} E_{\text{пр}.i} \cdot \xi_i \cdot \nu_i. \tag{3.2}$$

In most cases, the project significance index (ξ) is a company's internal characteristic. In cases whereby a project goes beyond the corporate boundaries and becomes a national project (for example, SuperJet 100), and whereby its goals coincide with national ones, the project significance index (ξ) should increase.

The high-speed rapid development evaluation method is based on the fact that a rapid development plan also includes steps that ensure ultimate improvement of an organization's state in its most important spheres. Each event should result in an innovation and, consequently, a quantum leap in the product, resources used, production, and design processes.

Different programs used by a single company may and should differ in the types of events and degree of innovativeness. The novelty of an event in each sphere can be measured by the greatness the product or resource quantum leap. An event (and, consequently, a program), which gives a company a stronger push toward high-quality product, greater potential and effective management and therefore shows a speedier development, is more innovative.

Indicators characterizing the qualitative state of an object at a certain moment of time (t) are shown in Table 3.2.

Structural changes (or the speed of rapid development) at moment $t_0 + \tau$ (where τ is the duration of the rapid development program) are calculated according to formulas:

$$W_1 = \Delta P^{\text{pr}} = P^{\text{pr}}_{\tau+t_0} - P^{\text{pr}}_{t_0}; W_2 = \Delta P^{\text{cons}} = P^{\text{cons}}_{\tau+t_0} - P^{\text{cons}}_{t_0};$$
$$W_3 = \Delta P^{\text{org}} = P^{\text{org}}_{\tau+t_0} - P^{\text{org}}_{t_0}; W_4 = \Delta P^{\text{other}} = P^{\text{other}}_{\tau+t_0} - P^{\text{other}}_{t_0}.$$

For the most part, prognosis and evaluation of the main indicators rely on big data analysis. In order for an organization to continue on the rapid development course, it is necessary to monitor the environment surrounding it. Therefore, the traditional big data analysis, which does not provide for real-time analysis of large data blocks, should get up to a higher level. This should enable real-time analysis of relatively

Table 3.2 Indicators showing an entity's quality at specific time (*t*)

Activity area of a company	Measure of state at specific time (*t*)	Formula used to calculate the final result at specific time (*t*)	Indicators
Product	Product quality	$P_t^{\text{prod}} = a_1D_1 + a_2D_2 + a_3D_3$	D_1—market share; D_2—innovative product's market share; D_3—share of a rapidly developed product
Potential	Quality of potential	$P_t^{\text{cons}} = b_{\text{P}}Q_{\text{IP}} + b_sQ_s$	Q_{IP}—quality of industrial potential; Q_s—quality of scientific potential
Organization	Quality of organization and management	$P_t^{\text{org}} = \text{c}_1\text{A} + \text{c}_2A_2 + \text{c}_3A_3$	A_1—use of advanced management methods; A_2—use of IT in production
Property and other directions	Change in ownership structure and diversification	$P_t^{\text{other}} = d_1C_1 + d_2C_2$	C_1—government's share in the property; C_2—diversification

Note: *a, b, c, d* are weighing coefficients defined by an expert-based or statistical method

large data (it is needed for low signal management) mainly with the use of computer memory. This will help an organization to influence events and tendencies at the time when they are observed.

The graphic image of a structural change that results from rapid development is shown in Fig. 3.13.

The final effectiveness index of a rapid development program is calculated according to the formula:

$$\boldsymbol{E}_{\text{rapid dev.}} = \mathbf{0.4\Delta}\boldsymbol{P}^{\text{pr}} + \mathbf{0.3\Delta}\boldsymbol{P}^{\text{cons}} + \mathbf{0.2\Delta}\boldsymbol{P}^{\text{org}} + \mathbf{0.1\Delta}\boldsymbol{P}^{\text{other}},$$

where weighing coefficients are based on the importance of the organization's area of activity, which is part of the current rapid development program.

Large and highly integrated organizations boast a great potential for rapid product development. They do because it is impossible for a company to handle rapid product development without cooperating with other organizations. Design and creation of high-tech products is carried out by several companies and their design and engineering divisions. These cooperating enterprises' ability to properly follow their individual technical design specifications and come up with a technically relevant output product will determine the high-tech project's general performance. That means meeting appropriate technical and economic characteristics, which can maximally satisfy customer and market needs and have consumer properties that can form a sales market for a new product.

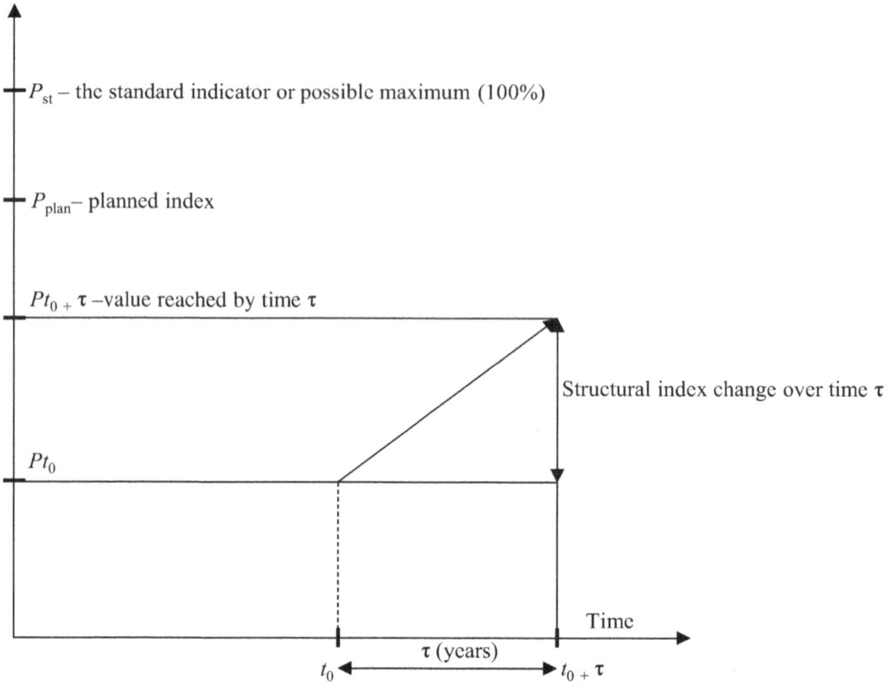

Fig. 3.13 Measuring structural changes

Assume a parent organization has an *n* amount of affiliate organizations. To evaluate the effectiveness of the company's overall development, it is possible to use a simple method and reevaluate the effectiveness while taking into account all projects run by affiliate companies. However, the significance of these affiliate groups' projects should be reconsidered from the perspective of the fulfillment of the company's goals. These goals may differ from those pursued by affiliate organizations.

The high-rate approach to evaluation of effectiveness, which is applied to organizations, implies counting up of all original indicators for an organization in general, treating it as a single economic entity. Actually, it is what defines the structure of rapid development, as all events and indexes should work and signal fulfillment of the goal pursued by all parts of the organization. However, practice demonstrates a possibility of calculating the rate of rapid development *(W)* for each affiliate separately. While natural indicators of development are comparable for affiliated organizations within a single industry (particularly, the comparison can be carried out as part of a technological audit), using a unified norm-setting standard can bring different organizations' indicators to one level (Fig. 3.14).

Mathematically, a rapid development rate is expressed in Fig. 3.14 by oblique angle α_i expressing an organization's development trajectory over time (τ) to the new quality index, i.e., the first derivative of innovative activity:

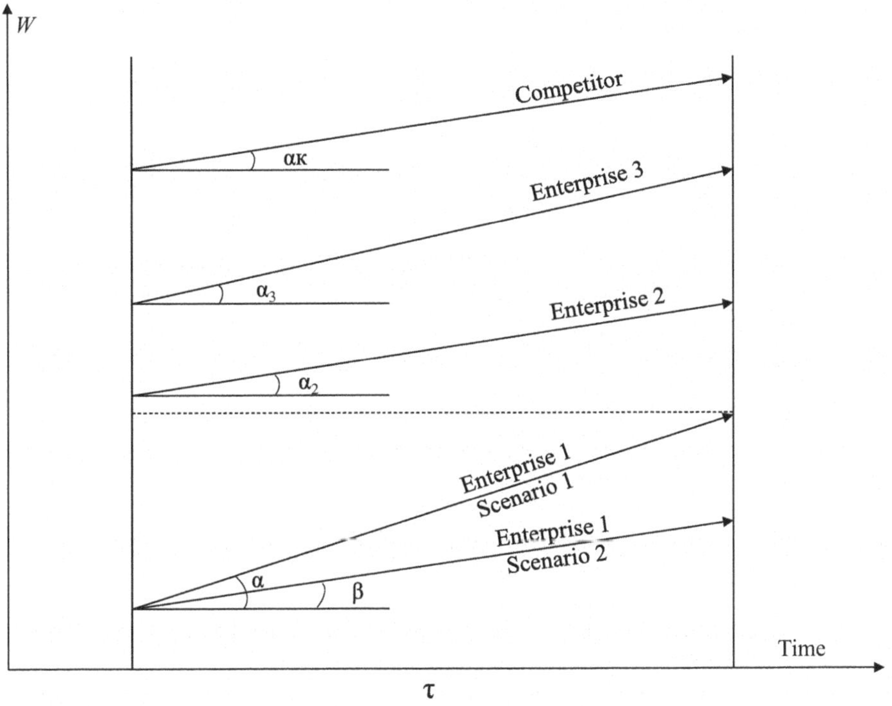

Fig. 3.14 Different rapid development rates for different advanced programs

$$\boldsymbol{W_\tau = W_t + \sin\alpha \cdot \tau}$$

Advanced product's quality indicators are closely connected to its production cost. Management of production cost should take place at all phases of the product's life cycle. Input data and information that is used for evaluation of production cost and labor intensity should be complemented, specified, and corrected at each design stage. The more specific a project is in all its parameters, the more precisely and thoroughly its prime cost and labor intensity will be evaluated. The modeling and economic analysis should be related to each other via processes, and their results should influence the designer's adjustments, allowances, etc.

Because it is necessary to evaluate the prime cost and expected price at the earliest stages of project execution, it is quite advisable to possess a set of economic tools and an information system, which helps to:

- Quantify a product's prime cost with account for types of materials used in it (providing that target technical characteristics have been achieved) and setting admissible run out accuracy, complexity of assembly, etc.
- Model the prime cost and labor intensity with varied allowances, adjustments, processing accuracy, materials, etc., given the necessity of achieving target technical characteristics and developing the most optimal product structure

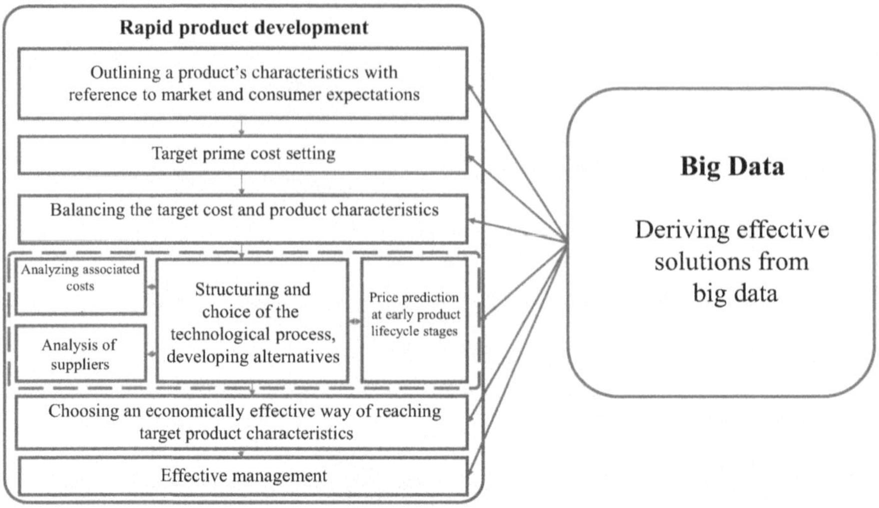

Fig. 3.15 The use of big data at all rapid product development stages

- Define a complete structure of the product's and its component modifications' prime cost with the use of specifications and flow routes
- Find and specify direct and production costs resulting from the use of various technological operations, processes, specific materials, and from meeting parameters (for example, processing accuracy)

Big data analysis is a powerful instrument, which can be effectively used at all stages of rapid product development and become a key driver of prime cost optimization (Fig. 3.15).

The versatility and effectiveness of the rapid comparison of possible rapid development scenarios is quite so evident. In order to compare rapid development programs followed by different companies, it needs to make a few assumptions. The rate of rapid development (or the intensity of work aimed at rapid development) should be matched to a company's size and its position in the market; in other words, it requires a look at its position in its segment relative to leaders and competitors. At certain stages of market development, a large company holding a substantial share in its segment may demonstrate low innovative activity and maintain growth through active use of older products. Large companies have a stronger inertia and have to exert greater effort (in terms of financial and resource investments) to gain speed in rapid development. This is a formal illustration of the fact that small companies operating within narrow segments can demonstrate maximal innovative activity and promptly bring an advanced product to market.

Therefore, to speed up the rapid development process, it could be advisable to stimulate highly innovative small affiliate organizations and use them as an additional driving force of the parent organization's development. Large companies, which operate confidently in traditional markets, tend to act conservatively within small time intervals and demonstrate explosive growth when releasing an ultimately new next-generation product. These innovative leaps are fueled by substantial investments in R&D and the new production. Also, most significant and some revolutionary innovations emerge in manufacturing processes when executing new projects that are aimed at satisfying advanced needs. As the task of creating an advanced product is executed, the entire productive–economic system will be changing, and a company will be able to make a breakthrough, as it has moved on to the next technological wave and taken over the leadership in the market.

A company's current intellectual potential and human capital, centers of scientific, technological and manufacturing competences, provide a basis for designing and creating a product with ultimately new features, competitive economic indicators, high customer appeal and values, modifiability, and adaptability to changing demand.

3.3 The Economy–Production–Economy Cycle as a Form of Creating Competitive Goods

The previous section focuses on creation of new types of goods for future needs. It states that predicting their technical and brand image should be based on a deep analysis of the economic environment, which includes both manufacturers and consumers. Also, it highlights the necessity of predicting a product's competitiveness and competitive price needed for outlining design and production processes while relying on these parameters. Thus, the external economic environment defines the portfolio of data, which needs to be used for launching internal production and design processes.

Because a highly innovative organization always aims for rapid development that consists in creating a new market or capturing part of an existing one, the result of this activity and the appearance of a new product influence the economic system and modifies it. The qualitative change of the market will depend on how competitive the product is going to be. The *economy–production–economy* cycle as a form of creating innovative products is shown in Fig. 3.16.

The structure of a product's significance for a manufacturer is shown in Fig. 3.17.

The main components of this significance are a product's prime cost and its ability to bring the company to rapid development. On the other hand, the product's significance for consumers is expressed in its market price and usability (Fig. 3.18).

First of all, a product's competitiveness is expressed in its salability in domestic and foreign markets with competitors' activity. A consumer often chooses a product, which satisfies a greater number of needs per price (price + consumption cost) than a similar one produced by a competitor.

Generally, competitiveness reflects the beneficial effect/price ratio. Capitalized products' consumption cost is usually much higher than their sales price. Consumption cost consists of general operation costs sustained throughout a product's life

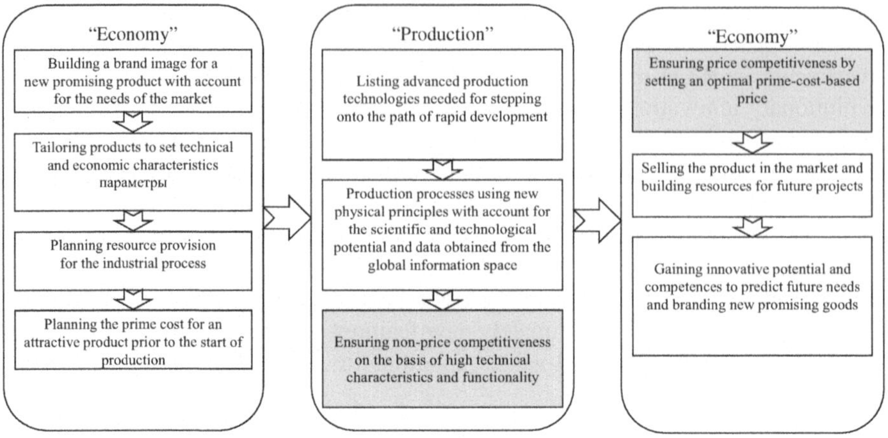

Fig. 3.16 The economy–production–economy cycle as a form of creating innovative products

cycle. Therefore, a product showing the lowest consumption cost throughout the operation period will be the most competitive one. It should be noted that actually a company sells not only the product itself, but also a complex of important services related to operation and maintenance of devices, equipment and materials consumed at all stages of the post-sale part of the life cycle. It is also noteworthy that a buyer can commit to long-term costs defined by the product's life.

The competition between similar products and services offered by different companies, as they differ in price, quality, and other characteristics, is referred to as *direct competition*. The concept is based on product differentiation, which is quite so common to civilized economies. Differentiation spreads not only to consumer goods, but also to production goods, and it is manifested in deeper specialization and a larger share of one-off production.

The relationship between direct competition and a product's competitiveness is quite evident. Most manufacturers (sellers), when entering a specific market, expect that there are products in the market similar to theirs and keep this fact in mind while ensuring their goods' competitiveness.

A competition between products, which satisfy just one type of need, is termed *functional competition*. It may occur even when producing goods that have one-of-a-kind characteristics. The relationship between functional competition and products' competitiveness is fairly concealed. Functional competition implies a "fight" between products relating to different industries yet satisfying similar needs.

Therefore, a seller embarking on a new market should be aware that the product is going to face competition both on the part of (functionally) similar and different ones, in cases whereby they are substitutional and meant to satisfy one and the same consumer need. According to M. Porter, manufacturers of substitutional goods make up one of the Five Forces.

The effectiveness of a managerial decision being made depends on the timeliness, integrity, and relevance of incoming information. Besides, the manager should

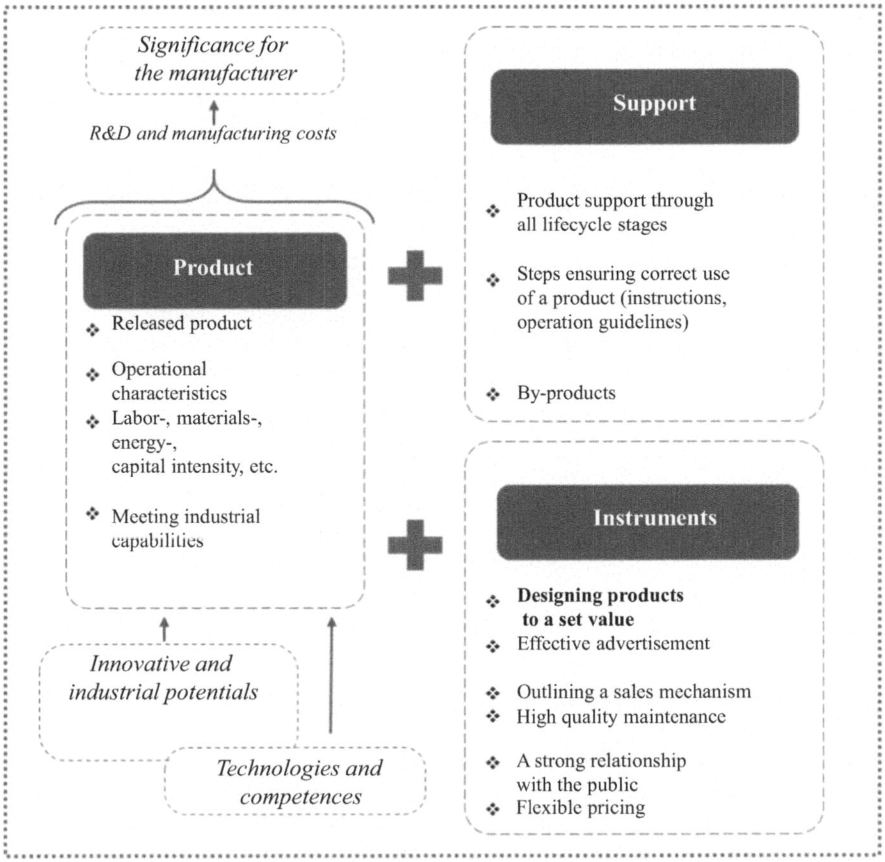

Fig. 3.17 A product's significance for the manufacturer

receive a compact and perceivable block of information, which should reflect a complete picture of the company's functioning with account for external and internal influences.

External influences are not created by a separate company; they are generated by acting market agents (competitors, raw material suppliers, materials, components, services), which exert indirect influence on a company's pricing, scientific/technical and production policy, as well as by the government that uses economic, political, and administrative tools to regulate state, private, and mixed-type companies and the industry in general (state laws, bylaws, directives, standards, regulations, political, and economic terms). These influences depend on the state's policy and may either limit (taxes, tariffs, customs barriers) or stimulate the creation of competitive goods, increase branch exports, gross output (subsidies, privileges, investments, etc.).

Internal influences, which determine an organization's and its product's competitiveness, include technical level of production, internal skills, strategy of development, financial policy, etc. With effective accounting, analysis, and management,

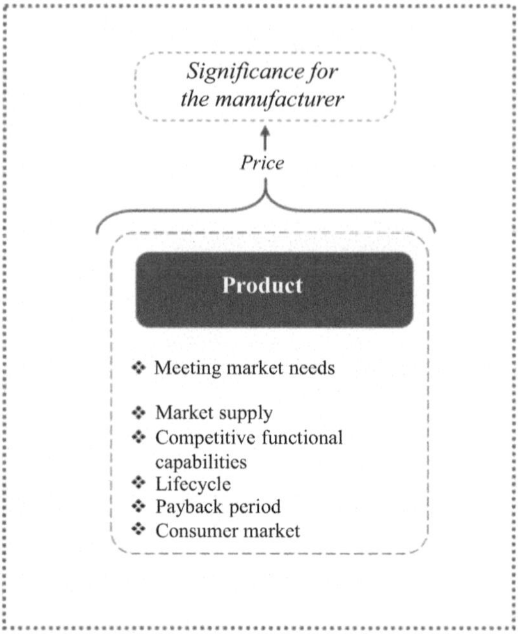

Fig. 3.18 Significance for consumers

these influences can help organizations to provide competitive advantages for their goods and themselves.

Today, there are new factors influencing organizations both from inside and from outside. These include digitization factors, which are presented in the general system of influences, which determine an organization's competitiveness (Fig. 3.19).

Once a new type of product appears in the market and consumer uncertainty increases, traditional management methods shown in Fig. 3.19 may lose effectiveness. In this case, products' parameters' failure to meet consumer needs will be a direct cause of slumping sales, increasing unsold output and, consequently, worsening financial state. To avoid a crisis and loss of opportunities for stable development, the organization should provide for the following things to manage, create, and implement competitive advantages:

- An advantage in sales markets over competitors' products
- A higher consumer performance through advertisement that will bring it to various markets, provide demand and stimulate sales
- Developing and launching production of ultimately new types of goods with a high customer performance, which will ensure sales in domestic and foreign markets
- A flexible corporate pricing policy

To do that, it is necessary to evaluate the product's competitiveness in the sales market in the following order:

- Analyze the market and select the most competitive product sold by a competitor

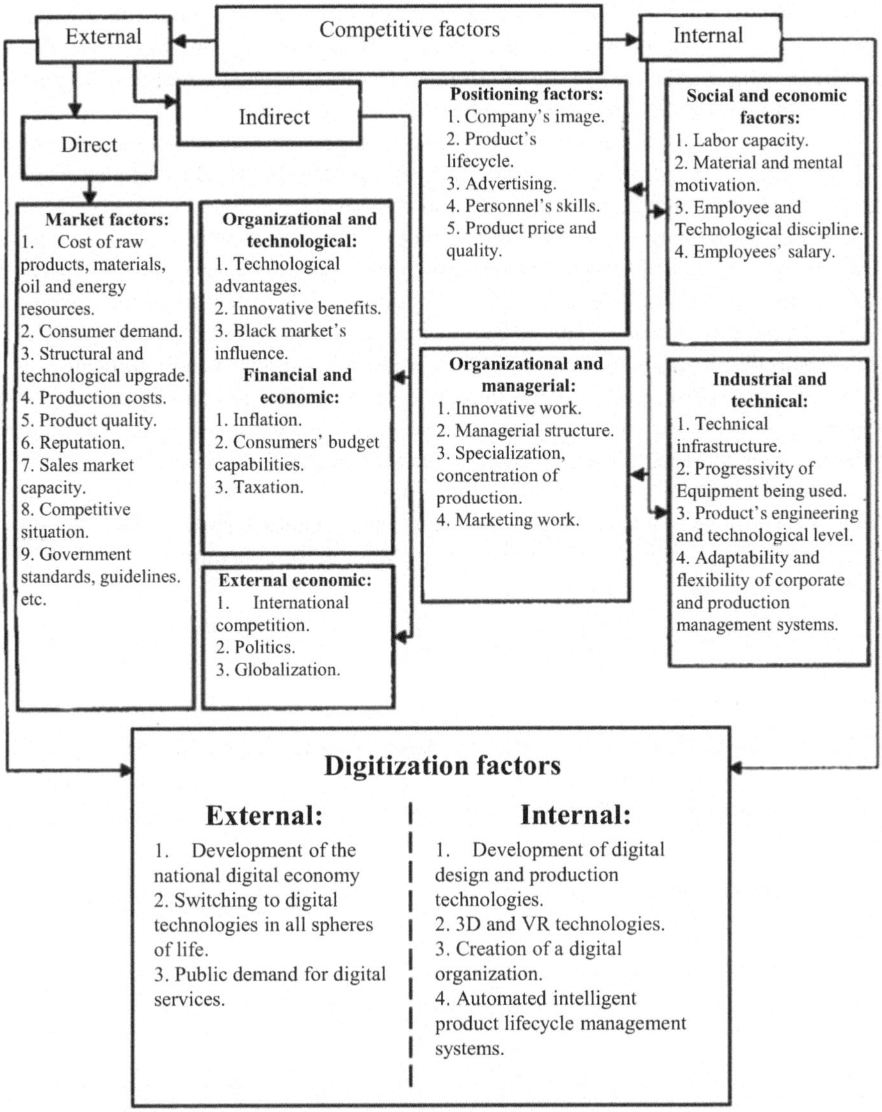

Fig. 3.19 Internal and external factors determining an organization's competitiveness

- Point out the most important characteristics of different types of products, which make them competitive in the market
- Calculate the product's integral competitiveness indicator

Data for recording and analyzing an organization's performance should be based in the main goal of its work, which is rapid product development. The amount of this data, which has a degree of significance, should be sufficient to enable the manager to make managerial decisions pursuing individual (private) purposes and the main goal. However, even the most optimal part of this data is huge and hard to analyze. For this reason, mathematical modeling methods are practiced widely, as they help to derive a broad indicator from calculated separate indexes.

The price is important in competitiveness management because most companies can only change it within set boundaries. Also, unlike other product characteristics, the price can be managed more quickly and effectively.

Characteristic of most products and services is an inverse price–competitiveness relationship. A reduced price means a higher competitiveness and vice versa. Meanwhile, qualitative assessment of the influence of the price on competitiveness is still a very difficult task. This makes it necessary to build mathematical models, which help define the influence of product prices on competitiveness indicators.

Another must-study issue is the influence of price changes on competitiveness, which depends on the time factor. Particularly, it is necessary to study the price's medium-term and long-term influence on competitiveness.

Assume the price/quality ratio is the main competitiveness indicator, quality being a composite index of the product's qualitative characteristics. The higher the quality indicator is, the more attractive the product is for consumers.

The model uses the following values:

K—product quality
S—price
Q—competitiveness

Because the competitiveness indicator depends on price dynamics, the model is not going to focus on statistical competitive, quality, and price indexes; instead, it is going to focus on how they change over time:

Q(t)—change of competitiveness
K(*t*)—change of quality
S(*t*)—change of price

A common formula for these functions is:

$$Q(t) = \frac{K(t)}{S(t)}.$$

To determine the influence of changing quality and price indexes on the competitiveness index, it needs to differentiate this ratio by time and plot a differential equation:

$$Q'(t) = \frac{K'(t)S(t) - K(t)S'(t)}{S^2(t)}.$$

It will be simpler in a case whereby the quality index is a constant value. In this case, the differential equation is:

$$Q'(t) = -\frac{K_o}{S^2(t)} S'(t);$$

$$K(t) \equiv K_o.$$

This equation reflects the influence of an instant price change on the competitiveness indicator. However, a deeper research on the medium-term or long-term influence of price change on competitiveness requires that these differential equations be modified to embrace the lagged effect and the residual effect.

To evaluate the medium-term influence of price change on a product's competitiveness, it is necessary to add lagged-effect addends to the right-hand side of the equation:

$$Q'(t) = \frac{K'(t)S(t) - K(t)S'(t)}{S^2(t)} + \sum_{k=1}^{N} f_k(Q(t - h_k), S'(t - h_k)),$$

where $h_k > o$ stands for lagged time.

This is a difference–differential equation reflecting the medium-term dynamics of the competitiveness indicator.

The point of this equation is that the dynamics of the competitiveness indicator depends not only on instant product price changes, but also past-time competitiveness indicators. From the medium-term perspective, that is because the competitiveness index is influenced not only by objective product characteristics, but also by consumers' subjective opinion about the product. As it is known, customers' opinion is somewhat inertial. Therefore, in case of an instant price change, the consumer opinion will change depending on what this opinion was like at the early stage of the medium-term period.

This model considers discrete time lags, because the consumer opinion changes significantly over a medium-term period, depending on the product price change. The addends in the abovementioned difference–differential equation contain lag variables and depend not only on past-time competitiveness indicators, but also on price values observed during those periods. This reflects consumers' common psychological reaction on product price changes.

A very simple example of a difference–differential equation describes the dynamics of competitiveness with the medium-time product price change dynamics:

$$Q'(t) = \frac{K'(t)S(t) - K(t)S'(t)}{S^2(t)} + \sum_{k=1}^{N} a_k Q(t - h_k) S'(t - h_k),$$

where a_k coefficients and negative and diminish in modulus with index k growing:

$$|a_1| \geq |a_2| \geq \cdots \geq |a_N|.$$

Evidently, all suggestions concerning the properties of f_k functions are fulfilled.

Next, the *long-term* influence of the price change on competitiveness should be considered. This case also implies generalization of the differential equation, which describes the dynamics of competitiveness with changing prices, as well as modification of differential equations with the use of functional and differential equations, which embrace past-time values. Discrete instants of time cannot embrace the influence of the history of indicators. Therefore, a distributed time lag should be considered. The equation is:

$$Q'(t) = \frac{K'(t)S(t) - K(t)S'(t)}{S^2(t)} + \int_0^t F(t-\tau, Q(t-\tau), S'(t-\tau))d\tau.$$

This is a functional–differential equation reflecting the long-term dynamics of the competitiveness indicator.

The economic concept of this equation is that the rate of competitiveness change depends on past-time competitiveness values and the product price change. Unlike the medium-term competitiveness dynamics equation, in this equation the time lag is distributed over the entire time interval. Definitely, the influence of the price change should diminish over time.

Another important matter is management of the competitiveness index through product price change. Management of competitiveness is paramount for all organizations. Although companies do their best to boost their competitiveness indexes, their competitiveness management opportunities are limited. Competitiveness is influenced by:

- Objective product characteristics
- Product price
- Objective characteristics of competitors' products
- Prices of competitive products
- Consumers' opinion

Most organizations are capable of changing their products' objective characteristics, prices, and consumers' opinion.

As a rule, changing objective characteristics poses serious technological hindrances and time expenditures. In most cases, innovations entail huge financial expenses.

For large companies, influencing consumers' opinion is a must-do procedure, and it is completed with the help of advertisement. However, influencing consumers' opinion is an unpredictably difficult task.

It is much easier to manage competitiveness by changing the price. It is possible to change it quickly, which is important in a competitive environment. Technically, changing a price is an easy task. However, the price is one of the most important drivers of competitiveness.

Creating high quality yet economically effective products is a crucial task, which can only be solved through the use of innovative technologies. They should reduce the absolute prime cost value of products being designed and released, as well as ensure a relative reduction of costs per unit of effectiveness.

One of the most important aspects of production is providing due output quality and high operational characteristics. High quality is provided through strict following of technical specifications concerning design and production technology. The quality of products depends on designer performance and technological support. However, quality criteria are not the only indexes of the technical base, because at the current stage of technology as a science, efficiency becomes part of it that equals integrity and quality matters. Only complex research on engineering and technology issues can produce correct and scientifically relevant solutions with accuracy, quality, integrity, and efficiency matters being treated as a single whole.

Therefore, reduction of the prime cost with due quality and competitiveness intact is a crucial part of introduction of innovative technologies. Analysis of certain industrial processes shows that it is paramount to automate operations that have the greatest influence on products' accuracy and quality indexes. Quality improvement, strict monitoring, and control over the use of materials greatly influence a product's prime cost.

Another way to reduce prime cost in engineering is error limit optimization for infrastructural elements, which provide for due quality with minimal costs.

Realization of these is possible with mathematical error limit modeling and engineering of technological processes, which rely on essential relationships between quality, size accuracy, cost of production, and production technology indicators. Calculation of error limit optimization is labor intensive and cannot be carried out without using computerized tools. Because economic effectiveness of a product is both an engineering/technological and economic objective of an organization, they should be treated as part of an entire task complex for integrated automated design, engineering, and production systems. Effective production is a major economic challenge. Output products' prime cost is a crucial qualitative characteristic of a company's work.

In mechanic engineering this characteristic depends on a lot of objective factors resulting from a product's infrastructural characteristics and organizational/technical manufacturing environment. Factors like the number of original parts, their complexity and accuracy, use and cost of materials and other factors, exert direct influence on the cost of production. These factors' influence largely depends on manufacturability, i.e., structures' embracing of optimal parts manufacturing and assembly conditions. Products with the same structural and operating characteristics may include several structural solutions, which differ greatly in prime cost and, consequently, in sales price, due to different manufacturability.

Also, there are organizational and technological influences on the cost of production. These include series production, line of production, use of standardized parts and units, technical performance and technological instrumentation, organizational and technical performance of all corporate auxiliary services, quality management of production, etc.

Economic aspects of tailoring a product to a specific cost (logical design phases), which illustrate the relationship between economy and production, are shown in Fig. 3.20.

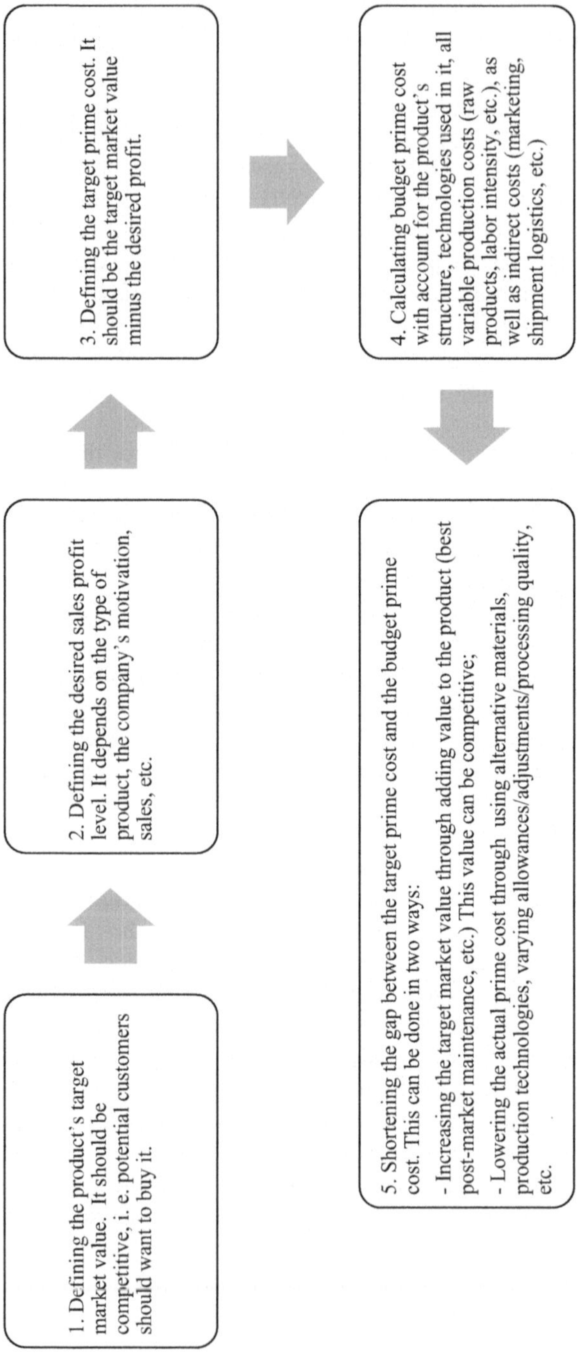

Fig. 3.20 Economic aspects of tailoring products to a specific cost (logical design phases)

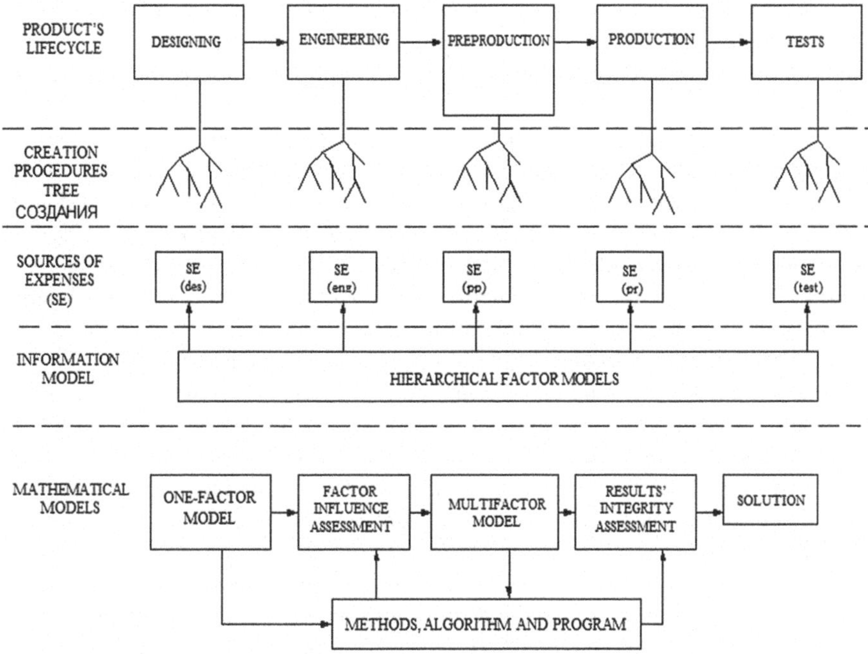

Fig. 3.21 General algorithm of prime cost change research used throughout a product's life cycle

Traditional statistical cost analysis and calculation methods can be quite effective providing there are large statistical samples (mass serial production), but they are not applicable to product design and engineering phases of one-off production. The latter case requires using more complex cost calculation models, which rely on big data analysis.

To analyze prime cost and find ways of reducing it, it is important to track the sources of production and sales costs, as well as the content, origin, and influence of prime cost drivers (Fig. 3.21).

When manufacturing a part, it is hardly possible to strictly follow dimensions set by the design specification. Consequently, there are always deviations from absolute dimensions and forms, and the challenge is to minimize errors. A design-basis justification for error limits can be based on the theory of dimensional chains. A dimensional chain is defined by independent parameters that are unified into a closed contour and act within it as interrelated values. For example, a group of single-diameter parts connected through holes can make a dimensional chain. In this case, minimization of deviations consists in minimization of clearance Δ, which depends on the diameter of the hole d_{hole} and shaft d_{shaft}. When building the "hole-shaft" link of the dimensional part, the clearance acts as a master link, and diameters d_{hole} and shaft d_{shaft}—as intermediate ones (augmenting or diminishing, depending on whether the master link Δ is increasing or diminishing).

All these factors produce different effects on a product's prime cost. Therefore, its cost is a function $C = f(x_1, x_2, \ldots, x_n)$ defined by independent parameters x_i, which describe these factors. The function of $C = f(x_i)$ expresses these parameters' influence on costs.

The number of influences is great, and many of them are qualitative and cannot be fully expressed in quantitative values; some of them are fairly vague.

Deriving the function of $C = f(x_i)$, $i = 0,1,\ldots, n$ from pure theoretical data is hardly possible. In this case, analysis of the past history of production plays an important role, as it provides actual data on costs and x_i indicators observed when carrying out similar operations or manufacturing a similar product. The availability of a sufficient number of analogous operations and/or products enables the use of statistical methods of predicting expenditures on science-intensive products.

A mathematical interpretation of prime cost will make it possible to manage an organization's or its product's competitiveness and, consequently, to take specific actions aimed at using innovative technologies. These technologies ensure high competitiveness, as they bring the product's characteristics into compliance with international standards or above those of similar products while maintaining a competitive price.

In turn, a competitive price can result from the use of innovative technologies, which boost labor efficiency, reduce the consumption of energy and materials, provide for uninterrupted and continuous use of equipment, etc. All these activities can reduce the product's prime cost and make it more competitive in the global and domestic markets.

Renovating products with account of the evolution of needs as a driver of competitiveness.

As a rule, competitive advantages are created at the engineering phase and through the technological setup, distribution, and promotion stages.

Product renovation is crucial part of boosting an organization's economic effectiveness (in the civilized world, automotive, aviation, engineering and other companies upgrade their products every 1 or 2 years). Therefore, it is necessary to reveal basic technical and economic conditions, which explain the practical necessity of making a specific product.

Effectiveness of all products, including new ones, is expressed in a ratio between the economic effect (profit) and total costs resulting from its engineering, manufacturing and operation. In a market environment, a company's progress tightly depends on the profit that results from economic activity; therefore, it should define the necessity of upgrading output products, estimate related costs and set optimal and economically relevant timeframes.

In practice, technical and economic conditions for upgrading products depend on a group of factors.

First, those are technical and maintenance factors, such as:

- Discrepancy between consumer demand (observed during a certain period) or expected technical or maintenance characteristics of a competitor product (TC_{comp}) that is readying for release, and the same characteristics of a phasing out product ($TC_{phaseout}$). In a formalized form, this condition can be expressed as $TC_{comp} > TC_{phaseout}$.

- Ratio between the market and consumer demand (TC_{comp}) and new products' characteristics (TC_{comp}), i.e., $TC_{new} \geq TC_{comp}$, observed during a certain period.

Second, there are economic drivers of renovation, which meet users and the manufacturer's economic interests and opportunities, such as:

- The price for the new product (P_{new}) should not exceed that of a competitor's product or consumers' budget capabilities P_{comp}, i.e., $P_{new} \leq P_{comp}$.
- The price for a phasing out product ($P_{phaseout}$), given the accepted wholesale discounting practice, should be higher than a possible consumer price for this product (P_{cons}), i.e., $P_{phaseout} > P_{cons}$, providing that P_{cons} tends to zero.
- Profit per unit of new product cost (PR_{new}) should exceed or equal profit per unit of average renovation cost within the organization (corporation) (PR_{av}), i.e., $PR_{new} \geq PR_{av}$.
- Profit per unit of cost phasing out product cost ($PR_{phaseout}$) should be lower than the average corporate profit (PR_{av}), i.e., $PR_{av} > PR_{phaseout}$ providing that $PR_{phaseout}$ tends to zero.

According to the formulated baseline standards of an organization (corporation), the production and distribution of the new type of goods will be economically relevant, and the goods will be competitive if the following technical and economic conditions are followed:

$TC_{new} \geq TC_{comp}$;
$P_{new} \leq P_{comp}$;
$PR_{new} \geq PR_{av}$.

In this case, a phaseout is expressed as follows:

$$\left.\begin{array}{c} TC_{comp} > TC_{phaseout}; \\ P_{phaseout} > P_{comp} \text{ with } P_{comp} \rightarrow 0; \\ PR_{av} > PR_{phaseout} \text{ with } PR_{phaseout} \rightarrow 0 \end{array}\right\}.$$

Technical and economic environment, in which a new product is launched, is just the beginning of justification of its renovation; they require further specification and clarification based on the product's specific significance, demand for it in a certain period of time, and the environment, in which it is used and operated. Given that every manufactured item operates throughout its rated life, during which it produces a particular standard effect E_{si} that is measured in dollars (i is a product's numerical order), and design, manufacturing and operation costs (standard costs S_{si}) have been calculated, the standard effectiveness of costs resulting from the design, manufacturing, and operation of a specific i-number product, is its standard effect per \$1 of standard costs.

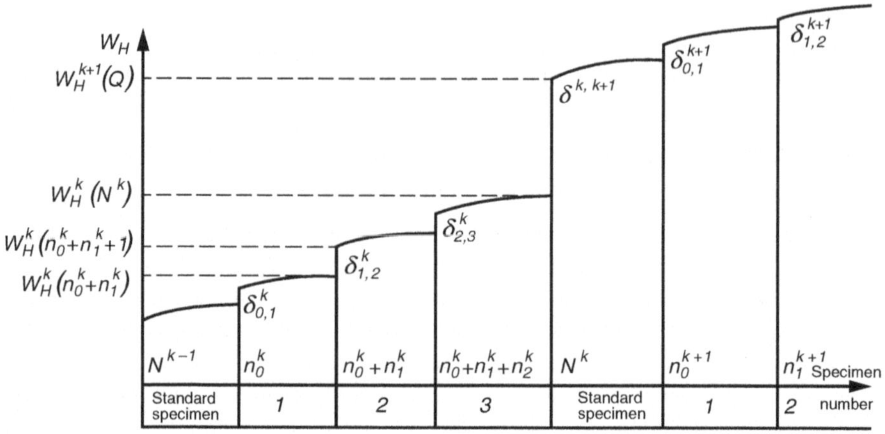

Fig. 3.22 Graphical model of product development: 1, 2, 3 are modifications of the standard product specimen

$$W_{si} = E_{si}/S_{si}.$$

Plotting on the *X*-axis numerical orders of the manufactured products, and on the *Y*-axis—their standard effectiveness, results in a diagram, which sheds light on how the standard effectiveness of costs changes as the mastered production grows technically and economically, new modifications of previously mastered products are created and new generations of particular types of goods arrive (Fig. 3.22).

The diagram presents the simplest model, which gives a quality-level idea of the laws of product type development. Figure 3.22 has the following legend:

n_0^k is the general number of manufactured *k*-generation standard product specimens;
n_j^k is the general number of manufactured *j*-modification standard *k*-generation specimens;
N^k is the general number of manufactured *k*-generation specimens:

$$N^k = \sum_{j \to 0} n_{Hi}^k;$$

W_{ni}^k is the standard effectiveness of the *i*-number *k*-generation specimen; $1 \leq i \leq N_k$;
$\delta_{j,j+1}^k$ is the difference between the initial standard effectiveness of the standard specimen (modification $j + 1$) and the achieved standard effectiveness of the previous standard specimen (modification *j*) (the standard specimen is designated as *zero modification*: $j \geq 0$):

$$\delta^k_{j,j+1} = W^k_{\mathrm{n}}\left(n^k_0 + n^k_1 + \ldots + n^k_{j+1}\right) - W^k_{\mathrm{n}}\left(n^k_0 + n^k_1 + \ldots + n^k_j\right);$$

$\Delta^{k,\ k+1}$ is the difference between the initial standard effectiveness of the new (k +1)-generation specimen and the achieved standard effectiveness of the latest-modification k-generation (previous-generation) standard specimen:

$$\Delta^{k,k+1} = W^{k+1}_n(0) - W^k_n\left(N^k\right).$$

In a market economy, an organization should master a new product when it is sure it is economically relevant: the economic effect, which has increased thanks to the rising standard effectiveness of the newly mastered specimen compared to the previous one, should in due (standard) time compensate for upgrade (production capacity creation) and preproduction costs.

That can be formally expressed in the following way:

$$\sum\nolimits_{i=i_0}^{i_0+P'_{\mathrm{new}}(t_{\mathrm{ok}})} W'_{ni}S'_{ni} - \sum\nolimits_{i=i_0}^{i_0+P_{\mathrm{old}}(t_{\mathrm{ok}})} W'_{ni}S'_{ni} \geq S_{pp},$$

where i_0 is the number of the product, which pioneers the release of a new specimen;

t_{payback} is the standard upgrade (production capacity creation) and preproduction payback period;

$P_{\mathrm{old}}(t_{\mathrm{payback}})$, $P'_{new}(t_{\mathrm{payback}})$ is the number of old and new items, which should be produced over t_{payback} to satisfy the demand for this type of product (hereinafter P_{o}, P_n);

S_{pp} is upgrade (production capacity creation) and preproduction costs.

The standard effectiveness and standard costs do not change much from product to product (unlike values $\delta^k_{j,j+1}$, $\Delta^{k,k+1}$), so they can be replaced by their own averages, which do not depend on i:

$$\left.\begin{aligned} W_{\mathrm{n}i} &= \overline{W_n} \approx \mathrm{const}; \\ W'_{ni} &= \overline{W'_n} \approx \mathrm{const}; \\ S_{ni} &= \overline{S_n} \approx \mathrm{const}; \\ S'_{ni} &= \overline{S_n}' \approx \mathrm{const} \end{aligned}\right\}.$$

With the assumptions, the inequation takes on the following shape:

$$P'_{\mathrm{new}}\overline{W_n}'\overline{S_n}' - P_{\mathrm{old}}\overline{W_n}\,\overline{S_n} \geq \mathrm{S_{pp}} \tag{3.3}$$

A comparison between values $\overline{S_n}'P'_{\mathrm{new}}\overline{S_n}\,P_{\mathrm{old}}$, which are the costs of satisfying the demand for old and new specimens over the payback period (t_{payback}), the following ratio will be economically relevant:

$$\overline{S_n}' P'_{\text{new}} \leq \overline{S}_n \, P_{\text{old}}. \tag{3.4}$$

Inequation (3.4) demonstrates that replacement of old products with similar advanced ones should not entail appreciation. The stability (lowering) of capital intensiveness with advanced equipment replacing outdated one is an example of how the requirement can be used in practice.

Inequation (3.4) can be expressed as follows:

$$\frac{\overline{S}_n \, P_{\text{old}}}{\overline{S_n}' P'_{\text{new}}} = 1 + \alpha, \quad 0 \leq \alpha < 1 \tag{3.5}$$

where

$$\alpha = \frac{\overline{S}_{\text{n}} P_{\text{old}} - \overline{S}'_{\text{n}} P'_{\text{new}}}{\overline{S}'_n P'_{\text{new}}} \tag{3.6}$$

Using inequations (3.5) and (3.6) in expression (3.3) results in:

$$\overline{W}'_n - \overline{W}_n \geq \frac{S_{\text{pp}}}{S'_n P'_{\text{new}}} + \frac{\overline{S}_n P_{\text{old}} - \overline{S}'_n P'_{\text{new}}}{\overline{S}'_n P'_{\text{new}}} \overline{W}_n. \tag{3.7}$$

The right side of inequation (3.7) is the minimal economically relevant increase of the effectiveness standard.

Assuming $W'_n - \overline{W}_n = \Delta W_n$ and designating a multitude of possible economically relevant values ΔW_n as $\{\Delta W_n\}_{\text{ec}}$, the right side of the inequation (3.4) is the achievable lower boundary of multitude $\{\Delta W_n\}_{\text{ec}}$:

$$\frac{S_{\text{pp}}}{S'_n P'_{new}} + \frac{\overline{S}_n P_{\text{old}} - \overline{S}_n P'_{new}}{S'_n P'_{new}} W_n = \inf \{\Delta W_n\}_{\text{ec}} = \Delta W_n$$

With $\overline{S}_n$, P_{old}, $\overline{W}_n$ being constant values and the fact that more sophisticated and advanced specimens entail higher costs (S_{pp}) and lower $\overline{S}_n$, P'_{new}, it can be concluded that switching to next-generation products should result in a much more pronounced increase in standard effectiveness than switching to next modifications, i.e.,

$$\Delta^{k,k+1} > \delta^{k}{}_{j,j+1}. \tag{3.8}$$

The correctness of inequation (3.8) confirms that in the new market environment, a quick transition from one generation to the next one is the key driver of companies' survivability, as it speeds up the renovation of each particular type of product.

The usefulness of mastering new product specimens and generations in different environments can be explained by a pricing policy only. A new product must be

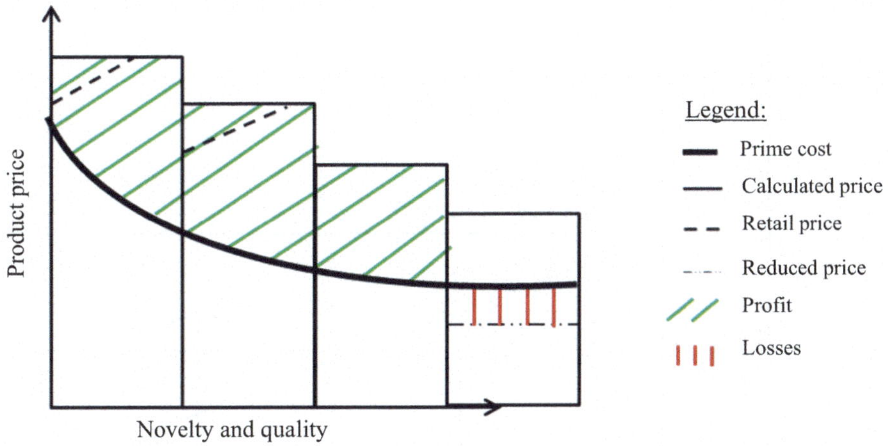

Fig. 3.23 The dependence of price for a product on its innovativeness and quality

more cost effective than its predecessor. Higher standard effectiveness of a new product should result in higher cost effectiveness. This is the only case whereby using an advanced product instead of an outdated one is economically relevant, because increased profit will help to quickly compensate for the lump-sum costs of its mastery.

The price/prime cost ratio changes depending on the new product's novelty and quality as shown in Fig. 3.23.

At the early stages of the mastery of a new specimen, the release price can be much lower than one that would motivate the company to produce it. This can attract customers. Later, as the demand increases, the price will grow until it reaches an optimal level.

Actual renovation processes depend on:

- Price for the new product (compared to the predecessor's)
- Value proposition (compared to the predecessor's)

The renovation process is quite visible from the perspective of a company's growing economic effectiveness (Fig. 3.24). Even if production and distribution of old and new products are equally profitable, it is necessary to thoroughly analyze the dynamics of the new product's prime cost and price. Replacement of the old product with the new one at time (t_1), when the cost of producing new goods (C_1) equal PRC_0, the company will gain a double-A1'1 profit over the t_1–t_4 period. Replacement at a later time—at t_3 and t_4, will not only reduce profit; the company will fail to release equipment, because by the time t_4 the cost of production of the new goods will be nearly twice lower than that of the predecessor.

P_0 and P_1 and the prices of the old and new products with the same standard sales profit; C_0 and $C_1$are the cost of production of old and new products respectively.

Timely renovation of products provides extra profit for the manufacturer on condition that the upgrade is preceded by a thorough economic analysis.

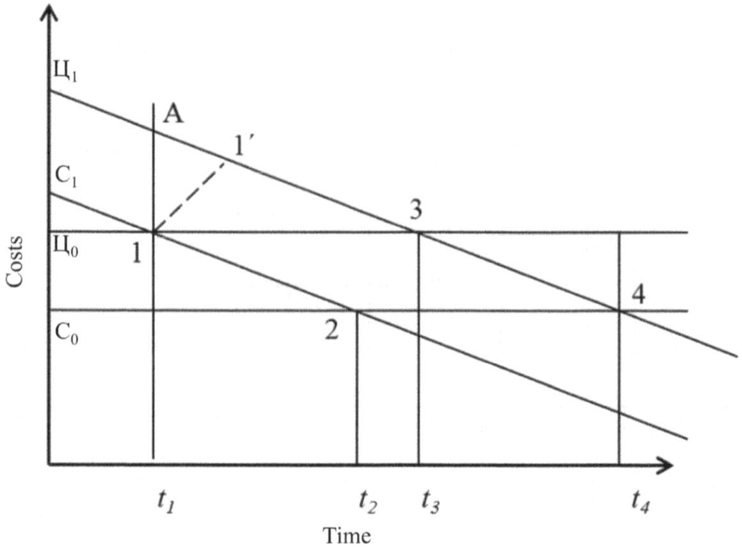

Fig. 3.24 Cost of old and new products

It is also noteworthy that some costs result from informational processes taking place prior to sales. A change in costs forces a company to change price not only because it wants to earn more (it does anyway), but also due to other common factors that occur during production and sales. An increase in price itself does not give any additional competitive advantages. Vice versa, it reduces them and forces an organization to develop steps in different spheres of its activity to lower costs. Therefore, it is necessary to analyze costs from time to time using a cost monitoring and control system (every company should have it) and do organizational and technical steps to prevent their increase and reduce them.

It should be noted that internal costs can be stabilized and reduced with the help of common methods that lower labor intensity and increase labor efficiency, stimulate resource saving, as well as other widely used measures. As a rule, internal production costs make up a 20–60% share of general costs depending on output products' complexity and the number of technological areas within an organization, which are needed for production of specific goods.

External costs, which include the cost of materials, components, energy resources, intermediaries' activities, the government's economic strategy, lowering prices, growing tariffs, taxes, subsides, etc., make up a 75%–40% of general costs and cannot be directly influenced by a the manufacturer; therefore, advanced companies, who strive to win a market, choose to predict changes in prices for supply services, tariffs, taxes and other external influences that can change production costs. The organization can use this prognosis data to avoid economic losses resulting from destructive external influences. This can help it stabilize prices and win the competition.

Chapter 4
Evaluation of an Organization's Ability to Tailoring Production to Set Parameters

4.1 The Model and Dynamic Evaluation of Innovative Potential with Rapidly Growing Competitive Innovative Solutions and Expanding Informational Space

A theoretical research of the rapid development process at an organization revealed the fact that management of the process should rely on an innovative approach aimed at executing rapid product development and manufacturing projects. A company's ability to achieve rapid development through execution of innovative projects depends on its innovative potential, which consists of material, financial, informational, scientific, technical, and other resources, as well as competences used in its innovative work, ability to properly react to negative influences that could impede the innovative process.

The result of an innovative process, which contributes to rapid development, is successful development of highly competitive products, which can form a new market or radically change an existing one. Organizations that sell advanced products, temporarily monopolizes the respective market, as the products begin to dominate it. As the theory of competitiveness states, a product's competitiveness is evaluated with reference to technical and economic characteristics of the product that dominates the market. Also, this refers to the organization producing it: it is the most competitive one within the market in question, and it shows maximum sales compared to its competitors.

High competitiveness of the product and the maker is an indicator of a high innovative potential, which helps it to market highly competitive products. Therefore, the concept of innovative potential, referring to the conditions helping a company to achieve rapid development, should be understood as a complex of drivers, which includes competitiveness, with account for interrelations between them and their dynamics. This complex approach to estimating innovative potential can help to develop a criterion of its sufficiency to achieving the goals of rapid development.

A. Tyulin, A. Chursin, *The New Economy of the Product Life Cycle*,
https://doi.org/10.1007/978-3-030-37814-1_4

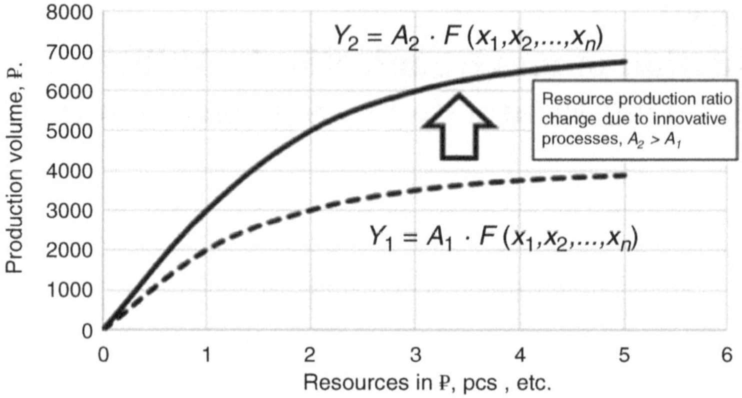

Fig. 4.1 A shift in production resulting from the use of incremental innovations

According to the theory of innovations developed by J. Schumpeter, there are two types of innovations [1]:

- The introduction of new processes, which are observed in improving technologies that create a basis for development, production, and upgrading of existing products
- The introduction of goods that generate new or modified goods

The first type of innovations helps companies maintain their own and products' competitiveness. The effect of introducing such innovations is an opportunity to produce goods in larger volumes with the amount of resources invested unchanged, or in the same volumes with lower resource investments (Fig. 4.1).

In terms of resource–production ratios, it means an increased opportunity to do so with the increasing technological coefficient A:

$$Y = A \cdot F(x_1, x_2, \ldots, x_n),$$

where A is the technological coefficient (it increases as innovations are introduced); F is the resource-production ratio; x_1, x_2, . . . , x_n are production drivers.

The majority of such markets are not dominated by any organization or product. The market of smartphones (Fig. 4.2) is a bright example, as it demonstrates a continuous multidirectional change of the share occupied by the biggest manufacturers. Supposedly, the main smartphone companies have similar innovative potentials. Newly emerging competitive solutions, which temporarily increase a company's share, are promptly mastered by its competitors. Also, the diagram demonstrates a gradual and nonstop decrease in other makers' shares. This means that these organizations do not have a sufficient innovative potential to compete with the leaders.

Innovations have helped smartphone manufacturers reduce prices through increased sales (Fig. 4.3).

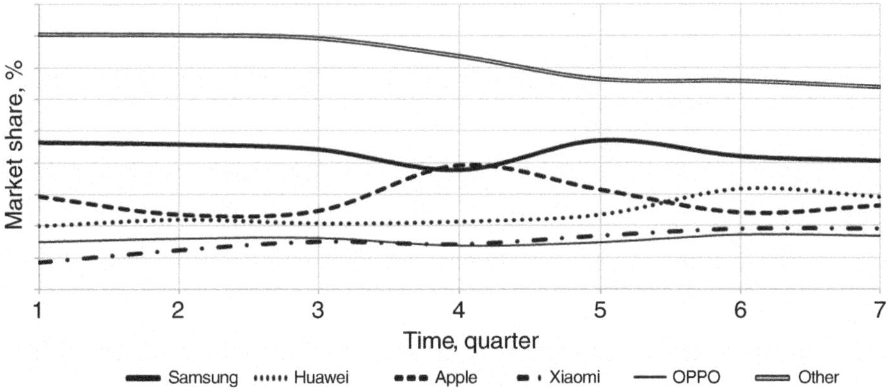

Fig. 4.2 Smartphone market dynamics (major companies' shares from the first quarter of 2017 through the third quarter of 2018)

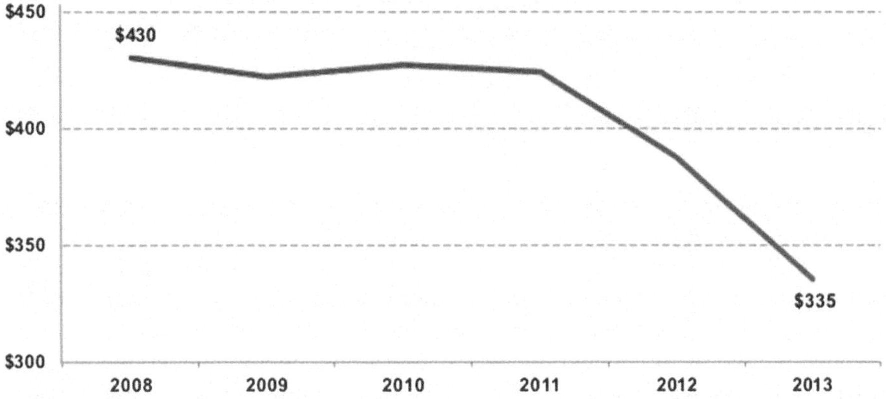

Fig. 4.3 The dynamics of an average price for a smartphone on a global market

Second-type innovations (innovations of products) aim for development of new types of products. These can be of several types:

- Products new to a specific company's markets. These products can be developed with the help of first-type innovations.
- Products new to the organization, as it has never produced them before.
- New market-oriented products, i.e., rapid development products.

It is absolutely necessary for any company to continue to move in all three directions to maintain stable economic growth and increase (maintain) competitiveness. The effectiveness of work done in each direction is evaluated according to a company's innovative potential. Thus, innovations that are needed for creation of traditional products rely on effective technologies of the past and successfully used competences. Innovations used for development and production of new goods and

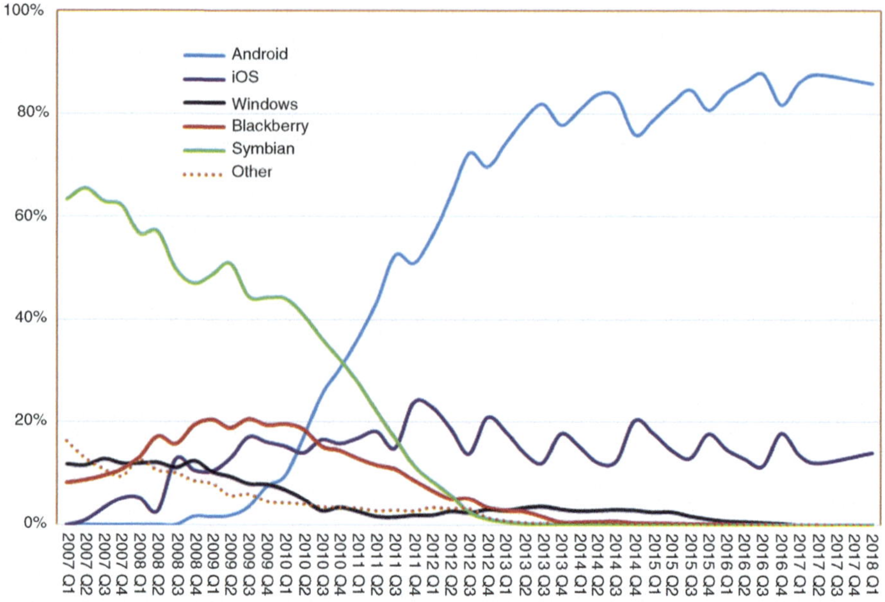

Fig. 4.4 The structure of the OS and mobile device market

those never made before may also rely on existing competences, as well as those acquired by way of transfer.

In turn, the development and introduction of breakthrough innovative technologies requires a substantial resource base and new one-of-a-kind competences. At pre-investment stages these projects rely on economically and mathematically modeled prognoses of demand for attractive projects, time to market, the duration of temporary monopolization, etc. These prognoses define an organization's ability to achieve rapid development. Along with the resource base and competitiveness, these prognoses define the sufficiency of a company's innovative potential.

The Android OS, which is used in smartphones and other devices, is an example of rapid development. Originating in 2008 (during the smartphone boom), in 2010 the OS captured a 10% share of the mobile device market, and in 2011 the share exceeded 50%. It has been dominating the market for quite a while: since 2013, the share of Android-operated devices has been above 80% (Fig. 4.4).

A closer look at the process of creation of Android reveals that it became possible thanks to the developer's (Google) high innovative potential. By the time the development started (2005), the company boasted high competitiveness within the software segment and one of the world's most powerful IT and software brands. By the time, the organization had gained a good amount of resources (financial, intellectual, software development competences). The decision to come up with and promote its own operating system in the smartphone market was based on the booming growth, which was predicted to continue for years to come, and the concentration of the most user-oriented services on the platform. The competence gap was filled through the acquisition of Android Inc., which had its own Android OS projects.

By the time Android appeared, the smartphone market had been shaped. Consequently, there were operating systems for mobile devices. Android's succeeded thanks to:

- Detailed prognoses of the growing smartphone market dynamics.
- The OS's technological potential achieved through involvement in the process of substantial corporate resources (the merger with Android Inc. alone went for $130 million).
- Ability to predict the development of other operating systems through innovative development and prevent their rapid development.
- An effective distribution model (the system itself is open source and can be used by other smartphone manufacturers. The operating income is realized from advertisement and sales of smartphone applications).
- An elaborate renovation strategy, which embraces prospective consumer expectations, competitors' activities, and drivers of the mobile device market.

If combined, these factors can produce a synergistic effect, which lies in the company's growing competitiveness and long stay in the rapid development mode within the mobile OS market.

To give a formal description of a company's potential for rapid development, it is necessary to present the economic and mathematical model that evaluates the sufficiency of a company's innovative potential in the face of quickly emerging competing innovative solutions and the expanding global information space. A specific feature of this model is prognosis of prospective consumer expectations, sales, and marketability, which largely reflects an organization's innovative potential.

Typical of high-tech segments of economy are oligopolies of science-intensive products. In these markets, several (usually about 10) manufacturers sell their products. Barriers to entry are high because of products' high technical and economic characteristics, which have appeared thanks to manufacturers' substantial innovative potentials. Each organization participating in this kind of market should follow its chosen competitive behavior pattern. An organization aiming for rapid development follows an advanced competitive behavior strategy. This strategy focuses on predicting competitors' activities and requires that the organization constantly monitors their behavior and analyzes their reaction to other players' activities.

The mathematical model, which describes the behavior of high-tech oligopoly market participants, is based on the Stackelberg game-theory model.

This game is nonsymmetric. The inequality results from differences in innovative potentials. As the Stackelberg formalism states, an organization boasting the highest innovative potential should take the lead in this game-theory model. The numerous participants (competing organizations) are aligned hierarchically.

The following example will focus on a game played by a group of participants. Those are companies marketing their products.

This game-theory approach to building a mathematical model implies that each participant has a set of strategic decisions. Decisions made by the leader, whose goal is reaching the rapid development status, are expressed as follows:

$$x_i \in X,$$

where X is a multitude of strategies developed by the organization aiming for rapid development.

The set of strategies followed by other participants is expressed in the following combination:

$$y_i \in Y,$$

where Y is the multitude of strategies followed by the rest of the participants.

In the process, participants choose some optimal strategies (x, y) based on their priorities. Gain functions are:

$$H_i : X \times Y \to R, \quad i = 0, 1.$$

In this market, the inequality means that the organization with a higher innovative potential and competitiveness has privileges over other organizations. Therefore, the game should be hierarchical. Here a few assumptions should be made. First, assume that the organization, which is heading toward rapid development with account for predicted consumer expectations and other players' activities, takes the lead. Second, other players' choice of strategic decisions depends on the first player's (leader's) choice of its strategy.

Based on this assumption, the most competitive organization comes up with (reserves) its own strategic decision, which is:

$$x^* \in X.$$

Once the leader has chosen its strategy, the rest of the participants make their moves. They choose the most optimal product sales strategy, which brings maximal profit:

$$y^* \in Y(x^*).$$

The gain (supposed profit) results from the solution of the following optimization task:

$$H_1(x^*, y^*) \to \max.$$

Assume that the first argument (x^*) is a fixed one.

This poses a common to the Stackelberg model challenge concerning the choice of an optimal decision for the first player—the organization heading toward rapid development. Within this game-theory setting, the organization should be the first one to choose its strategy while relying only on an advanced prognosis of other players' choice. The optimal response function (*P*) is expressed as follows:

$$P : X \to Y.$$

The optimal response function is:

$$P(x^*) = \arg\max \{H_1(x^*, y) : y \in Y(x^*)\}.$$

In these formalistic conditions, it is possible to use a mathematical model of the choice of an optimal decision for an organization heading for rapid development:

$$H_0(x^*, P(x^*)) \to \max .$$

A situation of equilibrium in the game-theory is analogous to an optimal decision. The equilibrium occurs when none of the players benefits from refusing to follow the equilibrium strategy (the Nash Equilibrium). However, players are nonsymmetrical in the given oligopolistic science-intensive market. Such situations reflect the Stackelberg game-theory formalism.

To implement this economic/mathematical model, it is advisable to use a simulation model, which can help to find optimal decisions determining the choice of respective strategies by organizations operating in the market. Also, the use of simulation models will help to embrace various occasional influences, which can affect sales.

The strategy choice is determined as follows:

$$H_0(x^*, \arg\max \{H_1(x^*, y) : y \in Y(x^*)\}) \to \max .$$

This optimization task is done with the help of some calculative procedures. Because this model embraces rapidly changing external factors, which are predicted as part of advanced competitive behavior, a simulation approach should be used. Occasional factors are integrated into the gain factor in the shape of the following argument:

$$H_0(x, y, \xi),$$
$$H_1(x, y, \xi).$$

Simulation modeling uses the following random values:

$$\xi_1, \xi_2, \ldots \xi_N \tilde{F}_{\xi_i},$$

where F_{ξ_i} are random distributions, which describe dynamic external factors affecting the marketing of products. These random distribution factors should be selected based on prediction techniques.

The process of choosing a strategic decision (which is determining equilibrium sales) can be exemplified in an individual oligopolistic case—a duopoly, where two organizations sell their products. One of them is a potential leader, as it strives for rapid development in the market.

As required by approach to determining the equilibrium price, which is commonly used in game-theory modeling, the price is a linear function of general supply—Q (the composite supply from both companies):

$$P(Q) = a - Q,$$

$$a = \text{const}.$$

Assume that c_0 and c_1 are marginal costs incurred by each of the companies. According to the Stackelberg game-theory formalism, the leading organization sets production output q_0. The second one determines production output q_1 while being aware of the q_0 production output. Taking into account the marginal costs, the profit of each organization will be:

$$H_0(q_1, q_2) = q_0(P(Q) - c_0) = q_0(a - q_0 - q_1 - c_0),$$
$$H_1(q_1, q_2) = q_1(P(Q) - c_1) = q_1(a - q_0 - q_1 - c_0).$$

The Stackelberg game-theory setup is a kind of dynamic game, which implies analysis of two stages. This setup makes it possible to use the backward induction method. At the first stage, an optimal output value (q_2^*) should be found, which will maximize the second company's profit with the known q_l output. In other words, q_2^* is the solution to the following optimization task:

$$H_1\left(q_0, q_1^*\right) = \max_{q_1} q_1(a - q_0 - q_1 - c_1).$$

The peak point is the solution to the following equation:

$$\frac{\partial H_1}{\partial q_1} = -2q_2 + a - q_0 - c_1 = 0.$$

Therefore,

$$q_1^* = q_1^*(q_0) = \frac{a - q_0 - c_1}{2}.$$

In this case, an organization striving for rapid development should be able to foresee that its choice of production output q_0 will predetermine the second organization's choice of production output q_1^*. Therefore, the first company should choose production output with reference to the following optimization task:

$$H_0(q_0, q_1^*) = \max_{q_0} q_0(a - q_0 - q_1^* - c_0).$$

This function's peak point is:

$$q_0^* = \frac{1}{2}(a + c_1 - 2c_0).$$

while

$$q_1^* = \frac{1}{4}(a + 2c_0 - 3c_1).$$

Because external factors are dynamic, gain functions may need a correction. An advanced external factor analysis technique is big data mining with big data contained in the ever-expanding global information space. This analysis can help to reveal the main economic climate change trends and predict directions of development of techniques and technologies, as well as other organizations' activities.

Because the Stackelberg model is dynamic, it can be used at any point of time to select an optimal competitive behavior strategy.

A graphic picture of a market described by this model is shown in Fig. 4.5.

Therefore, following an advanced competitive behavior pattern in a market helps to predict other player's activities and dynamically changing influences of the economic environment. This game-theory model has helped to foretell the market output (q_i) by each organization. A company heading for rapid development can choose a strategy, which will ensure that its products is going to dominate the market. Domination criteria may differ from nation to nation. For example, in Japan (a global high-tech leader) the statutory domination ceiling is 50% of the market volume. In the context of rapid development, exceeding this ceiling may signify a transition to rapid development in this particular market.

This simulation model reflecting companies' activities in an oligopolistic market can help to define the sufficiency of a company's innovative potential for reaching rapid development. Domination of a product in the market is a sign of a company's achieving of the rapid development status:

$$q_i > A,$$

where q_i is the company's share in the market (it is the solution to the game relying on the simulation model); A is the domination ceiling.

Another sign of domination and sufficiency of innovative potential for achieving rapid development is an organization's ability to independently define key market parameters, prevent entry by other players, and limit the work of those already

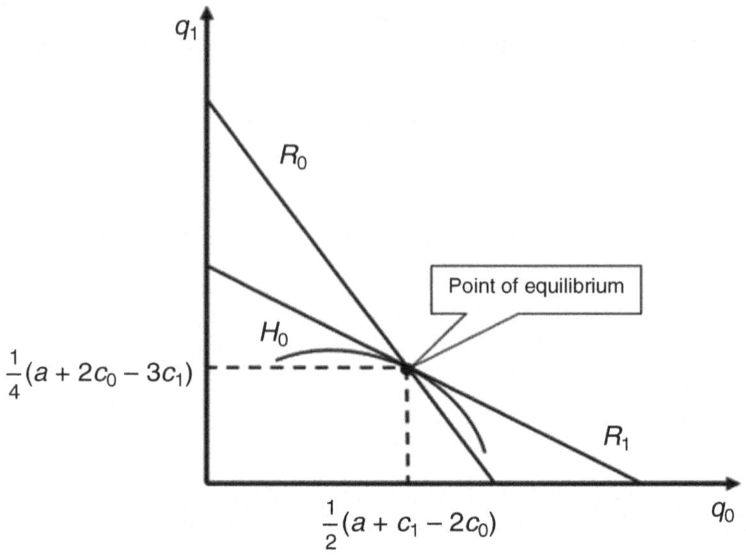

Fig. 4.5 The Stackelberg equilibrium (R_0 and R_1 are curves expressing reactions of each company to an output value offered by the other company)

participating in the market. The simulation model also helps to achieve that through selection of an optimal advanced competitive behavior pattern.

4.2 Modeling, Evaluation, and Prognosis of the Development of Unique Competences to Satisfy Prospective Needs and Ensure Their Transfer

It is possible for an organization to develop innovative potential sufficient for producing advanced goods through developing a management system, which can help to operate the process of creation of this potential. Innovative potential is a combination of all types of resources and influences, including industrial, technological, financial, economic, intellectual, scientific, and research, which are necessary for execution of innovative work. Each component of innovative potential relies on a knowledge set, and the knowledge should transform into companies' and their staffs' competences. In practice, some components of the innovative potential can reach high levels, while others may stay at a lower level. Spheres, which have not reached a sufficient innovative potential, require formation and development of new competences.

Each of the types of resources may have a different number of criteria, which are defined by different sources. As applied to an organization, these sources may include finance statements, production reports, personnel department data, personnel group polling data, etc. The system focuses only on criteria, the growth of which stimulates the growth of the company's innovative work. It sets an integrated index for each type of resources, and its value may suggest high, medium, or low innovative potential of the company with account for this type of resource or factor.

First, a company's innovative potential is defined by the following internal resources and factors:

- Financial and economic
- Intellectual
- Organization and management
- Scientific and research
- Production and technology
- Marketing
- Information and methodology

These resources may to a greater or lesser extent contribute to the formation of the company's general innovative potential.

Because innovative potential management methods entail a significant resource spend, it is necessary to possess tools that enable evaluation of the components of innovative potential. Based on the estimates concerning these components, it is possible to make managerial decisions to direct resources to boost a particular part of the innovative potential.

For instance, assessment of the scientific and research component can rely on the system of criteria shown in Table 4.1.

An integrated estimate (index) of the science and research element of the innovative potential makes up a weighted sum of evaluation criteria:

$$\mathrm{IP}_{\mathrm{R\&D}} = \sum_{i=1}^{13} w_i y_i;$$

$$\sum_{i=1}^{13} w_i = 1.$$

Also, these estimates can be applied to other components of innovative potential. According to these indexes, each component of innovative potential is categorized as follows:

- High level is assigned when the respective integrated index reaches: [0.65; 1].
- Medium level is assigned when the respective integrated index reaches: [0.5; 0.64].
- Low level is assigned when the respective integrated index reaches [0; 0.49].

With these indexes at hand, an integral element of the whole innovative potential (IP) can also be a weighted sum of its components.

Table 4.1 Evaluation criteria for the scientific and research component of a company's innovative potential [2]

Sr. no.	Criterion	Criterion's significance	
1	Intellectual property management	y_1	Ratio between corporate IPOs (intellectual property objects) and the general number of IPOs
		$y_2 = \frac{J'}{J}$	J' : number of corporate IPOs upgraded to the point of industrial and technological usability; J: general number of IPOs in the company
		$y_3 = \frac{\sum_{i=1}^{N} C_i + \sum_{j=1}^{K} C_j}{\sum_{r=1}^{R} C_r - \sum_{q=1}^{Q} C_q}$	N: number of innovative solutions developed without third parties' help K: number of acquired innovative solutions R: $N+K$ Q: number of innovative solutions not used in the company and implemented outside C_i: financial costs of independent development of the i innovative solution C_j: costs of the acquisition of the j innovative solution C_r: costs of development and acquisition of all innovative solutions C_q: income from sale of the q innovative solution
2	Intellectual property ratio	$y_4 = C_t/A_{nc}$	C_t: total number of IPOs A_{nc}—other noncirculating assets
3	Scientific research	y_5	Ratio of industrial company's agreements with scientific and technological organizations to the total number of agreements
		y_6	Ratio of corporate innovation-related one-off industries
4	R&D costs	$y_7 = N_{R\ \&\ D}/N$	Ratio between R&D personnel's income and other employees' income
		$y_8 = N_{R\ \&\ D}/Q$	Ratio between R&D costs and total sales
5	Non-R&D innovation costs	$y_9 = N_{nonR\ \&\ D}/Q$	Ratio between general innovation costs and total sales
6	Ratio of R&D and NPD personnel	$y_{10} = P_s/E_{av}$	P_s: number of R&D and NPD employees E_{av}: average number of employees hired by an organization
7	R&D and NPD property ratio	$y_{11} = Eq_{pt}/Eq_g$	Eq_{pt}: cost of pilot-test equipment, rubles Eq_g: general equipment cost
8	Level of new science-intensive technologies	y_{12}	Ratio between corporate advanced industrial technologies and the total amount of introduced technologies
9	Corporate innovative programs	y_{13}	= 0.25; company focusing on short-term plans only, as dictated by the situation, lacks an elaborate innovative program = 0.5; a company focusing on an annual plan as dictated by the situation, does not have enough resources to develop innovative

(continued)

Table 4.1 (continued)

Sr. no.	Criterion	Criterion's significance	
			programs = 0.75; in companies developing 2- to 3-year plans, innovative programs are accepted when the company has to do that by reason of market competition =1; in cases whereby a business plan relies on innovations as musts for reaching a company's goals
10	Emergence of innovations	y_{14}	= 0; in cases whereby no monitoring of new ideas is carried out because a company does not need to do that = 0.5; when a company chooses not to block employees' initiatives in this sphere; only potentially beneficial innovations are introduced = 0.75; an organization, which regularly collects and evaluates new concepts, uses them as a source of product and/or process innovations; in parallel to that, it studies innovation-related costs and risks = 1; providing that the managerial staff work toward finding opportunities for using new ideas offered by their own staff and external sources; the company uses new motivation and creativity evaluation systems; when new concepts are introduced, projects are checked for feasibility

Management of each of IP component consists in developing respective competences with the help of resource investments. An organization striving for market domination should launch a self-perpetuating process of competence improvement and effective use of resources. If this competence is hard to replicate for competitors, it can be deemed a key one.

Dividing competences into key and minor ones is conditional and depends on specific circumstances, market conjuncture and other influences. A once key competence may become minor after a while; and vice versa, in an individual matter, a minor competence can play a decisive role, as innovative processes are not quite so predictable.

An organization should be aware of as many of its competences as possible, retain an opportunity to promptly adjust priorities and be flexible in adapting to market standards. In this respect, competences should be understood in a broad sense as all knowledge, abilities, skills, and potential of a person or a group of persons, which enable them to carry out their work properly, with or without using respective types of tools and equipment, aimed at development, production, and promotion of globally competitive products and providing specific services.

With the digital transformation and quick emergence of new consumer expectations, companies should not only possess appropriate expertise and knowledge, but also have dynamic managerial capabilities and be able to quickly adapt to rapidly emerging new needs, update knowledge through a continuous learning system, maintain further cooperation, etc. This will help them to constantly improve their managerial knowledge base and at least par with their competitors.

As we can see, creating new competences and their development with the help of innovative technologies require huge resource investments. As development of advanced technologies is very costly, effective use of all types of resources takes on particular significance. While relying on key competence evaluation and advanced technology selection methods, it is possible to outline a program of work for a company and define the most important funding streams for the development of new competences and technologies. To do so, competence management mechanisms should embrace the necessity of their replication to enable creation of innovative technologies in a variety of directions of technical development. In other words, possessing a technology and its successful use in one particular direction does not mean it can be promptly introduced in another.

In the context of economic effectiveness, the transfer of competences and competence centers focusing on specific spheres, which encompass industrial resources and available resource-use competences, takes on a particular significance. The presence of such centers will help to promptly build an effective scientific and industrial structure of new projects out of separate components, thus minimizing the dissipation of resources around redundant divisions. This does not have to be confined to one organization only. Global practice widely uses temporary scientific and industrial structures (virtual organizations); only trusted leaders are invited to cooperate through contract agreements, outsourcing, and other legal routes.

The effectiveness of the competences being developed and innovative technologies springing from them should be evaluated from the perspective of products' technical and budget characteristics, which result from their use. At initial stages of the formation of advanced products' technical and economic image, economic and mathematical modeling appears to be the most effective evaluation tool.

To identify competences within a company, which need to be developed with the goal of stimulating the weakest components of the IP, it is necessary to evaluate the company's key competences. Mathematical modeling provides the most objective estimates, because they rule out the subjective factor, which is typical of expert judgment.

According to our research, evaluation can be carried out with the help of an economic/mathematical model (described in detail in its developers' paper [3]), which produces an aggregated value (index) characterizing the competence's validity and importance on the basis of some indicators, which are defined in keeping with a specific key competence description format. Key competences are graded through their comparison with the use of newly calculated estimates.

Thus, to make an aggregated estimate, quantitative values of each indicator, which reflect its particulars, need to be derived.

The formula that calculates a key competence's (EK) index is:

$$\text{EK} = \sum_{i=1}^{N} (w_i M_i l_i),$$

where N is the number of characteristics describing the key competence (for example, the presence of a highly professional and competent team with an effective scientific and research sector and advanced industrial base; the functional environment, in which the team does its scientific and industrial work; readiness of technologies developed thanks to the competence in question; the key competence's ability to spread to other industrial spheres; the presence of competing subjects possessing similar key competences; the company's strong and weak points compared to competitors, if any; availability of a scientific school that does research related to the key competence; having licenses, certificates, awards (especially international ones); the prospect of retaining key competences in the medium- and long-term perspective);

w_i—expresses weight coefficients that meet the ratio $\sum_{i=1}^{N} w_i = 1$; weight coefficients' values characterize respective parameters' relative contribution to the general assessment.

l_i is an estimate of a respective indicator according to our l scale: $0 \leq l_i \leq 1$.

M_i is an indicator's stability coefficient, which expresses the risk of a competence's elimination from the key competence list according to this indicator: $0 \leq M_i \leq 1$; in this case, the stability coefficient depends on the degree of risk of a drop of the estimate.

The estimate of a key competence's progress adopts values from interval [0; 1]. In cases whereby the index takes on a value that is close to *1,* most likely, the company does have a unique innovative solution. The solution becomes even more promising when a revolutionary technology is created.

Innovative technologies' marketability is defined by emerging products' cost and competitiveness, as well as by the possibility of their transfer to other industries. The product's set technical characteristics ensure that the chosen technology can help to optimize its cost based on the choice of design and technological solutions, use of digital, adaptive, and other advanced technologies. If using this technology does boost the product's characteristics while reducing (or retaining) its cost, the technology can be deemed effective. The following mathematical model can illustrate that.

This economic model should evaluate the effectiveness of innovative technologies, products, and services based on those, given the limited access to resources and with account of the level of key competences.

The model uses the following symbols:

$x(t) = (x_0(t), x_1(t), \ldots, x_n(t))$ is the row vector of product and service type set volumes that are based on traditional and advanced technologies and released over the t period of time, where $x_0(t)$ is the output of the main product released over the t period of time; $\{x_1(t), \ldots, x_n(t)\}$ is the group of output product and service type set volumes that are based on new technologies and released over the t period of time.

$p(t) = (p_0(t), p_1(t), \ldots, p_n(t))$ is the row vector of retail prices for products and services released over the t period of time.

$w(t) = (w_0(t), w_1(t), \ldots, w_n(t))$ is the row vector of costs of the resource volume per unit of product or resource released over the t period of time within a functioning interval $T = [1, 2, \ldots, N]$.

Resource–production ratios, which reflect the competence level-based proportion of resource used and output volumes, can be formally presented as:

$$X_l = \mathrm{EK}_l \cdot F(x_l(t)), \quad l = 0, 1, \ldots, n, \tag{4.1}$$

then $P(x_l(t))$, the gross profit on each type of product or service is:

$$P(x_l(t)) = p_l \cdot \mathrm{EK}_l \cdot F(x_l(t)) - [w_i \cdot x_l(t) + c_0], \quad l = 0, 1, \ldots, n, \tag{4.2}$$

where $\mathrm{TR}(x_l(t), p_l) = p_l \cdot \mathrm{EK}_l \cdot F(x_l(t))$ is the monetary earnings (turnover) from the volume of sold products.

$\mathrm{TC}(x_l(t), w_l) = w_l \cdot x_l(t) + c_0$ is the general production cost of product or Service 1, which include variable $w_l \cdot x_l(t)$ and constant c_0 production costs.

The gross surplus over N periods of time for all types of products (services) will be:

$$\Pi_{\sum}(x_l(t)) = \sum_{t=1}^{N} \sum_{l=0}^{n} \Pi(x_l(t)). \tag{4.3}$$

The intensity $q_l = \frac{dx_l(t)}{dp_l}$ of sales of the main products (services), depending on the sales price, is expressed as a differentiable function:

$$q_l = q(x_l(t), p_l), \quad l = 0, 1, \ldots, n, \tag{4.4}$$

The economic effectiveness of production of new types of products (services), in the simplest case, will reflect the condition:

$$\Pi_{\sum}(T) \geq \Pi_*, \tag{4.5}$$

where P_* is the minimal monetary gross surplus observed at the breakeven point $x_l^0(t)$ of production, i.e., when $P(x_l(t)) = 0$ and when an organization (or an industry) remains functional within an interval $T = [1, 2, \ldots, N]$.

As a criterion of economic effectiveness of the release of products (services) that are based on advanced technologies, a nondimensional value can be used:

$$\gamma = \frac{\Pi_{\sum}(T) - \Pi_*}{\Pi_*} \tag{4.6}$$

Economic effectiveness of innovation-based products and services should be evaluated according to:

- The release of new types of products (services) is economically effective if $\gamma \geq 1$.
- The release of new types of products (services) exerts a weak influence on economic effectiveness if $\gamma \in [0, \ 1]$.
- The release of new types of products (services) is economically ineffective if $\gamma \leq 0$.

The choice of an optimal structure of existing and new products (services) with limited monetary resources with account of demand functions q_l (4.1) is a multi-objective vector optimization problem:

$$\left(P(x_0(t)), P(x_1(t)), \ldots, P(x_n(t))\right) \rightarrow \max_{(x_0, x_1, \ldots, x_n)}, \tag{4.7}$$

on condition that:

(a) Gross surplus should cover all costs:

$$\sum_{t=1}^{N} \sum_{l=0}^{n} P(x_l(t)) \geq P_* \tag{4.8}$$

(b) Product (service) creation and sales costs should not exceed the set value i over the period T:

$$\sum_{t=1}^{N} \sum_{l=0}^{n} w_l x_l(t)) \leq C^*. \tag{4.9}$$

An optimal solution to task (4.1)–(4.9) will be one that will ensure maximally effective production of innovation-based goods (services).

Stable development of advanced production will not be possible without investments in new one-of-a-kind competences and innovative marketing of new products. These processes are both micro- and macroeconomic, which reflects the shift of competition toward unique key competences in a large number of sectors. Also, it produces an effect, which is observed in the fact that development of competences and resulting appearance of new technologies create a breeding ground for new consumer markets. It is a well-known fact that new markets emerge both due to demand for and supply of new types of goods. Competition in the high-tech and unique competence sphere has mechanisms that are different from those of the traditional industrial sector. These mechanisms boast a much quicker reaction to changes in competitors and new consumer expectations, and they force organizations to adjust their behavior more quickly and effectively.

High-tech production mechanisms create new innovative technologies as corporate competences grow. The laws of development of science-intensive organizations are conducive to a process when fundamental achievements give rise to applied projects, which create a basis for innovative technologies. Newly emerging innovative technologies manifest themselves in creation of new goods or services. Commercialization of innovative technologies forces high-tech organizations to create or contribute to creation of new markets to promote their new products. This has contributed to the formation of the *competence–innovation–market model,* which demonstrates that the progress of competences helps consumer markets to expand as new products, which satisfy prospective needs, appear. Market mechanisms contribute to rapid growth on investment in breakthrough technologies, which are used for production of these goods. As stated by the laws of innovative economy, these investments are hugely conducive to the progress of competences in this sphere.

Evidently, this explosive growth is transient and is observed only then there is a significant expansion of consumer markets. Marginal economic laws state and all consumer markets finally become saturated. The saturation is expressed in a drop in consumer market dynamic, which occurs at a certain moment. Once that happens, innovative technologies become ordinary. This model can be presented as a general consumer market evolution law based on the generation of new competences: creation of unique competences increases resources in high-tech organizations, boosting their innovative potential until it becomes sufficient for creation of a dominating product. As these products appear, needs for new amenities increase. Economic growth stimulates further demand for new technologies.

This law demonstrates a specific spiraling *competence–resource–product–need–product* cycle. Thanks to the *market growth–market decrease* cycle, the modern high-tech economy has (accumulates) a substantial stock of competences (intellectual resources), which predefine the emergence of a new market evolution cycles and competence progress. To describe these processes, it is necessary to use different mathematical approaches in the sphere of dynamic systems. This can be observed in a formalized version of this scheme.

This economic model can be described with the help of several dynamic parameters, which change over time. These parameters' continuous interdependence is key to their development. Here is a description of the main economic parameters, which will be analyzed within the model describing the relationship between the competence progress and appearance of consumer markets. Every next parameter is closely related to the previous one. The model describes the following parameters:

- Competence progress (describes the current average level of technological, organizational, managerial and other competences, which can be used by science-intensive industries in development of advanced technologies)
- Level of innovative potential (describes various components of the innovative potential)

- New products' level (reflects the degree of compliance with future needs and potential for market domination)
- Market development (describes creation and development of markets stimulated by the supply of new products and their usability)

To build a mathematical model, it is necessary to formalize the main concepts. It should be noted that the model uses constant processes with continuous time $t \in [0, T]$, $0 < T < \infty$.

The dynamic model that describes the relationship between competences and new markets is based on the following variables:

EK(t)—a qualitative estimate of competence levels
IP(t)—a qualitative indicator of innovative technologies
$\gamma(t)$—a qualitative indicator output products' effectiveness
$q(t)$—the product's market share (the relationship between innovative potential and release of dominating product is described in Sect. 4.1)

Here is a formal system of differential equations:

$$\begin{cases} \dot{\mathrm{EK}}(t) = F_1(\mathrm{EK}(t), \mathrm{IP}(t), \gamma(t), q(t)); \\ \dot{\mathrm{IP}}(t) = F_2(\mathrm{EK}(t), \mathrm{IP}(t), \gamma(t), q(t)); \\ \dot{\gamma}(t) = F_3(\mathrm{EK}(t), \mathrm{IP}(t), \gamma(t), q(t)); \\ \dot{q}(t) = F_4(\mathrm{EK}(t), \mathrm{IP}(t), \gamma(t), q(t)). \end{cases}$$

Numerical values of functions EK(t), IP(t), $\gamma(t)$, and $q(t)$ are values of integrated indexes that describe competence levels, progress of innovative potential, the effectiveness of an output product and its market share. The dynamics of these indexes can be described by the following linear model:

$$\begin{cases} \dot{\mathrm{EK}}(t) = A_{\mathrm{EK}}\mathrm{EK}(t) + A_{q,\mathrm{EK}}\, q(t); \\ \dot{\mathrm{IP}}(t) = A_{\mathrm{IP}}\mathrm{IP}(t) + A_{\mathrm{EK,IP}}\mathrm{EK}(t); \\ \dot{\gamma}(t) = A_{\gamma}\, \gamma(t) + A_{\mathrm{IP},\gamma}\mathrm{IP}(t); \\ \dot{q}(t) = A_q\, q(t) + A_{\gamma,q}\, \gamma(t), \end{cases}$$

where A are a group of coefficients that should be determined based on the statistics of the process being described.

This model describes a system of linear differential equations, in which every dynamic variable has a certain diffusion coefficient reflecting an expected decrease of all indicators resulting from a lack of additional management.

However, this model uses a cyclic dependence between these indicators, as every next one grows due to the influence of the previous one, and the last indicator—a market share of a product—influences the initial competence progress indicator via resources, which are used to raise the competence level.

This linear dynamic system, which should describe competences' mutual influence on the evolution of consumer markets, can only describe the period when these values grow. Therefore, to embrace cyclic events taking place along with the competence and consumer market development process, it is necessary to analyze nonlinear dynamic models as well.

Cyclic events consist of two clearly defined stages: a growth stage and a decline stage. They can be expressed through the following hysteretic function:

$$H[Y(t)] = \begin{cases} H_{\min}, & Y(t) > T_{\max}; \\ H_{\max}, & Y(t) > T_{\min}. \end{cases}$$

To describe the interdependence of the progress of competences and consumer market, the following dynamic model can be used:

$$\begin{cases} \dot{\mathrm{EK}}(t) = A_{\mathrm{EK}}\mathrm{EK}(t) + H_{\mathrm{EK}}[\mathrm{EK}(t)]A_{q,\mathrm{EK}}\, q(t); \\ \dot{\mathrm{IP}}(t) = A_{\mathrm{IP}}\mathrm{IP}(t) + H_{\mathrm{IP}}[\mathrm{IP}(t)]A_{\mathrm{EK,IP}}\mathrm{EK}(t); \\ \dot{\gamma}(t) = A_{\gamma}\, \gamma(t) + H_{\gamma}[\gamma(t)]A_{\mathrm{IP},\gamma}\mathrm{IP}(t); \\ \dot{q}(t) = A_q\, q(t) + H_q[q(t)]A_{\gamma,q}\, \gamma(t). \end{cases}$$

This model describes qualitative dynamics of values only, which characterize market volume's dependence on the progress of competences. A quantitative analysis of particular situations requires accurate statistical data, which help define coefficients, which are part of dynamic equations.

A simulation modeling, which is based on this model, gives solutions shown in Figs. 4.6 and 4.7.

The simulation modeling shows that this dynamic system, which describes the mutual relationship between competence growth and consumer market growth, shows various dynamics modes of the main values in the *competences–IP–production–markets* law. We can observe a distinct cyclicality in these values, which describes a spiraling growth of science-intensive markets. The growth is fueled by IP, which increases due to progressing competences.

This situation is often observed in the process of developing and introducing innovative technologies, which takes place during science-intensive production, when initially accumulated resources help to form IP sufficient for creation of new products and, consequently, new consumer markets. Afterwards, these markets grow rapidly, and the funding of new competence and IP development increases. This situation may repeat itself several times. It should be noted that the time lag, which naturally occurs between the introduction of new products and the funding of

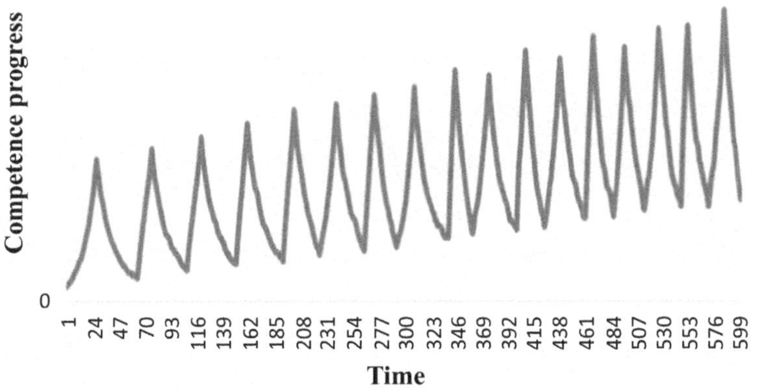

Fig. 4.6 Cyclic investments in key competences

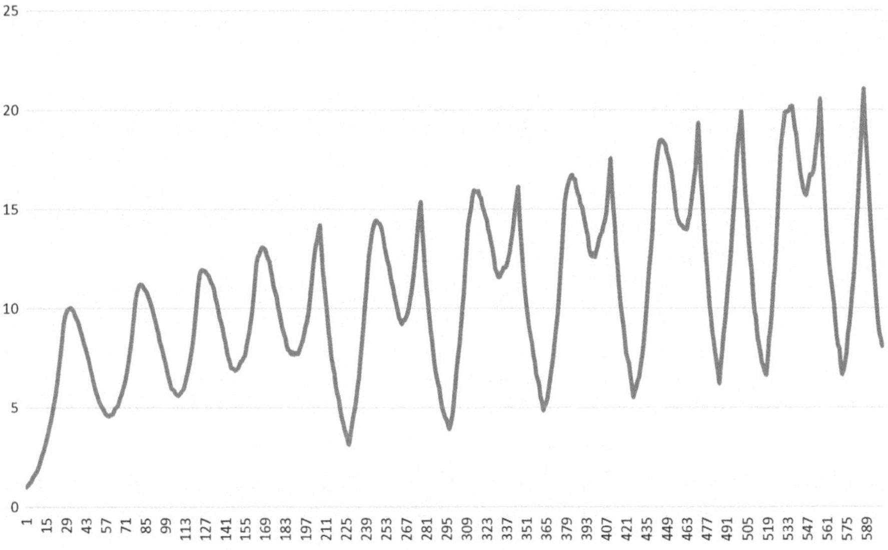

Fig. 4.7 A cyclic growth of a product's market share resulting from a cyclic progress of key competences due to resource spends

respective R&D, naturally desynchronizes the growth and decline periods to various extents in time. That is clearly visible in the unified diagrams (Fig. 4.8).

The level of a company's technological competences is partially defined by the effectiveness of the competence development management process. All major organizations use management tools, which regulate flows of information. These tools must be used to motivate collective use of innovative resources and transfer of knowledge both inside a company and outside of it. In other words, large companies can manage creation of new knowledge by way of controlling and transferring the

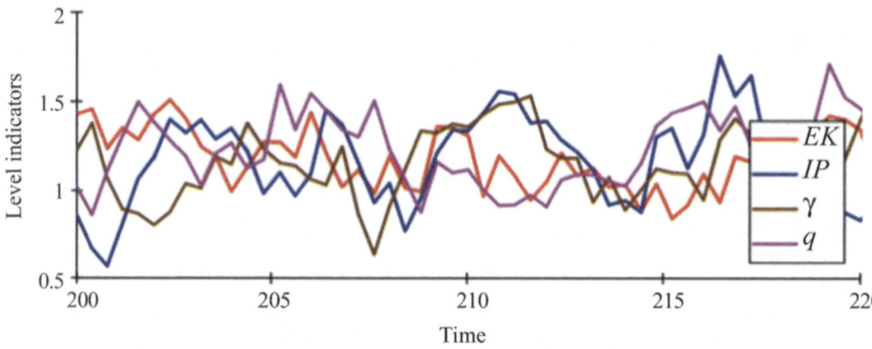

Fig. 4.8 Unified system solution on the interval $t \in [200, 230]$

existing knowledge. This concerns all types of knowledge: separate intellectual property assets, technologies, informal skills, experience, and competences. Therefore, exchange of information in high-tech organizations should be analyzed in three dimensions:

1. Exchange of knowledge between separate divisions of a company aimed at intensifying work in the sphere of innovations. Considering the fact that science-intensive work involves different companies using industry-specific technologies, one may talk of an internal interindustry knowledge exchange. This is only possible when industry-specific space is formed and functioning effectively.
2. The use of companies' scientific potential for external transfer of technologies to allied activities (including commercial ones), which are not related to the main activity. In this case, a structural elements of a company, as well as their scientific, research, and industrial potential should be regarded as a cross-industry catalyzer of innovative work, as it is practiced in other countries.
3. Using the international information space to analyze consumer expectations and existing innovative technologies, selecting the most promising and competitive ones and incorporating in an industry those contributing to a higher competitiveness of national science-intensive products and their leadership in global branch markets.

Therefore, creation and effective use of key competences and innovative technologies throughout a company's business processes is a prerequisite of development and manufacturing of dominating products. Customer expectation prediction tools can be helpful in determining directions for the most important competences:

- Technological: tailoring to a set price and creation of advanced products
- Organizational: transition to a digital organization form; transformation of key business processes; complex product life cycle management
- Managerial: taking effective decisions based on big data analysis

Newly created key competences define a product's competitiveness through innovative technologies, which are formed by these competences. It is possible to eliminate a competence's shortcomings through transfer of competences. Imperative for maintaining an effective innovative process within an organization is to have a substantial resource base, which can provide a basis for creation, development, and transfer of competences.

This mathematical model, which relates resource-driven competence development to the growth of IP and the market share of IP-driven advanced products, and which helps to build corporate resources, illustrates the process of reaching dominance in a market. Further operation of the IP aimed at enhancing (or retaining) the product market share should rely on the rapid growth of competences, which should result from the analysis of prospective consumer expectations.

4.3 Resource Provision Models for Future Products

The competence development and transfer processes described above, mastering of advanced technologies, as well as implementation of new advanced product creation projects through realization of the newly created IP, require a substantial resource base. In this respect, the effectiveness of rapid development depends on the quality of the established resource provision system for future projects, which focus on creation of highly competitive products.

Based on the management of resources and competences of persons, who make decisions concerning resource provision planning, are controlling activities that should regulate the formation and development of a company's IP, which is used in creation of new and advanced future projects. It is planning and resource use rationalization models that determine how effectively IP will be realized and whether or not an organization is capable of creating a product that can capture new markets and acquire advantage in rapid development over leading competitors.

Today, automated resource control systems are used widely. They help to adjust economic processes within organization in a maximally effective way. The most famous ones include MRP, MRP-II, ERP, CSRP, etc.

Meanwhile, the ongoing digitization of all spheres of economy necessitate development and creation of resource provision process management systems, which use advanced technologies (artificial intelligence, machine learning and neural networks), which have helped to direct the process of creation and management of the resource potential at the formation and effective use of IP, which provides for rapid development.

Ineffective and nonoptimal resource management will have negative consequences, which are lower productivity, breaches of deadlines, poor quality, increased costs, missed opportunities, and morals running low.

Today, as companies follow in the path of making highly competitive goods, they face a controversy while, on the one hand, they have to introduce innovative and technologically advanced solutions and, on the other hand, they have to provide resources for that. Eventually, the entire resource stock must be used and concentrated on the most important directions relating to development and manufacturing of future goods, which can put the manufacturer on the path of rapid development through projecting effective controlling action on its IP.

The technical and economic image of a future product and its manufacturing depends on a key element that is the company's IP, which sets appropriate resources for production, as they are part of this IP (Fig. 4.9).

Only in case there is a sufficient resource potential, such as the one shown in Fig. 4.9, an advanced competitive product can be created and launched onto the market on time. Thus, a sufficient resource potential will ensure effective IP fulfillment and stability of a future project throughout its life cycle.

Formation of a resource potential for fulfillment of attractive projects relies on the so-far accumulated scientific and technical assets, existing technological base, competences, as well as financial resources, which are derived from the sales of a competitive product (once the project reaches the point of return). The main resource provision sources for a new future project are shown in Fig. 4.10.

Providing resources for attractive projects means not only direct resourcing a particular project. Also, it is forming a set of expenses resulting from having to improve and upgrade existing production facilities.

Upgrades should rely on regular assessment of the technical level of production and its ability to meet international and main competitors' standards.

This comparison can help to reveal potential reserves for the effectiveness of production and develop a technical strategy aimed at implementation of such reserves while providing resources for these processes. Modernization of industrial facilities is aimed at boosting an organization's level of development, reduction of the prime cost and time of creation, and marketing of advanced products, which can help the organization to achieve rapid development.

As to providing resources for a project focusing on development and manufacturing of advanced products, it is necessary to choose one of resource provision models:

1. Resources for development, manufacturing, shipment, and storage of the product at the manufacturer's warehouse (in this case, the company does not incur any distribution or logistics costs).
2. Resources for development, manufacturing, maintenance, and customer education. In this case, the manufacturer should build a pool of resources to ensure shipment of the product to the end user (materials management and building an integrated logistic process, Fig. 4.11), development of tuition systems and training facilities, as well as for offering additional services, which can bring a substantial additional income to the organization.

However, all abovementioned resource provision models for future projects are inevitably implemented due to limited resources, and effective fulfillment of a company's potential will depend on the effectiveness of resource distribution.

In order to fulfill projects with a limited resource base, it would be paramount to select the most promising one, which can bring a company to the point of creation of a new market segment.

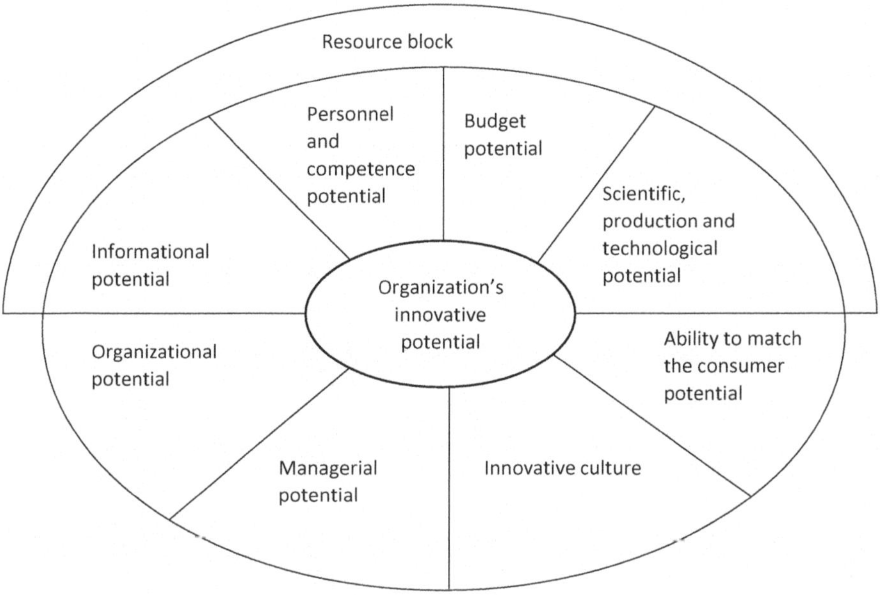

Fig. 4.9 The structure of a company's innovative potential

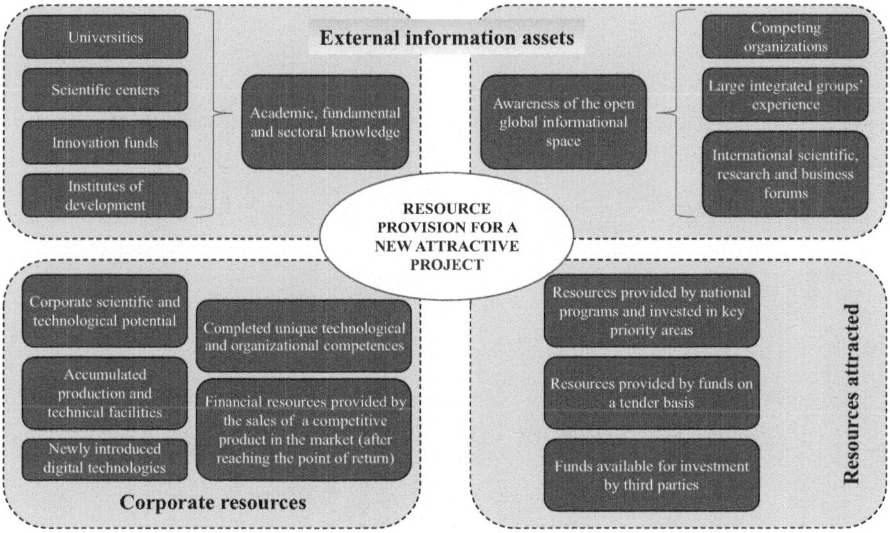

Fig. 4.10 Resource provision for an attractive future project

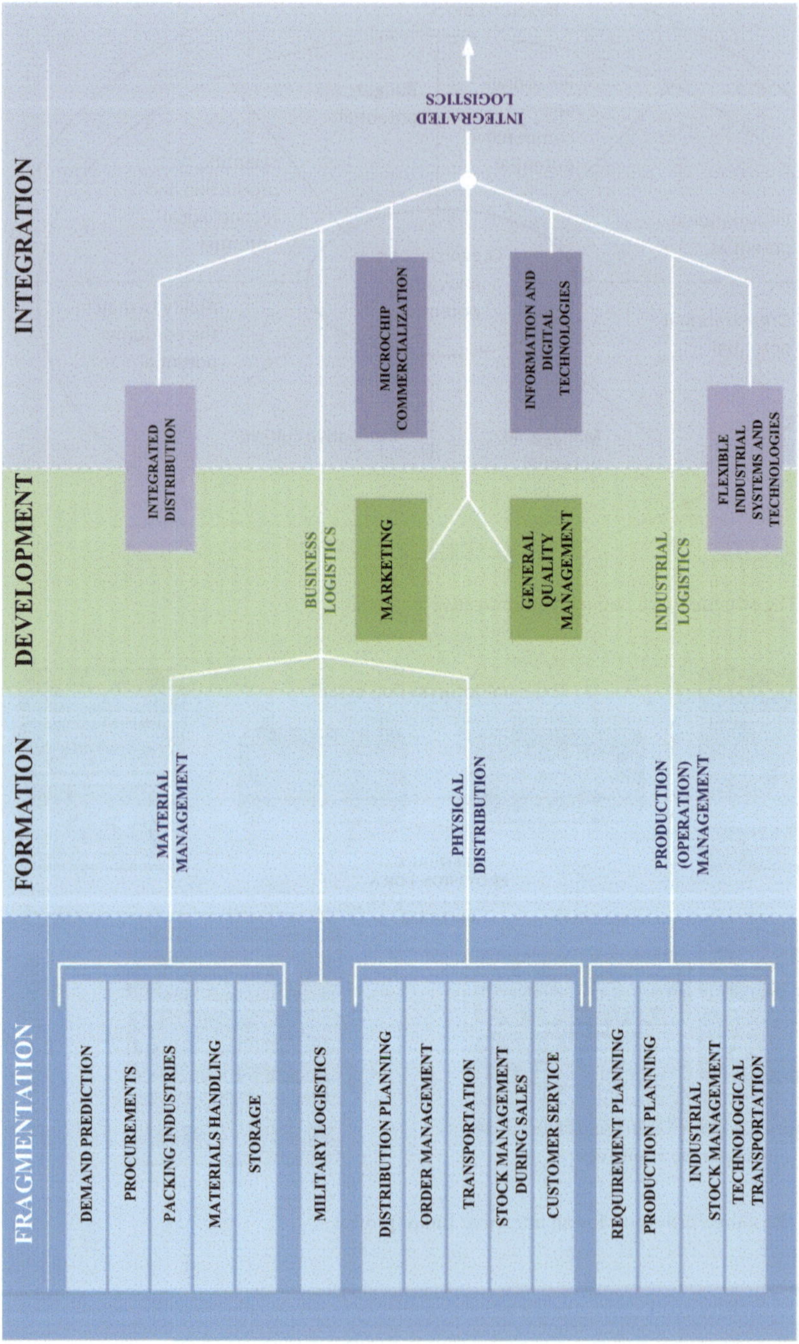

Fig. 4.11 Logistics as an integral management tool

From the multitude of innovative concepts it is necessary to choose ones, which can ensure high economic effectiveness and put an organization onto the path of rapid development. This poses a crucial issue of rational use of the formed resource potential; in other words, it should focus on the fulfillment of a promising project, which could eventually increase the efficiency of an organization's work and competitiveness.

The most attractive project is selected with the help of a decision-making model in uncertain environment. The key aspect of this model is that it is based on a clearly defined calculative procedure, which rules out any subjective factors when choosing future projects.

The most important benchmark for the model is the set of hypothetic promising projects, which are to be executed in an uncertain environment. The uncertainties result from macroeconomic instability, various types of risks and other internal and external factors. According to the model, these factors are known, and their probability distribution will not be counted, because the economic reality suggests that these factors are largely uncertain because of economic instability.

The most important part of selection of promising projects is knowing productivity increase ratios, which are fulfilled when choosing a project and the fulfilled uncertainty. These ratios should be calculated using respective methods.

With the help of this model of selection of attractive projects, which boost effectiveness (reaching a higher competitiveness and rapid development), it is possible to define:

- The ranking of each attractive project
- A productivity increase ratio for each attractive project

The benchmark for the model is:

(A) A multitude of promising projects that are viewed as a source of increased productivity and competitiveness. This multitude is formalized with a set of technical and economic characteristics.

The set of technical characteristics is expressed by vector P:

$$P = \begin{pmatrix} P_1 \\ P_2 \\ \dots \\ P_N \end{pmatrix}.$$

The number of economic characteristics is expressed by vector T:

$$T = \begin{pmatrix} T_1 \\ T_2 \\ \cdots \\ T_N \end{pmatrix}.$$

(B) The multitude of various factors of uncertainty are expressed by vector F:

$$F = \begin{pmatrix} F_1 \\ F_2 \\ \cdots \\ F_N \end{pmatrix}.$$

For example, vector F may include the following factors:

F_1—the dollar rising against the ruble
F_2—rising world prices for raw materials
F_3—rising world prices for components
F_4—rising world prices for equipment
F_5—rising internal electricity tariffs
F_6—rising transportation costs

(C) Benchmark unification matrixes:

$$A = \begin{pmatrix} a_{11} & \cdots & a_{1n} \\ \vdots & \ddots & \vdots \\ a_{m1} & \cdots & a_{mn} \end{pmatrix}.$$

(D) Weighting coefficients denoting the importance of attractive projects' economic indicators:

$$a_m, \quad m = 1, 2, \ldots, N_M,$$

Meeting the following conditions:

$$a_m > 0, \quad m = 1, 2, \ldots, N_M,$$

(E) Correcting coefficient for aggregated scenario factors

$$\gamma_i > 0, \quad i = 1, 2, \ldots, N_M.$$

To implement an algorithm for making decisions, which raise projects' effectiveness and competitiveness in uncertain environments, it is necessary to calculate a productivity increase ratio matrix for aggregated scenario uncertainties.

An algorithm for building aggregated scenario uncertainties looks as follows:

Step 1. Choosing an uncertainty factor:
F—a chosen uncertainty factor.

Step 2. Fixing the chosen interval of probabilistic meanings for the chosen uncertainty factor:

$$F^0 \in [\psi^a, \psi^\gamma].$$

Step 3. Once uncertainty factors are used up, the algorithm ends.

Step 4. Choosing the next uncertainty factor and moving over to Step 2.

An algorithm for building a productivity increase ratio matrix for attractive projects is describes below.

Step 1. Unify all benchmark data within a single scale with the help of a benchmark unification matrix.

Step 2. A piecewise interpolation function is built on paired values—technical characteristics $\Leftrightarrow$ project indicators:

$$F : R^n \to R^m$$

where R^n stands for technical characteristics; R^m stands for project indicators; F is the technical characteristics $\Leftrightarrow$ project indicators factors.

Step 3. Calculating the integrated index using weighing coefficients of the importance of economic indicators:

$$\text{IT}(T) = \frac{\sum_{m-1}^{N_M} a_m T_m}{\sum_{m=1}^{N_M} a_m},$$

where T^m are economic indicators of attractive projects; N^m is the number of indicators.

Step 4. Calculating productivity increase indexes:

$$a_k = Q(T^k) = \frac{\mathrm{IT}(T^k)}{\mathrm{IT}(T^*)},$$
$$T^* \in \{T^1, T^2, \ldots, T^{k-1}, T^{k+1}, \ldots, T^K\} : \|F(T^*) - F(T^k)\| \to \min \}$$

where $a^k = Q(T^k)$ is a company's productivity increase indicator; $\mathrm{IT}(T^*)$ is the general integrated index of a project T^*; $\mathrm{IT}(T^k)$ is an individual index of an attractive project T^* based on indicator T^k.

The mathematical expression $\|F(T^*) - F(T^k)\| \to \min$ means that technological indexes should follow project indicators mentioned in Step 1.

Step 5. Building a productivity index increase matrix using the formula:

$$\{a_{ij}\} = \gamma_j a_i.$$

Next comes the decision-making algorithm for selection of attractive future projects developing in uncertain environments. Assume there is a group of attractive projects, which is presented by multitude T:

$$T \subset R^n.$$

The selection consists in the fixing of a vector:

$$t \in T.$$

A project, which increases effectiveness and competitiveness, is chosen with the goal of maximizing the scalar-valued function $f(t, s)$, which depends not only on the decision made (vector t), but also on the uncertainty fulfillment vector.

The target function $f(t, s)$ stands for productivity increase ratios. Uncertainty factors can take values from the multitude:

$$S \subset R^n.$$

Assume that the uncertainty vector s has become fulfilled providing the following vector has been chosen randomly:

$$s \in S.$$

Thus, the method helps to solve a decision-making task in uncertain environments:

$$\langle T, S, f(t, s) \rangle.$$

The concept of choosing a decision is that the person making it does not control the choice of vector $y \in Y$. Many real tasks do not even have a known probability distribution of the uncertainty vector. In this case, it is necessary to take into account all possible scenarios of uncertainty.

A promising project's increasing effectiveness is evaluated with the help of the following function:

$$f : T \times S \to R.$$

When making decisions, it is advisable to build a *minimax regret function*.

For each scenario of the uncertainty

$$s^* \in S$$

function $\widetilde{f}(s^*)$ can be built according to the formula:

$$\widetilde{f}(s^*) = \max_{z \in T} f(t, s^*).$$

This function can help to define the highest value of the target function every time the uncertainty scenario is realized. The minimax regret function is introduced in the following way:

$$\Phi(t, s) = \max_{z \in T} f(z, s) - f(t, s^*),$$

where z is an attractive project.

The following pair is referred to as a solution to the task of attractive future projects, which help to improve effectiveness and competitiveness in uncertain environments:

$$\left(t^0, \Phi^0\right) \in T \times R^1,$$

This pair is defined as:

$$F^0 = \min_{t \in T} \max_{s \in S} F(t, s) = \max_{s \in S} F\left(t^0, s\right).$$

It uses function F, which can be calculated according to the formula:

$$F(i, j) = \max_{k=1, \ldots, m} a_{kj} - a_{ij},$$

where a_{ij} is the productivity increase ratio.

This model is intended for finite aggregate cases T and S.

Therefore, it is necessary to choose one attractive project out of a finite aggregate of set technologies

$$t_1, t_2, \ldots, t_m$$

when there is a finite uncertainty set:

$$s_1, s_2, \ldots, s_n.$$

Thus, the task comes down to the following mathematical setting.

There is an $m \times n$ matrix:

$$A = \begin{pmatrix} a_{11} & \cdots & a_{1n} \\ \vdots & \vdots\ \vdots\ \vdots & \vdots \\ a_{m1} & \cdots & a_{mn} \end{pmatrix}.$$

It is necessary to choose line t_i, which should be optimal in terms of increased effectiveness and competitiveness. The person, who makes the choice, is not informed on the fulfillment of uncertainty.

To solve this task, which formally can be expressed as an ordered triplet, an algorithm is needed:

$$\langle T, S, A \rangle$$

where an alternative $t_i \in T$ is chosen by a person, who makes the decision, which is the choice of the number of an attractive project:

$$i \in \{1, 2, \ldots, m\}.$$

Independent of and along with the choice of alternative $t_i \in T$, the uncertainty is fulfilled, and it is expressed in the choice of the number:

$$j \in \{1, 2, \ldots, n\}.$$

This results in the fulfillment of the productivity and competitiveness increase ratio, which is calculated as an element of matrix A: a_{ij}—the productivity increase ratio.

The algorithm of choosing the most attractive future project, which helps to improve productivity and competitiveness, is:

Step 1. According to the set matrix A:

$$A = \begin{pmatrix} a_{11} & \cdots & a_{1n} \\ \vdots & \vdots\ \vdots\ \vdots & \vdots \\ a_{m1} & \cdots & a_{mn} \end{pmatrix}.$$

we calculate:

$$\begin{aligned} a_1 &= \max_{k=1,\ldots,m} a_{k1}; \\ a_2 &= \max_{k=1,\ldots,m} a_{k2}; \\ &\ldots \\ a_n &= \max_{k=1,\ldots,m} a_{kn}. \end{aligned}$$

Step 2. Based on the values $a_1, a_2, \ldots, a_n$ calculated at Step 2, a new matrix B can be built:

$$B = \begin{pmatrix} a_1 - a_{11} & \cdots & a_n - a_{1n} \\ \vdots & \vdots\ \vdots\ \vdots & \vdots \\ a_1 - a_{m1} & \cdots & a_n - a_{mn} \end{pmatrix} = \begin{pmatrix} b_{11} & \cdots & b_{1n} \\ \vdots & \vdots\ \vdots\ \vdots & \vdots \\ b_{m1} & \cdots & b_{mn} \end{pmatrix}.$$

Step 3. According to matrix B resulting from Step 2, the following value is received:

$$\begin{aligned} \beta_1 &= \max_{j=1,\ldots,n} b_{1j}; \\ \beta_2 &= \max_{j=1,\ldots,n} b_{2j}; \\ &\ldots \\ \beta_m &= \max_{j=1,\ldots,n} b_{mj}. \end{aligned}$$

Step 4. The productivity and competitiveness increase ratio is calculated in the following way:

$$F^0 = \min\{\beta_1, \beta_2, \ldots, \beta_m\}.$$

Step 5. Finding an attractive project, which would be maximally conducive to a higher productivity and competitiveness, according to the formula:

$$i_0 : \beta_{i_0} = \min\{\beta_1, \beta_2, \ldots, \beta_m\}.$$

The algorithm described above can be used for the choice of attractive future projects, which can help to improve productivity and competitiveness, as well as to achieve rapid development. These projects are selected through ranking.

When the ranking procedure is completed, an organization decides which projects it can realize using the created innovative and resource potential.

The innovative and resource potentials, assessment of their sufficiency for fulfilling attractive future projects, which have been ranked according to their influence on the organization's competitiveness and ability to achieve rapid development, public demand for its products, as well as expected academic and scientific achievements and development of public and individual needs, while using advanced design methods can help to build the image of a product, which will put the organization onto the path of rapid development.

It is a critical task, because its products' image is the foundation of the company's competitiveness. The image should be created with the help of advanced digital methods and technologies based on a prospective market need and dynamics analysis, as well as with account for the company's existing innovative and resource potentials. These potentials characterize the organization's ability to fulfill the new image as a highly competitive output product. This is for further research.

References

1. Samuelson, P. A., & Nordhaus, W. D. (2015). *Economics* (p. 1360). New York: McGraw-Hill.
2. Chursin, A. A. (2010). *Innovations and investment in a company's work*. Mechanical Engineering (p. 469).
3. Tyulin, A., & Chursin, A. (2017). *Competence management and competitive product development: Concept and implications for practice*. Cham: Springer.

Chapter 5
A Product's Image as a Basis of Its Competitiveness

5.1 Assessment of a Corporate Microenvironment and Its Role in Creating a Brand Image of a New Competitive Product

Production of science-intensive goods is determined by their technical and economic images, which, in turn, determine a product's design and technologies used in it and influence the labor intensiveness and production costs. The formation of an innovative microenvironment within a company should be based on rational engineering with the use of advanced production technologies. With the rapidly growing digital economy and cyber-economy, the innovative microenvironment should rely on a digital organization's platform, which should be highly transformable, resource-effective, ergonomic, and be able to integrate customers and business partners into business and product brand imaging processes.

The shaping of a product's technical and economical image wholly relies on customers' individual preferences and real-time intelligent big data analysis. The technological basis of a digital organization's innovative microenvironment consists of cyber-physical systems, in which software components are connected to mechanical and electronic parts via a complex of data, as well as via the Internet of Things, in which physical objects—devices, sensors, and systems—can send and receive data via the Internet without any activity on man's part. Automated routine decision-making is enabled by a well-developed "thing-communication" system, which implies the aptitude of things to identify each other, evaluate conditions, exchange and process data. It excludes man from the process of interaction between things, making the interaction more autonomous, reliable, quick, systematic, and controllable. The main goal of the formation of a company's innovative microenvironment (Fig. 5.1) is helping it achieve rapid development and leadership in key areas. This can be achieved through transformation of business processes within companies in four key blocks: development, technologies, production, and personnel. Each of the blocks implies introduction of elements of the digital production concept known as Industry 4.0,

A. Tyulin, A. Chursin, *The New Economy of the Product Life Cycle*,
https://doi.org/10.1007/978-3-030-37814-1_5

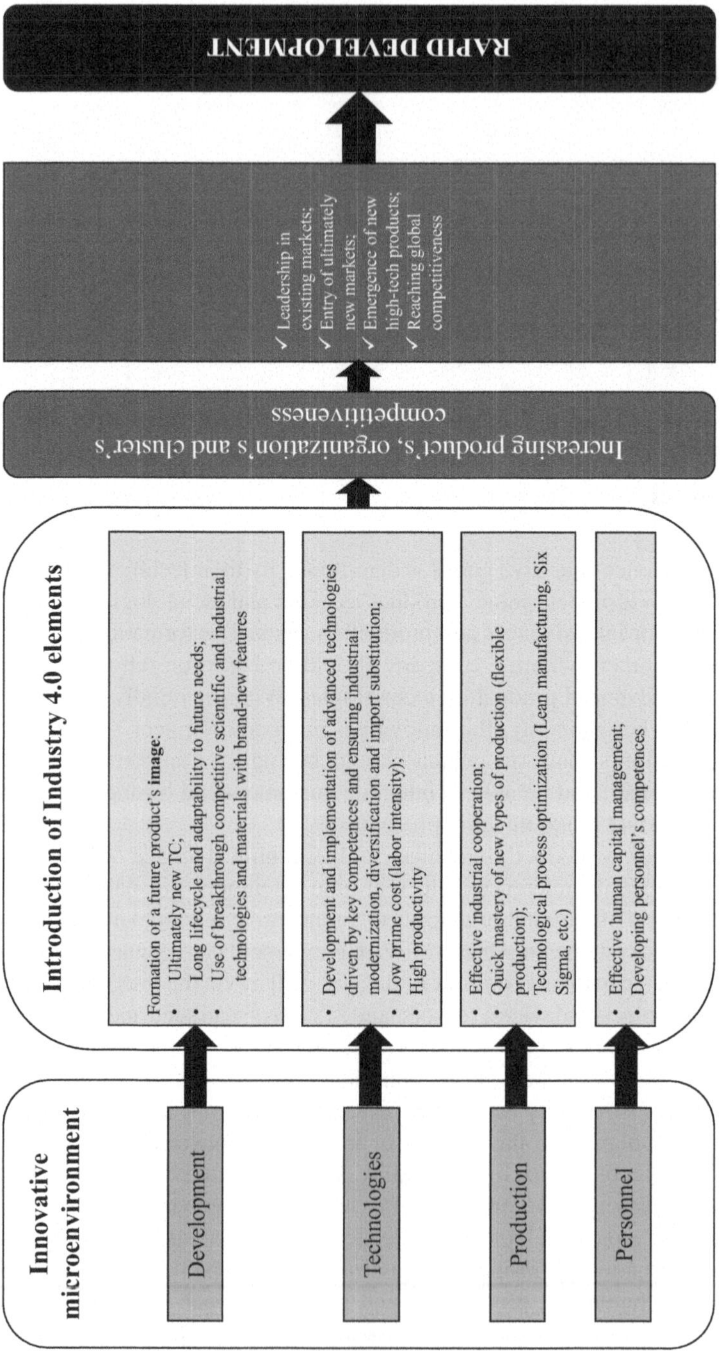

Fig. 5.1 A company's innovative microenvironment as the basis for its rapid development

which should greatly optimize organizational, economic, and industrial processes within an organization, increase its products' and the whole organization's/clusters' competitiveness. As a result, thanks to achieving leading positions in existing inter-branch markets, entering ultimately new markets and launching the production of new high-tech items, which set the tone for global competitiveness, the organization will enter the rapid development mode.

A crucial component of a corporate innovative environment as an ecosystem for development and production of new competitive products, are unique competences, based on which an organization develops an ability to build a technical and economic image of a future product, as well as information resources, which provide optimal relationships and interaction between economic objects and subjects when engineering and producing new goods. Development of cyber-economic elements and decision-making approaches, which are based on convergent global data analytics within an organization, increase economic processes' effectiveness through optimal management of relationships and interaction between subjects' and objects' subsystems (potential consumers, component suppliers, and resources) while building a future product's image.

The technical and economic image is a complex description of a product being developed, which reflects its most significant consumer characteristics, which include:

- List of main consumer parameters and their fulfillment
- Component descriptions (a component model)
- Description of the system's main utility functions (a functional model)
- Prime cost pre-calculations and price forecasts

Thus, a product's technical and economical image is a combination of technical and economic characteristics, which the product should have after the launch of production. This combination should be formed at the initial stage of the future product's life cycle, while generating the concept and giving it certain parameters and properties for it to materialize into a ready-made marketable product.

Development and implementation of breakthrough technologies, which are needed for outlining the image of a product that should become attractive for consumers, requires a substantial resource base, and economically unique competences. To make these processes effective, an organization should build an innovative microenvironment, in which its innovative potential driven by economic mechanisms can be transformed into a future advanced product. There are several things that characterize a corporate innovative microenvironment, which shapes a future product.

A modern company developing and producing highly competitive goods is a complex system with a multilayer structure, where layers' elements are interconnected vertically and horizontally. There are administrative, industrial, scientific, research, and other elements inside an organization. Each division performs a specific function and carries out its work using various types of resources (financial, manpower, etc.). A company's activity is influenced by internal and external factors. External influences are related to national economy and the industry, in which the company operates, consumer expectations within a market segment, which is a target one for the organization.

To prepare a company's scientific and industrial system for development and production of new types of goods, it is necessary to reveal and evaluate factors that characterize a company's ability to promote a new and demanded product. These factors form the internal innovative microenvironment, which can either help or impede the execution of breakthrough projects.

There is a system of indicators, which characterize a corporate microenvironment of an organization developing and producing highly competitive goods, as well as an economic toolkit for qualitative assessment of respective factors.

A company's current internal state can be estimated with the help of the following factor groups:

- Financial and economic
- Scientific and research
- Production and technology
- Availability of manpower
- Information and methodology
- Administration and management

The technical and economic image, which is being formed by the company, will be the foundation for the product, which is to be marketed and sold. Each particular product has a combination of technical and technological features, by which it is possible to evaluate the conformance of a company's microenvironment to standards, which consumers apply to its new product. A company's corporate microenvironment, which reflects its ability to image a highly competitive product, can be evaluated with the help of data reflecting the availability and use of specific production facilities, availability, and effectiveness of the main funds, use and consumption of energy, administrative and technical aspects of production, availability of manpower, and use of employees' work time. Another important source of information, which helps to assess an organization, is its financial documentation, which reflects the most important indicators of its work (balance sheet profit, profitability, turnover, operational profit, etc.).

Also, there are groups of factors characterizing a company's inner innovative microenvironment, and they can correlate with the specifics of a future product's image.

The first group characterizes an organization's financial and economic status. The factors reflect the availability and use of the company's budget resources. An analysis of financial and economic factor should evaluate the provision of its current industrial and economic work with respective types of resources, i.e., their effective use in the process of production. It is possible to evaluate a company's financial and economic status with the help of standard corporate economic indicators:

- Current liquidity ratio
- Inventory turnover period
- Current assets coverage ratio
- Financial leverage
- Equity-to-total assets ratio
- Debt-to-capital assets ratio

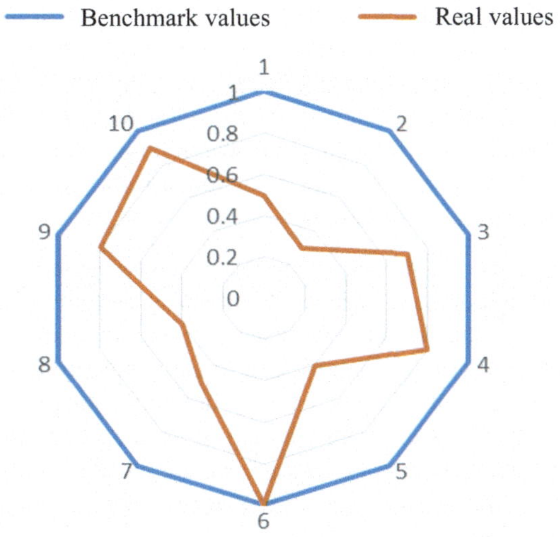

Fig. 5.2 Financial and economic values

- Asset turnover ratio
- Asset turnover period
- Return on assets
- Return on sales
- Profit margin

This is not a complete list of criteria, and it can be changed or supplemented in each specific case.

All these financial and economic characteristics are qualitative and can be calculated according to standard formulas based on the company's financial records. For example, the current liquidity ratio is expressed as a ratio between a company's turnover assets and short-term debt.

It is possible to demonstrate the evaluation of these indicators in the form of a diagram as shown in Fig. 5.2. This diagram can be used conveniently for comparing real values to benchmark ones.

Also, it is possible to use differences in financial and economic indicators. In this case, standardized vectors' components should be multiplied by the weights of respective indicators; after that, these vectors should be ranked through comparison with the benchmark ones according to selected metrics.

The following set of factors characterizes an organization's scientific and research sector. This group of factors is of particular interest, because in most cases launching a science-intensive project requires a number of scientific and research steps. Scientific and research factors characterize an organization's expertise and experience in developing science-intensive products. Its scientific and research potential is characterized by:

- Availability of scientific schools and individual specialists, who carry out research and development works in different spheres.

- Maintaining stable contacts with leading scientific organizations and centers, whose focuses are related to the company's advanced project.
- Possessing intellectual property and respective rights, patents for inventions, utility model patents, computer programs, trademarks, service marks, etc.
- A share of experimental and research property, machinery, and equipment related to technological innovations included in the total cost of all production process facilities and installations.

The latter is the only quantitative index. Patents, certificates, etc. make up a qualitative indicator, because the physical number of patents and certificates is not always reflective of an organization's innovative and technological potential. Other characteristics of the scientific and research potential should be qualitatively assessed based on a system of signs, which is tailored to the project stream. This system can be processed effectively with the help of an automated logic system. The result of the system's work is well-performed evaluation of a company, and it should reflect the correlation between its scientific and research potential and tasks that are solved as part of the project.

An important group of factors is one that characterizes a company's industrial and technological potential. These factors characterize the state of productive assets and the infrastructure, availability of technologies, free production facilities, which provide for development and release of new products, etc. A firm's industrial and technological potential can be analyzed according to the following factor groups:

- Availability of technologies needed for the formation of an advanced product's image.
- Availability of all required technological conversions.
- A company's ability to embrace new equipment and advanced technologies.
- Availability of required production facilities.
- Productive assets' state.
- Conformance of industrial equipment to (international) standards.
- Cooperation with partners in the sphere of industrial technologies.

These characteristics can be complemented by other ones, which conform to the future product's image.

Most of these characteristics can be assessed using a qualitative method as well. It is possible to check the presence of all needed technological conversions by checking each particular one. The *Availability of Required Production Facilities* characteristic also implies assessment of the number of facilities. The assessment demonstrates the sufficiency or insufficiency of available productive facilities. It is necessary to analyze all main industrial operations, which need to be completed to release a new product. Therefore, each technological conversion should be supplied with a table reflecting the lineup and industrial capacity of equipment, which is going to be used in production. Also, this table can reflect equipment's specific characteristics, such as functional depreciation, difficulty obtaining tools, and accessories. The task of assessing a company's productive and technological potential can be solved with the help of an automated logic system.

Closely related to industrial and technological factors is a group of factors characterizing a company's human resources. These factors describe the professional performance of specialists operating at various layers of the corporate hierarchy. Thus, the availability of appropriate workforce is evaluated in the same way as that of productive facilities. When necessary, it is possible to assess a company's needs to hire new specialists to fulfill a project. Another important characteristic of workforce is a potential for manpower development. Personnel's characteristics are also qualitative, and it is possible to draw conclusions with the help of a logic system.

Information and methodological factors characterize the automation of a technological process, use of automated communication systems, advanced design tools, etc. For example, using a digital production model in a company's work is an information and methodological factor.

Finally, administrative and organizational factors characterize the effectiveness of production administration and management. First, it is mandatory to analyze the effectiveness of interaction between a company's structural elements and to try to reveal strong and weak points. Second, an organization should be evaluated according to signs like *having a stable development strategy, introduction of innovative management mechanisms*, etc. Also, it is possible to carry out an analysis through the use of an automated logic system.

These are the main groups of indicators, which can help assess the level of production. Table 5.1 presents a rough outline of indicators, by which it is possible to make an assessment.

If a company does have an innovative microenvironment with these qualitative characteristics, its further focus will be prediction of the new product's brand image with account for scientific and technological tendencies and changing market and public needs. Predicting it on a national scale is a challenging task, because it requires a large number of actual figures and reference to a lot of factors. However, it is possible to make industry-wide predictions.

The most important methods used to evaluate the effectiveness of a future product's technical and economic image are simulation modeling and prognosis. While relying on these, a company can define consumer expectations and their transformation, as well as the level and tendencies of scientific and technological progress. Based on such predictions and with a sufficient innovative potential, an organization can start forming a technical and economic image of a future product.

5.2 Use of Communication Methods and Technologies in Market and Consumer Demand Analysis When Shaping a Product's Image

Primarily, the task of forming a future product's technical and economic image is related to the availability of unique competences in an organization, sufficient resourcing and innovative potential, which can be used for designing and

Table 5.1 Indicators used for evaluation of technical and economic level of production needed to shape a future product

Groups of signs	Characteristics
Financial and economic	Current liquidity ratio, inventory turnover period, current assets coverage ratio, financial leverage; equity-to-total assets ratio, debt-to-capital assets ratio, asset turnover ratio, asset turnover period, return on assets, return on sales, profit margin, etc.
Scientific and research	Presence of scientific schools and individual specialists; maintaining stable contacts with top scientific organizations and centers; possessing intellectual property and respective rights; a share of experimental and research property, machinery, and equipment related to technological innovations included in the total cost of all production process facilities and installations, etc.
Engineering and technology	A company's ability to embrace new equipment and advanced technologies; availability of required production facilities; productive assets' state; conformance of industrial equipment to (international) standards; cooperation with partners in the sphere of industrial technologies.
Workforce potential indicators	Professional background of specialists working at different corporate levels, availability of appropriate manpower resources, a necessity of hiring new specialists for projects, availability of personnel development opportunities, etc.
Information and methodology	Automation of technological processes, use of telecommunication systems, the digital production concept, etc.
Administrative and managerial	Effective interaction between corporate structural elements, using sustainable development strategies, availability of advanced management mechanisms, etc.

manufacturing of a new product boasting competitive advantages that attract customers.

In this case, a crucial role belongs to a company's intellectual potential, which is created by its staff, and their competences, which enable cost-effective and quick production. Shortening of the concept-to-market period is an increasingly important task, especially given product life cycles shortening due to the rapid technological progress.

Thus, building a future product's brand is preceded by creation of a scientific and technological platform, on which it is possible to develop and manufacture and advanced product given the available intellectual and industrial potential and the company profile.

However, this poses another task, which is identifying the tendencies and rate of the development of public needs, which should form within a market segment in a few years. Changes in public needs are explained by scientific and technological factors, the public's growing awareness of new achievements and scientific discoveries, which should improve the quality of life. Right now, the development of self-driving vehicles and drone aircraft is under way. Not unlikely, flying drone taxi and personal flying vehicles will appear in future, which could reduce traffic congestion in large cities; besides, people will not need to attend driving schools.

Self-cleaning clothes, which may appear in future, could reduce water consumption. Today, there are some prerequisites: Chinese scientists have developed a kind of fabric, which can self-clean in sunlight. The technology relies on the use of titan dioxide nanoparticles. It is not complete yet, because the effect requires a high ultraviolet dose. In medicine, microscopic robotic devices may be invented, which are supposed to be introduced into a human body and restore damaged cells.

These examples illustrate that the environment, in which man and the society lives, changes over time. Consequently, the outer environment and progressing technologies generate demand for new products with images that are very different from existing ones'.

Therefore, shaping a new product's brand image is related, on the one hand, to studying the transformation of the outer environment and, on the other hand, to future products' brand image trends. It is a major task, as it requires collection and analysis of a large amount of information concerning markets, technological progress, scientific discoveries, modern people's, public and state's needs; none of that can be fulfilled without using the global information space and advanced communication methods and technologies.

Today, there are lots of communication methods and technologies, which function thanks to the use of global information resources.

These technologies open up new horizons in branding, engineering, preproduction, and production of future goods with the use of contemporary digital methods.

Firstly, this refers to cloud big data processing and analysis solutions and methods, which have great advantages over traditional analytical data collection and processing methods.

The importance of big data depends not only on how much data an organization has, but also on how it uses it: the more effectively it applies accumulated (collected) and processed big data, the greater a potential for economic growth it develops through creating a highly competitive advanced product based on a multidirectional market and consumer need analysis and current innovative potential.

The use of advanced big data processing methods and technologies produces a number of beneficial effects (Fig. 5.3), which is the goal of industrial companies wanting to gain a good share in the existing market niche or create new markets.

Digital big data processing methods and technologies help to carry out analysis and make prognoses, based on which it is possible to make future product imaging decisions with account for consumer preferences. This will help to achieve high competitiveness and outperform leading competitors. Today, organizations should outline their strategies, including portfolio development strategies, while relying on analyzed big data, to compete, introduce novelties, and set consumer attractiveness.

Digital big data processing methods help companies to create a new basis for growth and use of big and diverse information for predictive technical servicing, endowing products with unique consumer properties tailored to current market standards, proactive risk management, etc.

Thus, today's businesses that are striving for leadership in the market and economic sustainability create their own digital and computer platforms and use

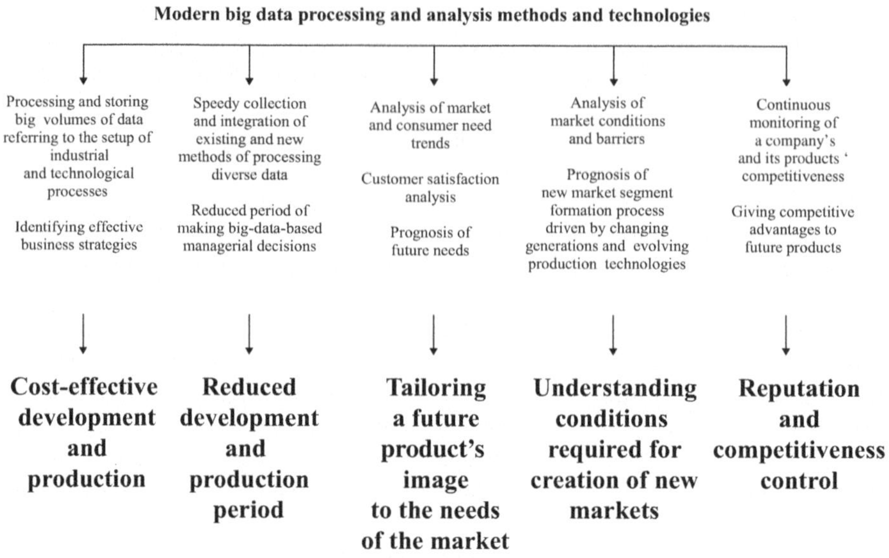

Fig. 5.3 Benefits of using digital big data processing methods and technologies

cutting-edge digital methods and technologies to integrate two processes: consumer market analysis and shaping the image of a future product. These platforms rely on creation and modular unification of product imaging and designing systems that use corporate information resources, technologies, and competences.

Meanwhile, the use of these information methods and technologies in the process of building a future product's image will be possible as a succession of steps is carried out (Fig. 5.4).

As part of building an advanced product's image, at the initial stage it is necessary to collect and analyze information about the current and future market demand for different types of products, with the use of cutting-edge information methods and technologies. The biggest part of it is identifying a dominant product, and its characteristics, which make it valuable for consumers.

At the second stage, it is necessary to predict the development of the market to understand whether or not the now dominating product is going to be on demand with permanent upgrades and improvements of its consumer characteristics due to technical and technological progress.

To evaluate the technical and technological progress, a mathematical model for evaluation and prediction of market capacity in relation to the currently dominant product can be used.

There are lots of potential segments where a future product can be solved:

$$G_1, \; G_2, \; \ldots, \; G_N.$$

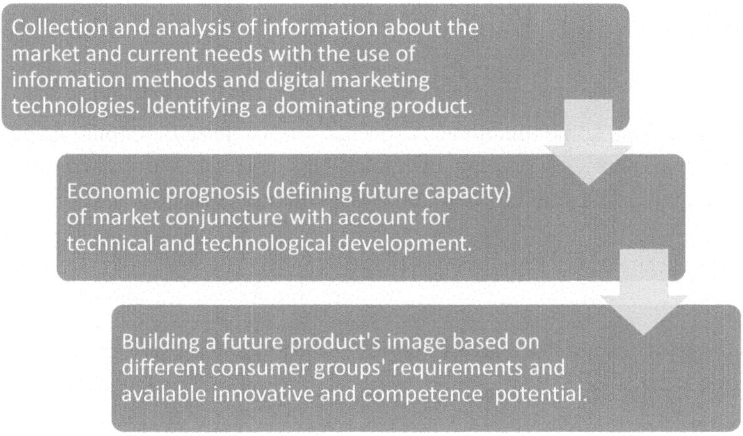

Fig. 5.4 Building a future product's image

Each market can be described by the following parameters at a fixed period of time:

A_n A product purchase volume
C_n Capacity of the market used by products of future
P_n A potential for the growth of a future market

The purchase volume describes the current state of a market. Measuring units of A_n are financial units. The capacity of the market C_n describes a maximum possible extension of the market in future, while measuring units describe financial units. The potential of the growth of the market P_n describes a possible increase in the market volume, which depends on the current market capacity. Mathematically, this parameter is a random value. It is measured in percentage.

First, there is a model describing the dynamics of purchase volume in different market segments. Assume that at the beginning, it uses set purchase volume, market capacity, and growth potential parameters for each particular consumer of goods/ services. Later on, these parameters are recalculated according to changing external parameters, which mainly include the weighted average price for the product being sold:

$$R_t, \quad t = 1, 2, \ldots,$$

where t denotes the time interval, within which buyers' parameters change.

So, customers' parameters change at every time point according to the following rule:

$$A^{t+1} = A^t + f(C^t, R^t);$$

$$C^{t+1} = C^t + g\left(A^{t+1}\right) + C^t P^t;$$

$$P^{t+1} = h(P^t, R^t),$$

where f, g, and h are some functions. A specific type of these functions can be chosen based on customers' geographic characteristics and the current conjuncture of market segments. The general shape of these functions can be:

$$f(C^t, R^t) = k_1 \cdot \sqrt{C^t} \cdot \frac{k_2}{1 + k_3 R^t};$$

$$g(A^{t+1}) = k_4 \cdot \sqrt{A^{t+1}};$$

$$h(P^t, R^t) = P^t - k_5 \sqrt{R^t},$$

where k_1, k_2, k_3, k_4, and k_5 are some positive coefficients. It does not specify the type of the random value P_n; instead, it describes a theoretical mean value.

To complete a calculation according to this model, consecutive computations should be made with reference to future products' predicted value.

Because the model embraces a number of market segments, it is necessary to take into account that they can be interrelated; for example, successful sales in one market can boost the market potential within another segment.

The model of interrelated markets can be described with a market interdependence diagram. Its top nodes will represent segments with advanced products, and oriented branches will reflect the interdependences between these markets. A proper calculation can be made with a help of a (loaded) diagram, where each branch is supplied with a coefficient of influence on the potential. Also, this weighing coefficient will also change over time, and it can be designated as:

$$Y_{ij}^{t+1} = k_6 \sqrt{A_i^t} \cdot C_j^t,$$

where i is the number of the node, from which a branch goes, and j is the number of the node, toward which a branch goes.

With reference to this coefficient, the market capacity dynamics equation will take on the following form:

$$C^{t+1} = C^t - g\left(A^{t+1}\right) + C^t \cdot P^t + \sum Y^t,$$

where all branches that go into the n-node are summed.

This model helps to quantitatively assess and predict markets that present products that boast great competitive advantages and a high customer appeal.

At the third stage, once market predictions are made, a future product's image should be built for different customer categories based on their needs and values. At this point, the key task is satisfying customers' needs.

While solving the task, three main points should be followed:

1. Meeting national and global technological tendencies when building a future product's image.

   In case whereby the state is the consumer, it is possible to track needs along with the ongoing transformation of economic activity models within a country, which can be fulfilled through scientific, technical, and technological development (Fig. 5.5).

   For example, in many countries there is a tendency of using self-driving vehicles, also as part of social governmental programs (public service vehicles).

   For this purpose, intelligent multiple camera, multi-radar, and multisensory vehicle driving systems are being developed that ensure accurate object detection (to detect people, vehicles, and other objects) even during rain, mist, at night, or in case of blinding. They require the use of respective technical and technological solutions that keep high-precision intelligent systems functioning.

2. Shaping a product's image should be aimed at winning leading market positions. In this case, a product should be shaped in such a way as to technically and economically outperform the currently dominating product.

   This requires a transformation of project management principles that are applied for creation of advanced products and use information methods and integrated databases (Fig. 5.6).

   Therefore, expected market standards should be considered and highly competitive technologies should be developed during the product image planning and shaping phase.

3. The shaping of a future product's image depends on changes in products' technical and economic characteristics in destination markets, which will be observed in future.

   This case refers to predicting changes in public needs and a manufacturer's potential, which are observed after 3–5 years, when an advanced product enters the market. Shaping the image of a product, which should outperform the currently dominating one, does not guarantee rapid development. It will do when the future product's image shows better characteristics than the image of a competitors' product and when the product itself gains the ability to satisfy public and individual customer needs.

Effective following of these three statements is possible with the use of information methods and technologies while shaping a future product's image as part of a unified corporate digital platform's work.

Some existing information methods are used and offered by Siemens in maintaining a unified digital product life cycle management platform (within a virtual organization).

In order to create and develop a virtual corporate and business process management platform, Siemens offers the following Industry 4.0 solutions:

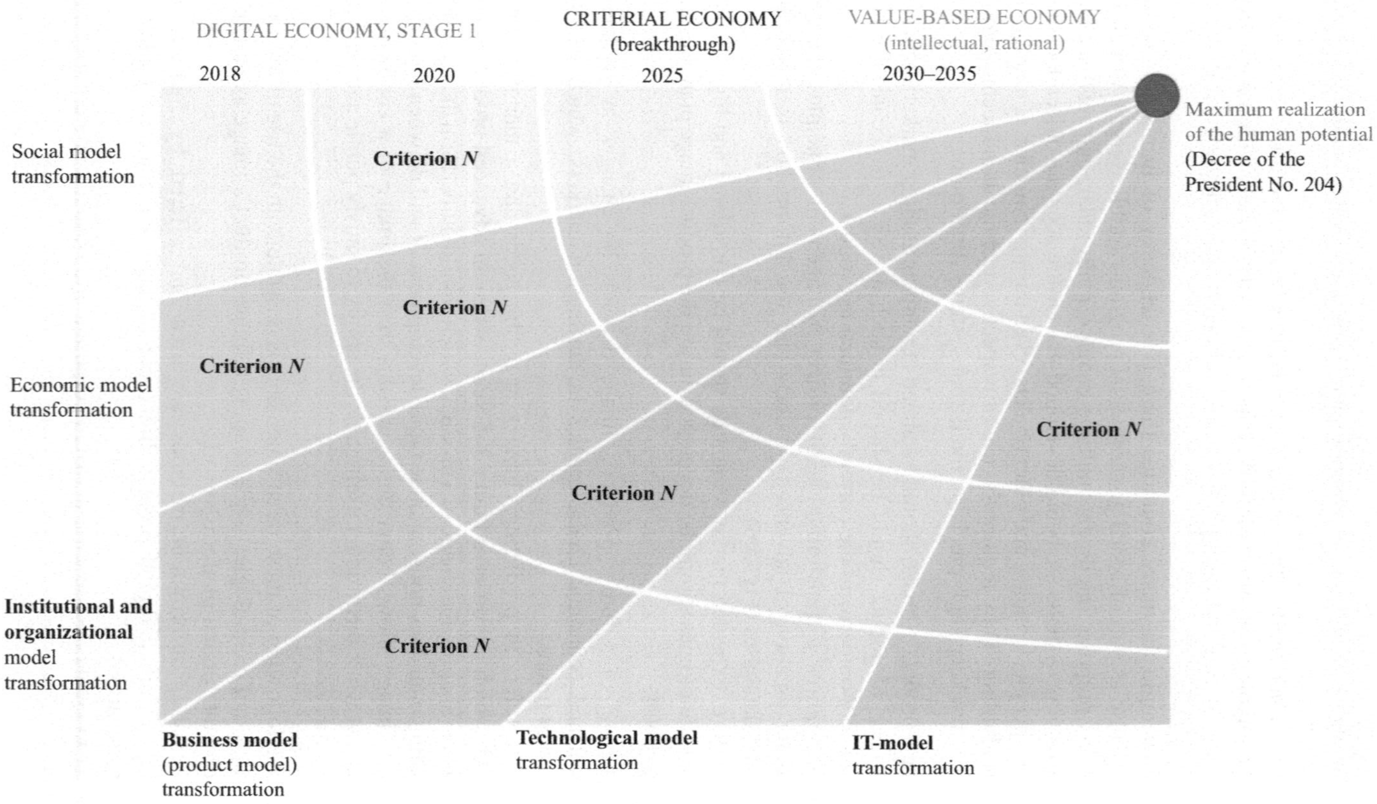

Fig. 5.5 Transformation of the state's needs and economic models

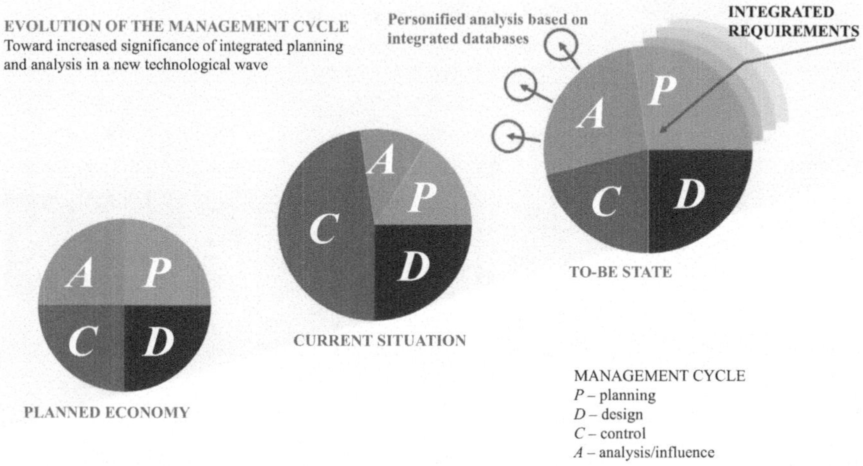

Fig. 5.6 Evolution of the management cycle

- The MindSphere platform, which provides access to big data and data processing tools.
- COMOS (for uninterrupted processes) and TeamCenter (for discrete processes) platforms, which enable companies to implement a model-based approach.
- The SIMATIC platform, a number of platforms used in drive technology, as well as measurement and automated tools in widely used cyber-physical systems.

The chain of production of an advanced item at an industrial enterprise includes the following phases (Fig. 5.7):

- Design
- Production planning
- Preproduction and engineering
- Production process
- Technical maintenance and repair

Each phase uses its own set of components, toolkit, and process installations and solves specific tasks, depending on the type of production. At all phases, a company cooperates with external and internal suppliers, which is a specific goal of optimization. It is what happens in the real world.

Digitization of business processes creates a reflection of this chain in the virtual world, which consists of "digital twins" of real-life objects and processes. Later, through advanced use of these twins, a hypothetic economic effect is calculated, which can be achieved through creation of a product based on its developed and modeled digital image.

Building a digital life cycle management platform could help to test future products on digital models at the image shaping and design stages: it will be possible to check units' mechanical compatibility and planned workload, reveal overheating

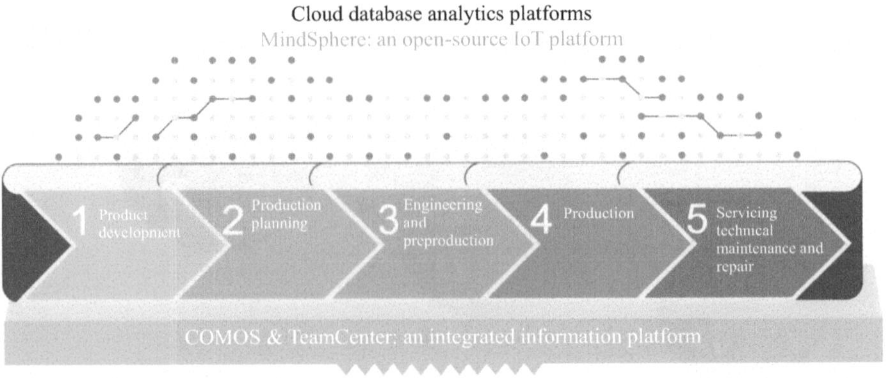

Fig. 5.7 A chain of product creation

areas, carry out electromagnetic tests, detect internal code errors, etc., without building a physical prototype.

With the help of digital models of mechanisms, process installations, and personnel, it is possible to replicate the work of a whole production site and department, as well as related logistics chains; this could help to calculate the length of time needed for building a unit, prevent mechanisms from traumatizing personnel, adjust material flows, and determine an optimal raw product stock that should ensure sustainable economic growth.

Also, creating an advanced product poses the challenging task of outlining the design, preproduction, and production processes in such a way that the output product retains its competitive advantages notwithstanding competitors' progressing technologies and product lines. To solve this task, it is necessary to use advanced design-to-cost methods, build digital twins that could help to model the commissioning process at the design stage and modify the product image to avoid emergencies and flaws. Using artificial intelligence in preproduction can make the production process more effective, minimize costs, and improve technical characteristics, which can help the output product to dominate the existing market or create a new one.

Thus, using information methods and technologies can significantly reduce the concept-to-market period and helps to predict products' competitiveness based on market analytics and revealed long-term needs at the product image formation stage. Further use of advanced technologies can help an organization retain leadership in a segment through upgrading and modernizing its product, which is already being marketed, with the help of new unique competences, scientific discoveries, and new technical and technological solutions.

Generally, market and future needs analytics, which accompany the shaping of a product, relies on advanced intelligent managerial decision-making technologies. The goal of using big data analytics and artificial intelligence is optimization of managerial decisions concerning products' technical and economic images. The most effective organizational and information means of supporting intelligent decision-making is transformation into a virtual organization and incorporating instruments into its platform. A corporate platform contains the company's unified knowledge base, which makes the core of effective intelligent decision-making methods, as well as methods of analyzing the global information space, which can help develop new technical and technological directions, enrich and update knowledge bases.

Using digital product design management infrastructure is the organizational and information basis for competitiveness management at the stage of shaping a product's technical and economic image.

5.3 Competitiveness Management of Science-Intensive Products When Shaping Its Technical and Economic Image

At the stage of building a product's technical and economic image of a future product, it is important to shape its competitive advantages. When solving this task, a company should take into account the market trends, public needs, specifics, and technical and functional characteristics of products represented in a particular segment. On the other hand, the company should combine innovative technical solutions and competences, which could help to create an image, which, by the time a product enters the market and begins to sell, will have gained great advantages over competitors' products.

Industrial enterprises face the challenging task of defining future products' unique consumer properties when shaping its technical and economic image, which should help it to achieve market domination in future. It is necessary to create a technical and economic image, which can give the product all the competitive advantages that are behind the project.

The hardest part of building a competitive technical and economic image is a lengthy concept-to-market period (3–20 years, depending on the complexity of a science-intensive product). Over this period, new products with new features appear in the market, new competitors emerge, new scientific discoveries are made; techniques, technologies, and public/state needs evolve. These factors gradually reduce the product's competitiveness unless it is constantly improved with the help of new scientific and technical reaches, acquired or developed competences.

It is important to pre-check and constantly monitor the situation in science, technology and the market, based on data derived from the global information space with the use of modern and advanced methods and technologies applied when collecting, processing, and analyzing diverse data. This is required for making timely and on-the-spot decisions aimed at improving a product's technical and economic image and stimulating a demand for it.

To provide competitive advantages for the product, its technical and economic shape should be changed and modified throughout the life cycle embracing its development and manufacture. This helps to improve both technical and economic parameters and to ensure that the product that is being tailored to the developed image outperforms its counterparts and dominates its market niche.

In the modern environment, this problem is increasingly tough, because innovations become outdated very fast, and all manufacturers strive for sustainable growth, retaining market positions, dynamic updating and extension of their product portfolios with the help of its current competences and take steps create a uniquely appealing image of a product, which they are planning to manufacture.

Thus, the biggest difficulty of reaching competitiveness when building a product's technical and economic image is in predicting technical and economic characteristics, which, when the image is materialized into and sold as a final product, should create significant advantages and help the product reach dominating market positions.

One way to solve it is to build competitiveness evaluation models when forming a product's technical and economic image and, as well as manage the improvement process based on these models, prior to confirmation of the terms of reference.

The modern economic literature describes a lot of approaches to the quantitative evaluation of products' competitiveness, which depend on price–quality ratios (technical characteristics); however, they are applicable only to products, which have already entered the production stage, and at a certain point of time; they cannot be used for evaluation of a product's technical and economic image. This indicator can help to define whether the product being created now with the goal of further production can dominate existing markets or create new ones, which should satisfy the public's new needs.

Therefore, a new market can be created only if the product does have the characteristics that are created at the stage of technical and economic shaping and capable of satisfying the growing needs of the public and business. Naturally, an advanced product that is based on an elaborate image, defines needs for new benefits and stimulates the appearance of new markets and economic growth, will eventually dominate the market.

Competitiveness evaluation, which is carried out when building the technical and economic image, is closely related to indicators of the innovative potential and key competences (including those related to the fundamental science), which are behind this image. Based on this evaluation, it is possible to define the effect resulting from the implementation of the technical and economic image as a finished advanced product.

Assume there is an item designed and pre-produced according to its image; it is in for further technical and economic improvement aimed at achieving a higher level of competitiveness along with an analysis of this parameter.

To analyze competitiveness along with building a future product's technical and economic image while taking steps to improve its parameters, which this product is going to have, the following mathematical model can be applied.

Assume that the time interval is $t \in [0, T^*]$.

T is the moment when improvement of the product's technical and economic image begins:

$$T \in (0, T^*),$$

where T^* is the time-frame boundary.

This moment reflects steps, which are going to change the level of competitiveness when shaping a product's technical and economic image.

Vector $Q = Q(t)$ within the time frame reflects competitiveness in the process of shaping the product's technical and economic image:

$$Q(t) = \begin{pmatrix} Q_1(t) \\ Q_2(t) \\ \dots \\ Q_n(t) \end{pmatrix}, n \geq 2,$$

where $Q_i(t) > 0$ defines the level of competitiveness when shaping the i-type product.

Assume that at the starting point ($t = 0$) the set competitiveness level is:

$$Q(0) = \begin{pmatrix} Q_1(0) \\ Q_2(0) \\ \vdots \\ Q_n(0) \end{pmatrix}.$$

This vector caters to a system of regular differential equations:

$$Q(t) = A(t)Q(t).$$

The influence of activities that are aimed at improvement of the product's technical and economic image on its competitiveness is evaluated with the help of the following vector function:

$$B(t, Q(t)) = \begin{pmatrix} b_1(t, Q(t)) \\ \vdots \\ b_2(t, Q(t)) \end{pmatrix}.$$

The original equation is:

$$Q(t) = A(t)Q(t) + B(t, Q(t)).$$

Functions $b_i(t, \ Q(t))$ describe the influence of activities aimed at boosting competitiveness and carried out when shaping the product's technical and economic image. Assume that

$$b_i(t, Q(t)) = \begin{cases} \beta_i(Q(t)), t \in \Omega; \\ \quad 0, \qquad t \notin \Omega. \end{cases}$$

Here $\beta_i > 0$ is a ratio denoting the degree of improvement of the product's technical and economic image, and

$$\Omega = \bigcup_{i=1}^{p} [t_i, t_i + \Delta_i]$$

reflects the division of the time frame into smaller intervals. There are two ways to define t_i and Δ_i as constant values:

$$t_i = \text{const};$$
$$\Delta_i = \text{const}$$

or as independent identically distributed random variables:

$$t_i = \xi_i;$$
$$\Delta_i = \eta_i.$$

It is necessary to ensure that the intervals do not overlap.

Steps aimed at improving the image of product $b_i(t)$ may differ. For example, $b_i(t)$ can be step function; in this case, image improvement will result in a jump in competitiveness.

In cases whereby image improvement activities are regular and are carried out with an unchanging intensity, functions $b_i(t, Q(t))$ are:

$$b_i(t, Q(t)) = \begin{cases} \beta_i, t > T; \\ 0, t < T, \end{cases}$$

where $\beta_i = \text{const}$.

Taking regular steps to improve the product's image ensures a stable competitiveness and helps to maintain it at a planned level.

This approach allows to consider different types of stimulating functions, which cater to various economic situations. For example, it is possible to take stimulating measures that imply introduction of new technological solutions: improving a product's image to tailor it to customers' needs, introducing a new functionality, improving technical characteristics using advanced technologies. However, it takes a bit of time to take a particular step. Over the time, competitiveness may decrease, while stimulating measures can raise it. Therefore, this process tends to be cyclic. These properties can be expressed through the introduction of new functions $b_i(t, Q(t))$:

$$b_i(t, Q(t)) = b_i(t) = \xi_i \sin(\eta_i t + \mu_i) + \varphi_i Q_i(t) + \phi_i,$$

where $\xi_i > 0$ is the value characterizing the acceleration or slowing down of the change of the competitiveness level when shaping the product's image; $\eta_i > 0$ is the value characterizing the regularity of product image updates; μ_i is time translation,

$\varphi_i > 0$ and $\phi_i > 0$ are two sets of numbers, which define weighing coefficients that regulate the calculation of the current competitiveness level.

This calculation model makes it possible to evaluate the dynamics of competitiveness when building a product's technical and economic image.

Thus, based on this economic/mathematical model, a competitiveness evaluation method has been developed, and it is to be used when shaping a product's technical and economic image. The computational algorithm relies on a mathematical apparatus of differential equation systems.

The model explains and proves the necessity of permanent improvement of a product's image, from which a new and highly demanded product can be created, which can ensure the manufacturer's domination of the market.

Meanwhile, the technical and economic image should be created in such a way as to keep a product's competitive properties high until the project reaches the point of return and brings a profit, which should be invested in a new advanced project.

Therefore, the technical and economic image of a product, which is expected to enjoy a demand, should be shaped as shown in Fig. 5.8.

Based on the scheme shown in Fig. 5.8, a number of recommendations for industrialists, who are supposed to shape a new advanced product's technical and economic image:

- New products' technical and economic image is built through research of similar competing products, which are present in the market, predicting the development of science, techniques, and technologies, which shape advanced products, and based on the formed corporate microenvironment, which will ensure that the future product will be tailored to the image.
- Carrying out competitiveness management work while building the image with the help of economic and mathematical tools, as well as advanced methods and technologies of collection and analytics of data derived from the global information space and concerning market conjuncture, competence development, and newly emerging technologies, which can be used for improvement of technical and economic characteristics of a future product.
- During the formation of a product's technical and economic image, competitiveness management is carried out with the help of a developed instrument—an economical/mathematical model, which helps to fulfill dynamic evaluation. In case whereby indicators resulting from the a model-based calculation are too low, an organization should reconsider and modify its product's technical and economic image.
- "Low signal competitiveness management." This tool has a similar algorithm, which is described in a monographic work [1]. Essentially, an analysis of market conjuncture and positions of competitors having more advanced specimens are carried out at all stages of a product's life cycle, as they can rob the future product of its market share. Should such threats emerge, it is necessary to correct the product's image with the help of the company's current potential.

In case whereby the corporate potential is too low to create an advanced product, it is necessary to obtain missing competences from boundary partners, cooperation

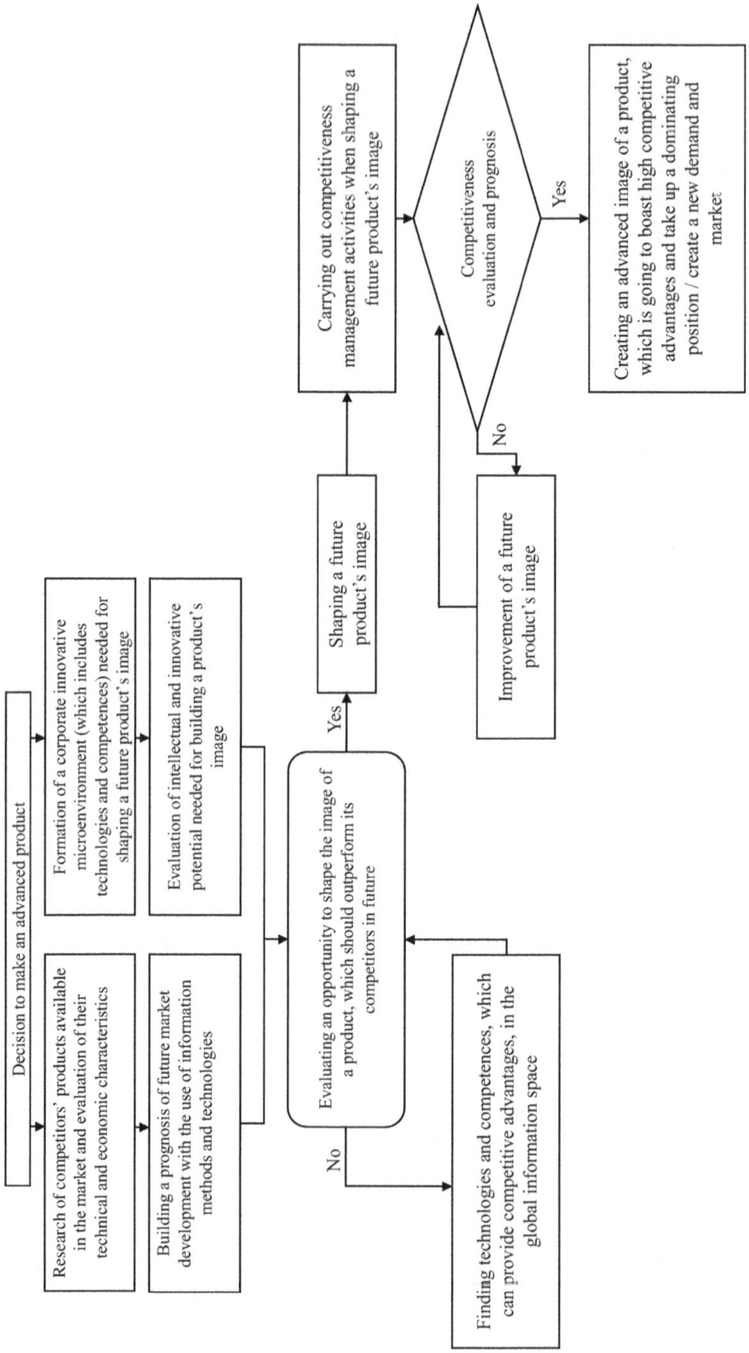

Fig. 5.8 Creating a product's technical and economic image

with whom can produce a visible effect in creating a unique image and, later, the advanced product itself.

At the stage of technical and economic image formation, as well as at other life cycle stages, competitiveness management also relies on the use of a system that monitors the appearance in the global production and science of competences and innovative technologies, which can help improve future products. In this case, it is necessary to provide additional resources needed for developing new competences and introduction of technologies in the process of design, engineering, preproduction, and production, to boost competitiveness when shaping a future product's technical and economic image.

In the process of competitiveness management, at the image building stage, the product's price, technical and functional characteristics are formed, which should make it price- and non-price-competitive in future. Later, to retain these price and functional parameters, the design process is tailored to a set price with the use of intelligent automated design management systems.

Reference

1. Chursin, A., & Makarov, Y. (2015). *Management of competitiveness: Theory and practice*. Cham: Springer.

Chapter 6
Economic Aspects of Developing Science-Intensive Products

6.1 Principles of Building an Intelligent Automated Product Life Cycle Management System

Creating a technical and economic image of a product, which should gain substantial competitive advantages over similar products that are present in a market, is just the starting point of making an advanced product; to preserve its uniqueness and attractiveness and to prevent loss of competitiveness over time, due to progressing markets, technologies, public needs, etc., it is necessary to manage its life cycle properly.

Product life cycle management is a challenging task, as it requires the use of makers' competences; therefore, there is a tendency of transferring this task from humans to computational technologies.

Computing technologies have become an integral part of effective implementation of various processes and functions previously carried out by personnel.

Today, automated process management systems are booming. So-called PLM (Product Life Cycle Management) systems are being created, which act as information management systems and are capable of integrating data, processes, business systems and, finally, the work of a big company's employees (Fig. 6.1) PLM-systems enable proper and cost-effective management of this information throughout a product's life cycle—from development, design, and production through maintenance and disposal.

A PLM system can be treated as an information strategy and as a corporate one.

Being an information strategy, a PLM system creates an integrated data structure through consolidation of different systems. As a corporate strategy, it helps large companies operate as a single team focusing on management of development, production, support, and phaseout processes while using cutting-edge solutions and gap analysis. A PLM system helps to make integrated and information-based decisions at every stage of a product's life cycle (Fig. 6.2).

A. Tyulin, A. Chursin, *The New Economy of the Product Life Cycle*,
https://doi.org/10.1007/978-3-030-37814-1_6

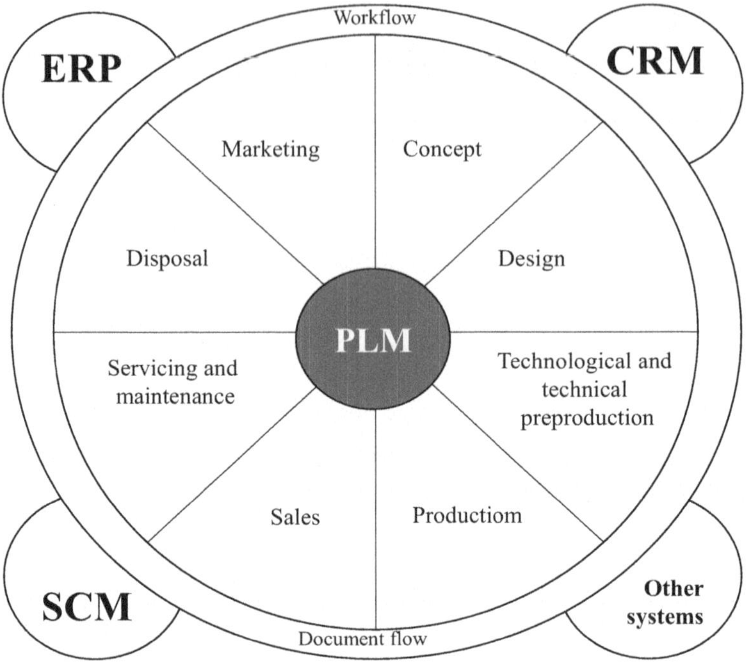

Fig. 6.1 Tasks solved by PLM systems

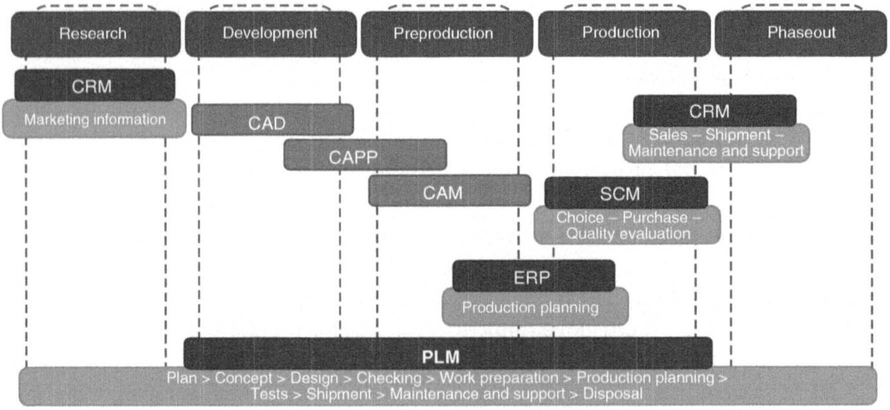

Fig. 6.2 A PLM system as an information strategy

PLM solutions create an integrated digital platform to ensure:

- Optimized cooperation ties throughout a product's life cycle and between companies.
- Creation of an integrated accounting system to provide data when necessary and to share with people the right data at the right time and in the right context.

- Maximum usability of an entire product portfolio.
- Gaining maximum profit through building new business models and reusing already mastered processes.

The greatest benefits of using PLM systems at each stage are shown in Fig. 6.3.

Systems embracing separate product life cycle stages are rapidly developing. For instance, in the product design sphere, automated design and engineering with the use of advanced digital 3D-modeling is an effective method.

Modern design and engineering systems can help create multivariant parametric and dynamic models, check their properties and behavior on a PC prior to making a prototype. This can substantially reduce the occurrence of design errors and speed up project implementation. It is during the design and engineering phase that, based on a product's technical and economic image, geometrical models of the product's parts and the entire product are created, and they are going to play a decisive role throughout the product's life cycle.

Today, all existing design and engineering software can be divided into three groups (Fig. 6.4).

In advanced end-to-end design systems and automated process management systems feature artificial intelligence elements. Their functioning is closely related to creation of intellectual knowledge bases, dynamic real-time resource planning within an organization, quick and flexible response to changes in the market environment.

Many companies demonstrate success in building and use of design and engineering infrastructures; particularly, TSMC, a Taiwanese chipmaker, has finished its design and engineering infrastructure for its 5 nm technological process as part of the Open Innovation Platform project, which includes technological files and design toolkits. During the technological process, silicon chips passed a number of reliability tests. Now it is possible to start making single crystal 5-nm systems for mobile and highly productive next-generation solutions that are tailored to rapidly growing 5G and artificial intelligence markets.

Further development of intelligent product life cycle management should focus on creation of large business systems that unify separate subsystems. These, in turn, pursue specific goals and a universal one—maintaining and improving advanced products' competitive advantages over the period from building a product's technical and economic image, its marketing and up until reaching the point of return.

Thus, a large business-oriented intelligent product life cycle management system should be created, which should make a product competitive over time.

This intelligent system should comprise a number of automated subsystems operating within an organization, and the result of their work should directly determine the product's competitiveness.

When building an intelligent management system, it should be remembered that it represents a large and complex business system consisting of different functional management systems that regulate specific business processes; these make up a whole system, which creates competitive advantages, ensures sustainable growth and therefore stable profit.

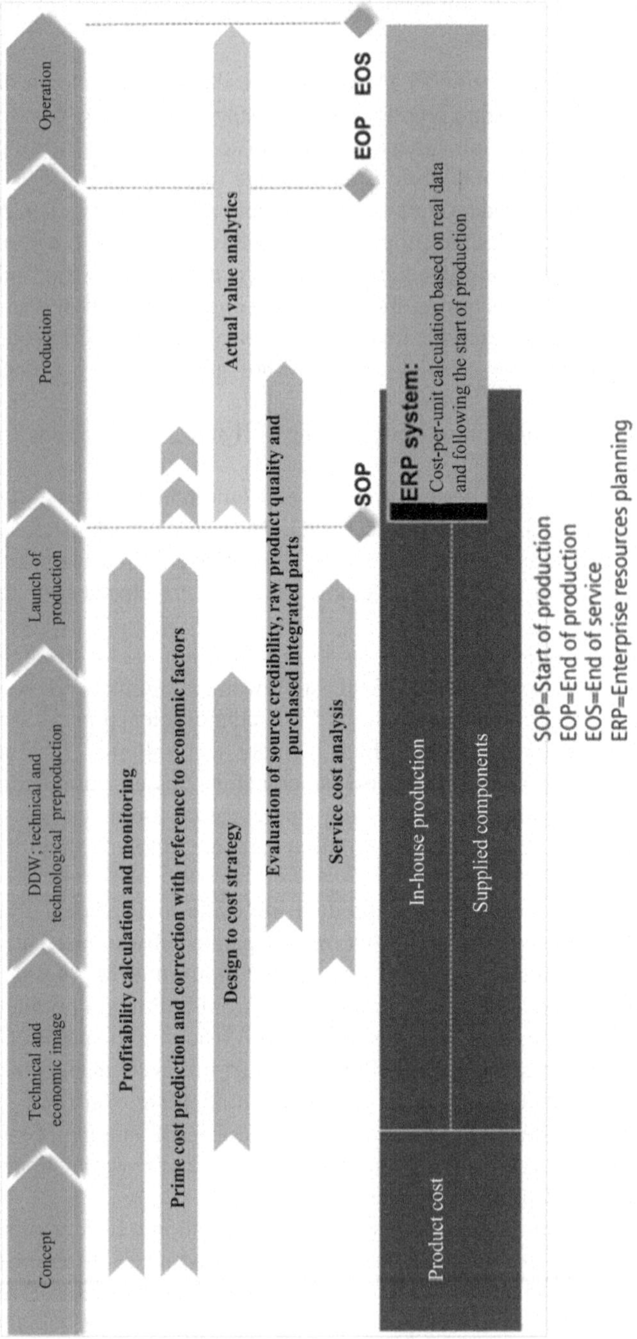

Fig. 6.3 Benefits of using PLM systems at all life cycle phases

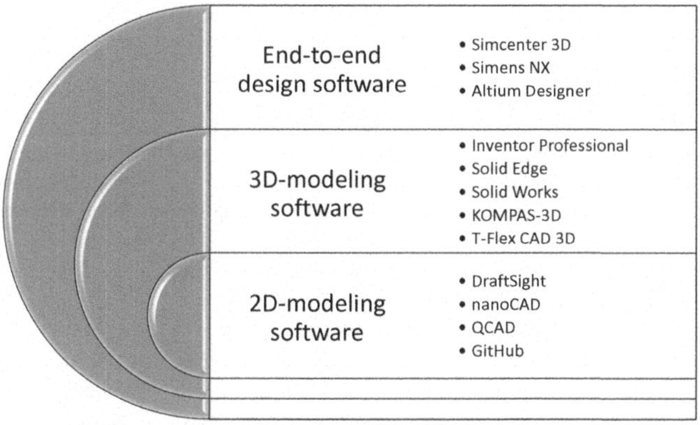

Fig. 6.4 Types of automated design and engineering software

The main subsystems, which should be integrated into a whole life cycle management system, their functional role and role in a project's competitiveness, are shown in Table 6.1.

Like all automated large-scale business management systems, a life cycle management system should have:

- Connection with the outer environment: Information and documentation exchange between its components to ensure acceptance and coordinated fulfillment of a managerial decision. There should be a sufficient amount of high-quality information at the right place and at the right time.
- Appropriate resources: Manpower, computing tools, corporate information, and novelties.
- Customer feedback, opinions, ideas concerning the introduction of new features, and other information coming in from the consumer sector.

The real-time competitiveness of a produced item is a criterion of the system's effectiveness.

The dynamic nature of the market, dramatic and often unpredictable changes in competitiveness parameters, often necessitates prompt readjustment of a company's work. An intelligent life cycle management system works well for these processes, because it has properties allowing it to automatically adapt to changing parameters and maintains the most effective, from its perspective, functional mode; it is versatile, flexible, and highly adaptive to changes in the outer environment.

This business-oriented intelligent life cycle management system, which includes the subsystems mentioned in Table 6.1, integrates them into a unified computing and analytics complex, should consider the influence of outer factors on each of these subsystems' competitiveness; it should not only control the functioning of each separate subsystem, but also embrace interconnections between them.

Meanwhile, intelligent life cycle management systems should be built according to the principles described in Table 6.2.

Table 6.1 Subsystems, which make up a large intelligent business system

№ п/п	System name	Life cycle stage	Intelligent element	Role in creating competitive advantages
1.	Product image shaping system	Shaping a technical and economic image	Methods and technologies used in processing and digital analytics of big data derived from the global information space and used for creating competitive advantages.	Shaping the image of a product, which is supposed to have competitive advantages.
2.	Automated product design subsystems (CAD, CAPP)	Design and engineering	Automated configuring of future products' technical and economic characteristics based on its technical and economic image. Using automated test result analytics and prognosis techniques and error detection at the design stage.	Creating competitive advantages at the design stage.
3.	Preproduction management subsystem (CAM)	Preproduction	Automated analytics and selection of the most advanced process technologies, which can provide substantial competitive advantages for a product and shorten the concept-to-market period. Choosing appropriate competences for a future product from the previously built digital database. Product testing and obtaining recommendations from the digital subsystem concerning prototype improvement to create a higher quality product in future.	Providing competitive advantages by choosing advanced technologies and competences for production.
4.	Process management subsystem (SCM)	Production	LEGO-building of the production process through optimal selection of operations by an automated expert system, which analyzes previous advanced production experiences.	Creation of competitive advantages with the help of a well-organized production process, stimulating profitability.
5.	Quality control subsystem	Production	Timely detection of process disruption occurrences, prediction of risks resulting from these occurrences, generation of recommendations by the	Creation of competitive advantages during the production stage and tailoring the product to market quality standards.

(continued)

Table 6.1 (continued)

№ п/ п	System name	Life cycle stage	Intelligent element	Role in creating competitive advantages
			subsystem, which contain a minimum number of technological operations specified by a pre-prepared process chart.	
6.	Sales management subsystem (CRM)	Marketing products	Using intelligent advertisement and customer communication methods, automated order management systems and optimal logistics schemes. Intelligent fault evaluation and testing systems, which should provide a multivariant or maximally optimal repair and maintenance solution.	Creation of competitive advantages at the stage of order receipt and processing; packaging and preparation for customer delivery, loading onto vehicles and transportation to the point of sale or destination and payment. Creation of competitive advantages during sales and communication with customers.
7.	Resource management subsystem (ERP)	All stages	Use of automated expert systems for defining the most optimal lineup and amount of all types of resources, which are needed for maintaining a product's life cycle.	Creation of competitive advantages through optimization of warehouse stocks, tax minimization, prevention of ineffective deflection of working assets, providing the required amount of materials, raw products, and components.
8.	Marketing process management subsystem	All stages	Evaluation of a company's potential, data mining, and prediction of market demand for several years ahead. Continuous need-change analysis. Receipt, processing, and mining of data derived from the global information space. Recommendations concerning all-stages product improvement to preserve and create competitive advantages.	Creation of competitive advantages through adjustment of the product portfolio policy, demand, sales management, and market research.

Table 6.2 The main principles of building intelligent product life cycle management systems

No.	Principle (characteristic)	Functions determining characteristics[a]
1.	Operation speed	Time between a change in the system's state and submitting analyzed information about the change to the company (division) manager: $T = f_T(S_{T,1}, \ldots, S_{T,N_T})$, where $s_{T,i}$ are parameters of the system's operation speed; N_T is the number of parameters: $s_{T,i}(i = 1, \ldots, N_T)$
2.	Accuracy of evaluation	Error $\varepsilon = f_\varepsilon(s_{\varepsilon,1}, \ldots, s_{\varepsilon,N_\varepsilon})$, where $s_{\varepsilon,I}$ are parameters reflecting the system's accuracy; N_ε is the number of parameters: $s_{\varepsilon,i}(i = 1, \ldots, N_\varepsilon)$
3.	Flexibility (adaptability)	Indicators characterizing the system's adaptability to changes in the outer environment: flexibility $\lambda = f_\lambda(s_{\alpha,1}, \ldots, s_{\alpha,N_\alpha})$, where $s_{\alpha,I}$ are parameters characterizing the system's flexibility; N_α is the number of parameters: $s_{\alpha,i}(i = 1, \ldots, N_\alpha)$
4.	Ability to predict	Variations in time of individual and generalized competitiveness indicators, as well of variations of the time period, within which these indicators change
5.	Multivariance of managerial decisions submitted to the manager	A group of individual and generalized indicators provided by the system and describing its state and changes, which influence the product's and company's competitiveness, as well as individual, generalized, and integrated product competitiveness indicators. Multivariance of managerial decisions should imply execution of measures aimed at improving integrated indicators
6.	Recommendations ("ready-made recipes"), for solving management tasks	Providing step-by-step problem-solving algorithms throughout products' life cycle stages
7.	Quick signaling of competitiveness declines	Signaling a decline in the future product's competitiveness and its causes, which are related to market development, public needs, scientific and technical progress, competence development, etc.

[a]As an example

Figure 6.5 describes the general structure of an intelligent product (project) life cycle management system.

In this scheme, the most important factors are information obtained from different sources of the global information space, machine learning, and artificial intelligence tools, which are used to process and analyze this information

Here is a general scheme of using this intelligent system for analytics and monitoring life cycle management processes.

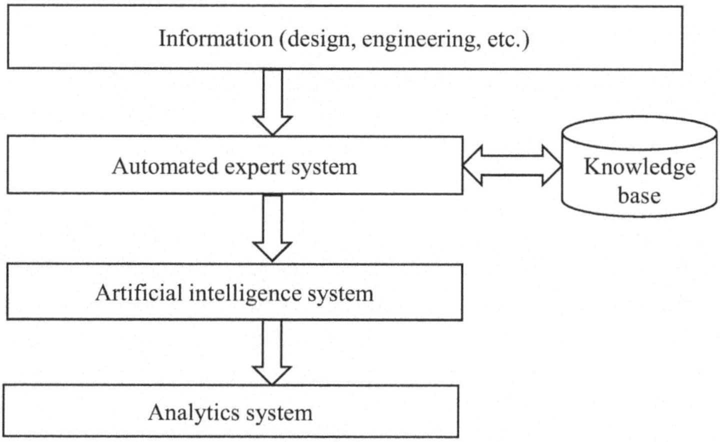

Fig. 6.5 Structure of an intelligent product life cycle management system

Step 1. Obtaining a future product's technical characteristics with reference to the project documentation and the technical and economic image.
Step 2. Calculation of the required amount of resources.
Step 3. Analyzing the optimality of available resources with the help of intelligent data analytics tools.
Step 4. Outlining offers related to an optimal range of technologies and competences involved in production based on data provided by the information and analytics system.
Step 5. Deciding upon the choice of technologies and competences with the use of an automated expert system and providing the system's calculation protocols.
Step 6. Continuous process monitoring.
Step 7. Continuous control over the results of the monitoring with the use of data mining and intelligent data analytics.
Step 8. Completion of the project and documentation, which should be confirmed by an expert system.

Although this pattern may vary depending on project specifics, its fundamental concept lies in the fact that traditional project analytics and monitoring methods are complemented by advanced artificial intelligence tools, which reduce the human factor's influence on managerial decision-making.

To solve the problem of building an intelligent life cycle management system, it is necessary to build a mathematical model of the automated project execution monitoring process.

The monitoring and analytics of project execution are key elements of an intelligent product life cycle management process. The monitoring is carried out with the help of advanced digital methods. The mathematical model relies on objective data obtained through big data processing and analytics.

Assume that project execution is evaluated in the discrete mode:

$$T = \{0, 1, 2, \ldots, N\},$$

where a period (for example, a month) is used as a time-slot.

Because all major projects are executed according to a specific schedule, it can be assumed that there is a plan, which covers the entire period:

$$P = \{P_0, P_1, P_2, \ldots, P_N\}.$$

Values P_k reflect the state of a project, which is planned for time point t_k. Definitely, these values can be multidimensional, which is not a significant factor for this model. It is the numeric value reflecting the difference between specific states that is important. Along with planned values, it can be assumed, a project can take on objective states, which can be expressed as:

$$S = \{S_0, S_1, S_2, \ldots, S_N\}.$$

These are unknown values, and their evaluation is the model's main point. As noted above, values P_k and S_k are hard to operate, because a project's state is always extremely complex; therefore, the model relies on the following composite valuation function:

$$\Delta S_k = F(t_k, P_k, S_k).$$

The result of using this function is an estimate of the discrepancy between the actual and planned indicators. Assume, this estimate is always positive:

$$\Delta S_k \geq 0, \ \ k = 0, 1, 2 \ldots, N.$$

It can be assumed that these values are random. To analyze and monitor the project execution process, a cumulative random value should be used, which is based on the formula:

$$R^n = \sum_{k=1}^{n} \Delta S_k, \ \ n = 1, 2, \ldots, N.$$

Therefore, value R^n is the accumulated discrepancy between planned and actual (objective) project state values at time point t_n.

Because the process appears to be random, a probabilistic space can be modeled, within which the production process is taking place. It is possible to use a natural filtration pattern. Assume that

$$\Phi_T = \{F_k : k \in T\},$$

where $F_k = \sigma(\Delta S_m : m \leq k)$ is a σ-algebra generated by random values prior to moment k. Consequently, there is a flow of σ-algebras generated by the process:

$$F_1 \subset F_2 \subset F_3 \subset \cdots \subset F_N.$$

Structurally, the random process R^n conforms with the natural filtration, i.e., the random value R^n is measurable relative to σ-algebra F_n.

From the perspective of information, σ-algebra F_n is information that is available at time point t_n.

It is important that the random process R^n is submartingale relative natural filtration. In fact, it appears that the process conforms to the chosen filtration. Next, it can be assumed (a natural economic assumption) that the process has finite expectations:

$$E[|R^n|] < \infty, n = 1, 2, \ldots, N.$$

Because R^n is the sum of positive random values, there is a submartingale characteristic:

$$E[|R^n|F_m] > R^m, m < k.$$

To analyze the random process R^n, which is a submartingale, the Doob decomposition theorem can be used. More specifically, process P^n can be expressed in the following way:

$$R^n = M^n + A^n,$$

where M^n is a marginal with respect to filtration F_T, i.e.

$$E[M^n|F_m] = M^m, m \leq n,$$

and the random process A^n is predictable; in other words, A^n is measurable relative to σ-algebra F_{n-1}. The predictable process A^n is the compensator of the process R^n.

The idea of the predictable process of random values is that the random value A^n is a function of random values:

$$A^n = \varphi(R^1, R^2, \ldots, R^{n-1}).$$

The martingale property of the random process M^n helps evaluate the trend of the random process R^n, which can characterize the project execution process.

However, direct measurement of random values

$$\Delta S_k \geq 0, \quad k = 0, 1, 2 \ldots, N$$

is difficult, because it is not always possible to obtain adequate information about a project' state.

There is a formal definition. Assume there are values observed at the same time points:

$$B = \{B_1, B_2, \ldots, B_N\}.$$

We are not going to specify B_k values, but their combination will create a flow of σ-algebras:

$$G_k = \sigma(g(B_m) : m \leq k),$$

where $g = g(t_k, B_k, P_k)$, $k = 1, 2, \ldots, N$ are random functions, which reflect the discrepancy between the actual (B_k) and the planned project state (P_k). Random values $g(B_k)$ are going to conform with the built filtration. Unlike S_k values, these random values can be physically observed.

Natural informational suggestions are that for the σ-algebra flows in question it is:

$$G_k \subset F_k, \quad k = 1, 2, \ldots, N.$$

The ratio suggests that information contained in filtration $\{G_k\}$ has a smaller volume than the information contained in filtration $\{F_k\}$, because observations provide but a fraction of factual information about a project's state. However, this supposition helps obtain new random values based on a conditional mathematical expectation. This can be done with the help of the Radon–Nikodym derivative.

Assume that in the original probabilistic space a probability measure P was used. Then, for each multitude A_k measurable with respect to the σ-algebra G_k and a random value ΔS_k, it is possible to calculate a new probability measure Q according to the formula:

$$Q(A_k) = \int_{A_k} \Delta S_k(\omega) \mathrm{d}P(\omega).$$

This new measure comes from the absolutely continuous Lebesgue integral.

Now it is possible to define the conditional mathematical expectation as a Radon–Nikodym derivative according to the formula:

$$E[\Delta S_k | G_k] = \frac{\mathrm{d}Q}{\mathrm{d}P}(\omega).$$

By definition, the conditional mathematical expectation of the random value ΔS_k with respect to the σ-algebra G_k, is a random value, although its values are an optimal prognosis with respect to the random values B_k.

With the help of a conditional mathematical expectation with respect to the σ-algebra of events, it is possible to evaluate the conformity of a project's current state with planned ones.

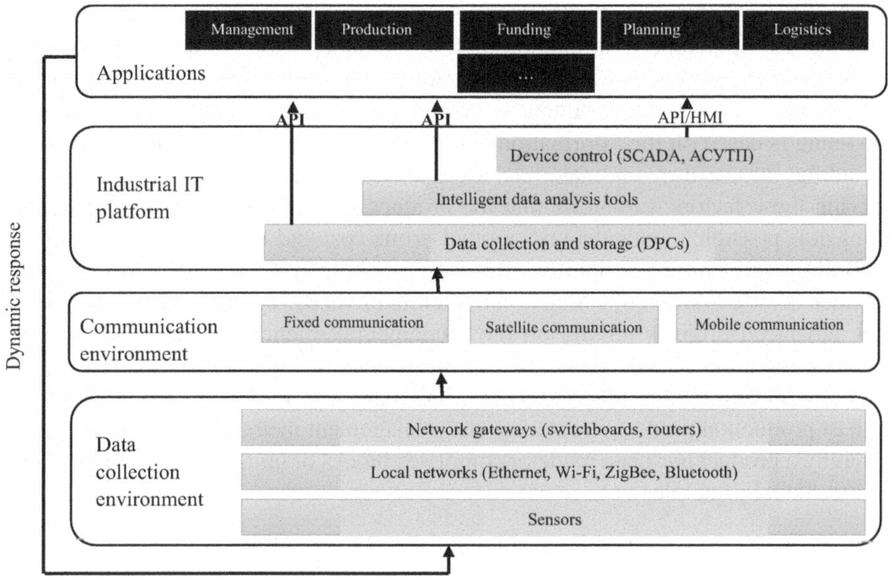

Fig. 6.6 An example of IIoT infrastructure

This mathematical model can be used as a foundation for an intelligent product (project) life cycle management system that is based on machine learning.

The Industrial Internet of Things is the technological basis for such intelligent product life cycle management systems. The Industrial Internet is a concept used by the computational network, industries and complex physical machines that are integrated with intelligent systems. Setting up such networks can rearrange economic and industrial processes, as eliminates the necessity of using manpower in some activities and operations and thus contributes to economic growth.

On a digital organization's platform, IIoT ensures communication within its infrastructure: digital management, computer-aided design and engineering, digital production, and interaction with the market.

A possible IIoT infrastructure with references to the positions of intelligent systems, which are described in this chapter, is shown in Fig. 6.6.

Introduction of an IIoT infrastructure into a company's platform can help it promptly respond to changes in design and technology standards.

Therefore, a digital product (project) life cycle management system should be built based on the aforementioned principles and the developed mathematical model. The primary function of this intelligent product life cycle management system is ensuring a future product's high competitiveness through boosting the economic processes, which accompany industrial ones, reducing industrial processes' duration, reducing prototype testing costs, minimizing the human factor and therefore errors, when making managerial decisions.

First, these tasks are solved through implementing the design-to-cost approach with the use of advanced simulation and 3D-modeling tools. Simulation modeling can help to choose geometric parameters/adjustments/allowances, which ensure an optimal prime cost and labor intensity. Simulation of prime cost and labor intensity modeling is based on the information about factors, which influence the prime cost, including information on materials used and corporate industrial facilities. By varying these factors, providing that the product meets the set technical characteristics, it is possible to calculate its optimal prime cost and labor intensity.

Building an automated intelligent decision-making system, while shaping a new product, is the evolution of simulation modeling methods. To implement this system, new advanced projects relying on the fundamental science, artificial intelligence, machine learning, and decisions based on analysis of big data blocks concerning factors influencing the prime cost. One of the main such factors is preproduction, as well as production process structuring and management methods, especially in cases whereby a product has been made or comprises blocks or units that use new physical and technical principles.

6.2 Advanced Digital Design, Modeling, and Production Methods

The previous chapters focus on an advanced product's technical and economic image and evaluation of its potential competitiveness. Further, life cycle stages of a science-intensive product include design procedures, as well as activities aimed at launching its production with account of the company's potential. The main procedures, which form the algorithm of managerial decisions concerning development and production of advanced goods, are (Fig. 6.7):

- Evaluation of a company's scientific potential, competences, and industrial capacity.
- Evaluation of the influence of technologies (existing and future) on the product's competitiveness. Evaluation of existing corporate competences and the necessity of developing new ones to master innovative technologies.
- Defining an acceptable prime cost per unit for the target market and development (concept-to-market) time.
- Specifying competitiveness estimates for the product in the target market.
- Simulation modeling to define optimal competitiveness with account for rapidly changing factors and risks.

At these stages, various advanced digital methods are used to boost the effectiveness of the main design, engineering, and production processes. The current state of machinery production is characterized by extensive use of scientific knowledge, projects, and science-intensive technologies. These industries rely on short-life cycle products, which can satisfy the growing public needs. Given the close ties within the

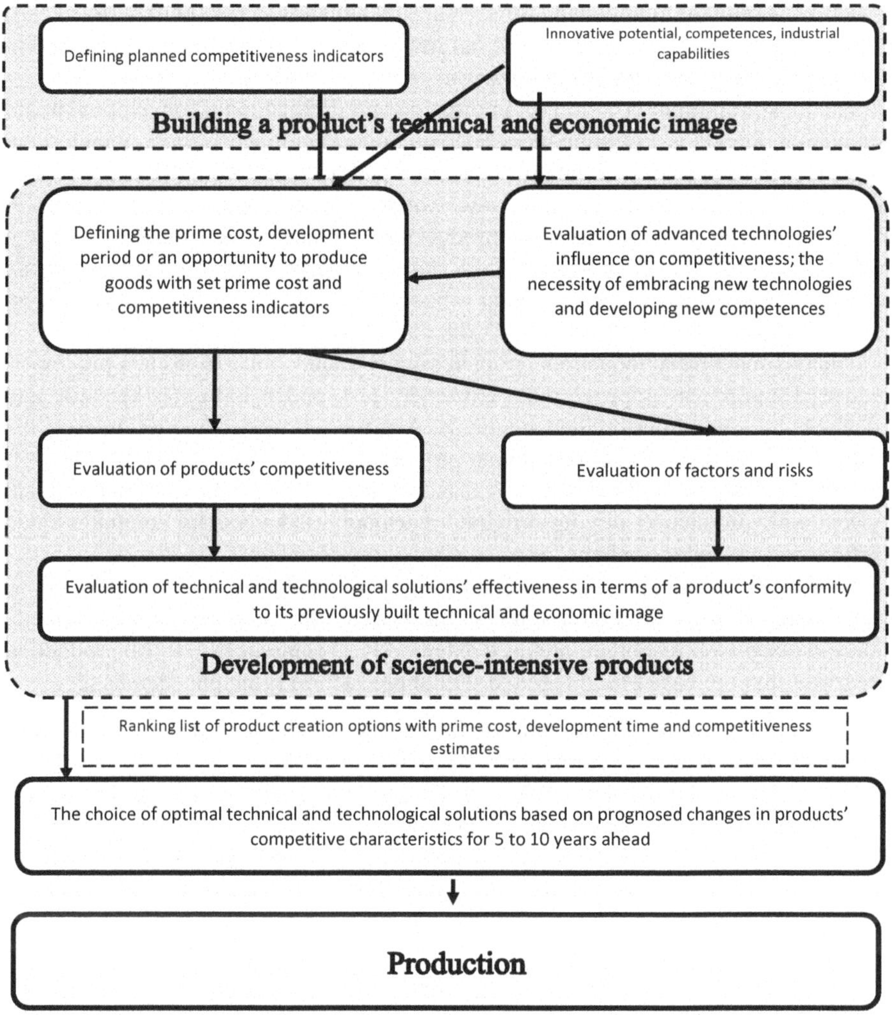

Fig. 6.7 The algorithm of product engineering and production

individual–organization and *equipment–technology* systems, production methods are increasingly conducive to not only further improvement of basic technologies, but also to creation of ultimately new science-intensive technologies, which help to solve increasingly challenging modern industrial tasks.

In case of introducing expensive equipment (for example, CNC machines), return on investment should be achieved by keeping it loaded and thus ensuring its nonstop operation. This raises tooling management, workpiece input, conveying, chip extraction, and some other support and maintenance standards. Practice shows that partial automation produces a limited effect; to provide a maximum effect, an entire company should be automated. This creates a situation when partial (localized) automation is not effective; therefore, an important part of designing a production system is defining the level of automation.

The most important criterion of automation and digitization of development and production process is providing the needed quality and competitiveness, which serve as general indicators of a company's scientific and technical status. The concept of management of an industrial company, which is based on permanent efforts to eliminate all types of losses, improve product quality, reach a higher labor efficiency, and cost optimization, is known as *lean technology*. There are several systems using lean technologies and exerting a great influence on output product quality: the 5S (and 6S, its advanced version), 8D, COTI, and TPM.

Today's machinery industries are increasingly eager to use high-tech processes, industrial equipment is progressing extremely fast, and it is highly versatile and adaptable to new output product standards. As new versatile industries emerge, the concept of integrated digital production progresses.

Integrated digital industries widely use computer-aided production management system with production relying on CNC machinery and versatile computer-aided crafting and assembly systems.

Nowadays, automated data processing and automated design systems (ADS), as well as technological preproduction systems, which fully integrate material, production and manpower resources, are used extensively. It is possible to fulfill production processes that are based on integrated digital production principles thanks to:

- Equipment featuring digital management tools, which combine flexible production systems with fully computerized production facilities.
- Using automated systems when designing and carrying out technological preproduction and management of production processes.
- Using systems intended for digital processing of large data in the sphere of economic computations, planning, marketing, management, technological preproduction, etc.

The most important components of integrated digital production are:

- Flexible crafting and assembly systems driven by computerized and intelligent tools.
- Highly effective technological processes.
- Systems ensuring timely shipment of workpieces.
- Automated supply and provision systems.
- Computerized automated drafting (CAD), planning (CAP), manufacturing (CAM), quality (CAQ), assembly (CAA), and product life cycle management (PLM) systems.

It should be noted that companies that use highly automated production facilities, which rely on integrated digital production principles, are increasingly eager to avoid using human labor in the process and confine people's functions to mere observation. This approach has been implemented at several "dehumanized" enterprises, which pioneered in Japan and Germany.

The concept of integrated digital production can be adapted to specifics of a particular company, because it is not always possible to ensure a desired complexity and automation of a process without using the "human factor" and a hierarchical organizational structure's support. The digitization process generates different production management concepts, which include lean production. The lean production concept simplifies corporate structures, eliminates whole hierarchical levels to improve information flows, and raise cost-effectiveness. This creates highly effective flexible industries, which promptly react to changes in the economic conjuncture.

Characteristic of lean production is reintroduction of the "human factor," its creative potential and advanced concepts into the sphere of creation of material assets and making key decisions. A specific trait of lean production is high process productivity, which ensures conformity between and integration of all elements involved in the process. The design and engineering stage involves all design and technological services simultaneously; component and consumer material suppliers and customers also join the design process. Effective functioning of quickly acting information exchange facilities is decisive.

Figure 6.8 describes an integrated digital production process with specified information flows and processes taking place within such a system. Its elements include:

1. Digital technological process development.
2. Engineering infrastructure and technology analytics.
3. Engineering data management.
4. Computer-aided design of accessories and preparation of management programs.
5. Automated CNC equipment.
6. Computerized equipment for engineering tasks, networks, and terminals.
7. Computer-aided equipment and output measurement process control.

One of the most important characteristics of a modern competitive organization is an opportunity to produce a broad spectrum of product modifications in small-sized portions over a brief period of time. Technical standards that define these modifications are formed by specific market segments. One possible solution that ensures the competitiveness of Russian science-intensive mechanical engineering in the new scientific and industrial system is the use of flexible new-generation production systems.

A flexible industrial system (FIS) is a computer-aided high-tech equipment complex consisting of flexible combination of flexible production modules (FPM) and (or) flexible production cells, automated high-tech preproduction, and functional support system. This system launches an automated readjustment function every time the program switches to goods, the production of which is limited by the equipment's technological potential. Generally, the FPS is a set of technological units, which are connected by a transportation system and unified into an automated industrial complex (including diagnostic and measurement facilities), which is capable of producing goods of a certain category in small- and medium-sized portions.

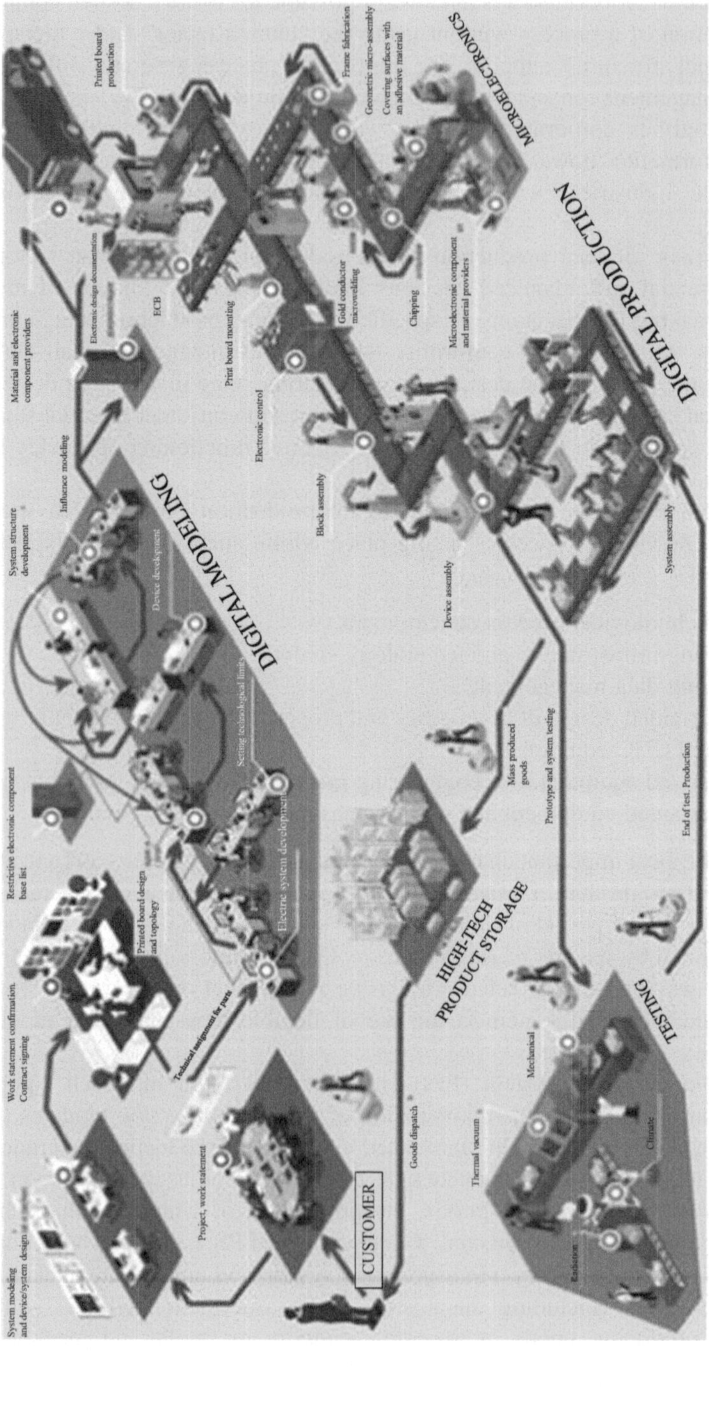

Fig. 6.8 Integrated digital production

An essential scenario of the evolution of traditional approaches to creation of industrial systems with rapidly progressing digital and cyber-economy is a transition to a digital corporate model. The process of building a digital ecosystem within an industrial company should be based on rational structuring with account for using advanced production technologies. An industrial system should be arranged in such a way that a company could effectively develop a new product and promptly adjust the production process to a maximally cost-effective manufacturing strategy.

An organization, which has introduced a digital production management system, has a number of specific features (Fig. 6.9) and advantages, such as no paperwork (automated in-process control of technological parameters); optimal machine loading, downtime and bottleneck management; equipment time in service, maintenance control, checking, natural lean production infrastructure, etc.

Digital production (intelligent organization, smart plant) is renowned for its transformability and adaptability, resource effectiveness and ergonomics, as well as integration of all economic subjects (consumers, investors, partners, etc.) into business- and cost-making processes. Digital production relies on individual preferences and science and technological tendencies through wide use and intelligent real-time analysis of big data derived from the global information space.

The technological basis of digital production is presented by cyber-physical systems, within which software components are interconnected with mechanical and electronic infrastructural elements via an architecture of data, as well as the Internet of Things, in which physical objects—devices, sensors, and systems—can send and receive data via the Internet without using manpower.

Automated making of routine decisions is enabled by a well-developed *communication-between-things* system, which implies a digital company's infrastructural object's ability to identify each other, characterize their status, and exchange and process data. The routine decision-making system can exclude man from the interaction between things, thus making the interaction autonomous, reliable, quick, consistent, and controllable. Incorporated into the Industry 4.0 system, the Internet of Things gives the production process a number of advantages:

- Flexible production is achieved through elimination of tough "pipeline" decisions, which enables mass acceptance and execution of personified orders, easier introduction of new solutions into the manufacturing process, and free outsourcing.
- Adaptive production is achieved through all-layer control and thanks to its functioning on a unified technological platform.
- Effective production is achieved through reduction of human-factor-related costs resulting from errors, downtime, and poor labor efficiency.

Cyber-physical systems are distributed intelligent systems—microsystems or microelectromechanical systems, which comprise electronic, mechanical, optical, and other, primarily digital components. Most of them also feature recognition and data processing functions and are connected to communication networks. They can execute perception, cognition, and other activities that are increasingly human. Cyber-physical systems' intelligent potential tends to become visible in the process

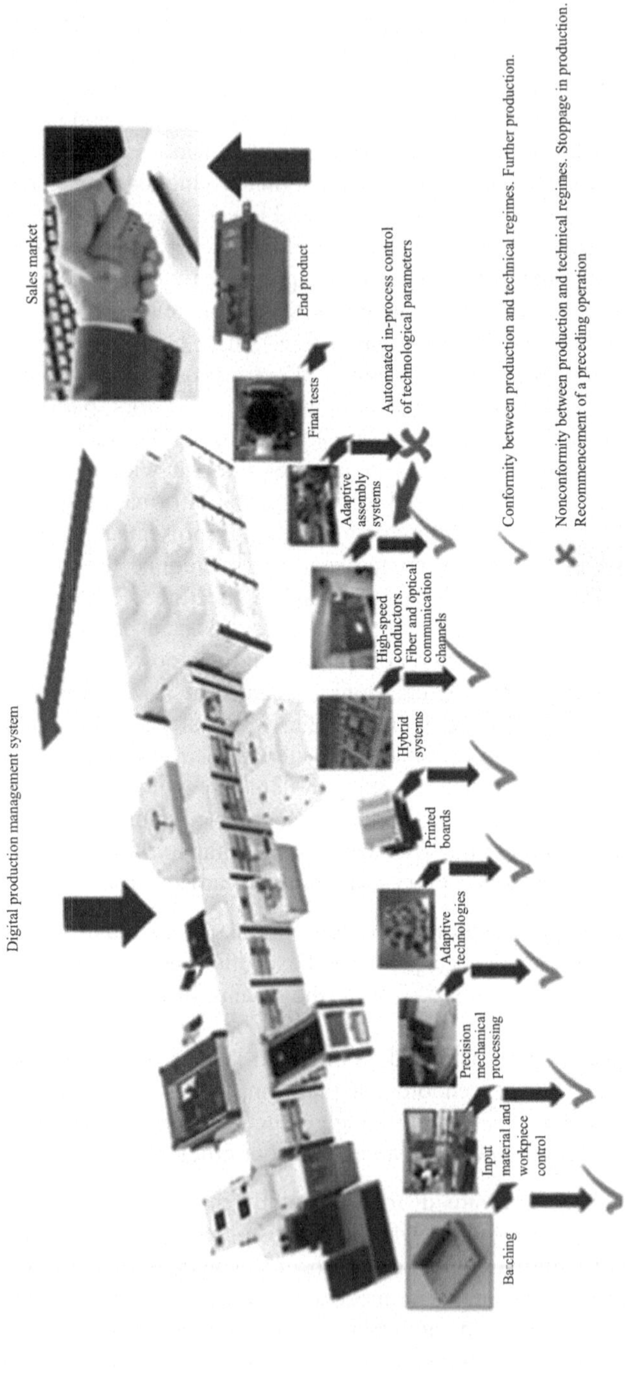

Fig. 6.9 Specific of digital production

of interaction between distributed system (the interaction should be more or less flexible).

Digital transformation of the manufacturing process requires development of new strategic approaches to introduction of technologies, competence building, and product marketing. In a digital economy, introduction of Industry 4.0 technologies into digital production evolves into cyber-economy, which appears to be the economic basis of digital production. It is a system of economic relationships, which functions within the information space and maintains optimal ties and interaction between economic subjects and objects, exchange, and distribution of material amenities. Cyber-economy offers a system of management of corporate resources, which is based on a digital approach; these resources form a basis for development and introduction of new technologies. Thus, cyber-economic mechanisms are to run a corporate technological platform and ensure a release of market-dominating products. Digital design and modeling are the key elements of this platform.

The cutting-edge digital design and modeling paradigm is based on the use of complex multidisciplinary mathematical models, which are highly adequate to real materials, structural elements, and physical/mechanic processes (including technological and industrial ones), which are expressed in mathematical/physical equations, primarily—in 3D unsteady-state nonlinear differential equations with individual variables.

These mathematical (smart) models accumulate all knowledge used when designing, manufacturing, and use/operation of a product, structural element, machine, installation, and technical or cyber-physical system:

1. Fundamental laws and sciences (mathematical physics, theory of vibrations, elasticity, plasticity, fracture mechanics, mechanics of composites and composite structures, contact interaction, dynamics and strength of machines, computational mechanics, aerohydrodynamics, heat-mass exchange, electromagnetics, acoustics, technological mechanics, etc.)
2. Geometrical (CAD) and computational finite-elemental (CAE) full-scale models of real objects and physical/mathematical processes
3. Complete data on materials, of which a product is made, including data on material's behavior resulting from thermal, electromagnetic, and other influences, high-speed deformation, vibrational, impact-, low-, and high-cycle loading
4. Information about operating conditions (normal, improper operation, emergencies, etc.), including information, which helps maintain a unit's specified behavior under certain conditions (so called programmable behavior)
5. Data on production and assembly technologies applied to separate elements and whole units
6. Other characteristics and parameters

Today, computer-aided technologies have made it possible to substantially raise physical models' adequacy, and the new digital design and modeling paradigm has raised mathematical models' adequacy and, consequently, the correctness of numerical results (Fig. 6.10). This allows to fully dismiss random engineering, as new products are usually built according to already operating prototypes.

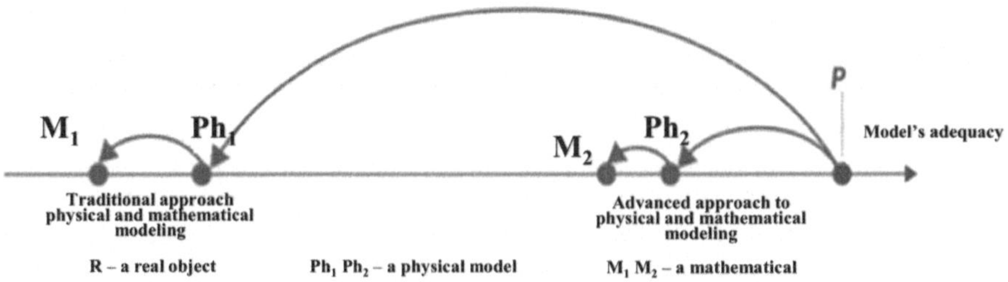

Fig. 6.10 Comparison of traditional and digital modeling approaches

Actually, at this stage the responsibility for a modernized product to previous generations of engineers, many of which had more time and financial resources for the development of products that are already in use. In addition, many engineers keep wondering *how a modernized product is going to behave and how it is going to work in particular operating modes*. This is the reason why there are few next-generation and/or ultimately new units/machines/devices, which could have been globally competitive, demanded, personified, and even custom-made, i.e., initially tailored to constantly toughening consumer and global market requirements.

The use of the new digital design and modeling paradigm has made it possible to avoid the regular situation, when the growing number of changes (resulting from old or new errors and missed data) and, consequently, rising costs affect the entire development life cycle—from the design stage through massed production (late changes increase a company's expenses). Finally, it is quite possible to introduce the majority of changes during the design phase and thus minimize the total costs, reduce expenses, and create science-intensive and high-tech new-generation products as soon as possible (Fig. 6.11).

To illustrate[1] the point of the "smart" mathematical model, it is possible to use the automotive industry as an example, as it is the most science-intensive, rapidly developing and competitive one with global production amounting to about 100 million vehicles per year; besides, technologies and materials used in released automobiles are instantly studied and copied by competitors. A modern car should meet a huge number of target characteristics and indicators, including performance (comfort, ergonomics, aesthetics, etc.), as well as active and passive car safety, aerodynamics, manufacturability, etc.

The most complete and challenging type of quality and safety test is a crash test. To gain global competitiveness, each vehicle should pass a series of certification and benchmark tests. It should be noted that a crash test is extremely costly; the only thing that can help minimize costs and speed up a car's release is a virtual test (it is noteworthy that the world's top carmakers have radically changed the ratio between full-scale and virtual tests: from 100 to 100 in 2007 down to 5 to 10,000 in 2017!)

[1] http://fea.ru/news/6721

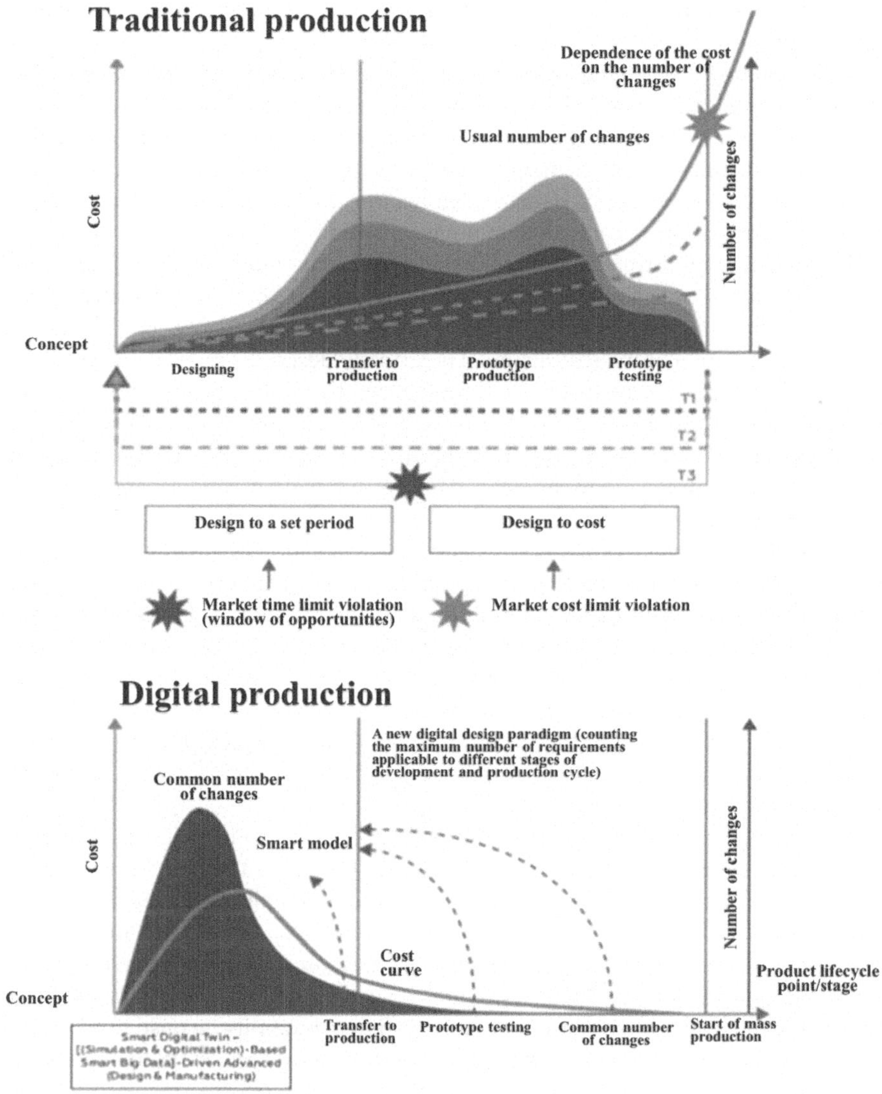

Fig. 6.11 Comparison between the traditional and digital approaches to production

Results of thousands and tens of thousands automated virtual tests are used to create a "smart" model, which allows, for example, to program the destruction of 5000–8000 of car bodies' weld points given various possible crash scenarios in such a way as to achieve sufficiently high passive safety (nonlinear deformation and body elements' destruction ensure survival and minimize damage to drivers and passengers). This unsteady and nonlinear behavior of a car is detected during a full-scale crash test, and a highly detailed "smart" model actually programs each structural part's behavior in operating and crash conditions.

A virtual crash test is a multidisciplinary crown jewel, which comprises nearly all sciences ranging from material science and engineering, mechanics, to manufacturing technologies and, definitely, all physical processes embraced by aerodynamics, vibration, dynamics, strength, fatigue, and all types of nonlinearity (geometrical, physical, contact interaction, damage summation, localized damage, etc.), and a whole spectrum of optimization technologies is used.

To reduce a unit's weight, prime cost, improve structural, vibrational, acoustic, performance, and other indicators in a car body, about 200 different materials are used, including metals, alloys, polymers, composites, and, finally, metamaterials, which have an optimal microstructure. To correctly describe physical and mechanical processes taking place inside a unit influenced by various factors (for example, dynamic), one should know a broad array of parameters and characteristics, including elastoplastic strain curves observed at different strain rates, fracture initiation criteria, fracture propagation models, damage summation models applicable to different materials, etc.)

Apart from materials' properties, an adequate evaluation of a particular car body elements' behavior while building the "smart" model, it is important to consider technologies used when manufacturing these elements, such as "intelligent molding," "intelligent forging," assessment and management of the preceding strain and deformation state, local thin-outs, warping, etc., following technological processes. These factors can seriously affect a whole unit's behavior. Also, virtual manufacturability of parts is evaluated to ensure whether or not it is possible to make them using a specific method, ensure due strength, workmanship and assembly, and many other characteristics.

Also, it is important to evaluate transitions between car body elements. These are connected at welding points, seams, and adhesive bindings. The body of a premium car can have more than 7000 welding points, a total length of welded seams may exceed 6 m, plus various types of adhesive bindings—glass, structural, semi-structural, and expanding. Each element demonstrates its own behavior model under different influences. The positioning of welding points greatly influences the whole body's behavior (strength, vibrations, fatigue, durability, acoustics, etc.), and it is extremely important that crash tests demonstrate programmed destruction areas, as at specified time points (at specified milliseconds) strictly defined elements located in strictly defined areas should be destroyed and, most important, they should be dynamically destroyed so as to keep passengers safe.

An automobile's structure includes a large number (up to 100) of different mechanisms, such as the motor, suspension mount, hood, boot lid, window lifters; some modifications feature a folding roof or sliding doors. A "smart" model contains data about each unit of the mechanism, its kinetic, dynamic, and structural characteristics, which help evaluate the functioning; this data is expressed in the form of separate mathematical models that are described in unsteady-state nonlinear partial differential equations.

In order to carry out virtual crash tests, it is necessary to have complete virtual replicas of all testing equipment units, all crash test rigs, which are used in full-scale crash tests. They makeup virtual test site with anthropomorphic crash test dummies that allow to accurately simulate a human body's biomechanical dynamic behavior observed under various unsteady influences and assess injury criteria; there are male, female, and child dummy families featuring over 10,000 measurement sensors, more than 20 certification and rank test barriers and 20 ram testers for different parts of the human body for additional estimates, as well as for pedestrian crash tests.

Impact interaction with barriers is a high-speed dynamic process lasting about 200–250 ms. The integration step that is used for numerical solution of problems is 1 ms. The total number of integration steps exceeds 200,000. This information makes up big data (more than $2 \cdot 10^{12}$ parameters) at the "smart" model's input. A completed virtual test complements the data volume resulting in Smart Big Data: a supercomputer modeling of a 200 ms process yields an output data volume, which contains more than 10^{14} parameters. From several tens of millions nodes—about $\sim (1\ldots3) \cdot 10^7$ nodes, more than 50 parameters are recorded, which include transition, speed, acceleration, deformation, strain, etc. As a result, there are $5 \cdot 10^8$ curves, which give a comprehensive description of the "smart" model's behavior.

To develop a "smart" model, a multilayer matrix of target indicators and resource limits (temporal, financial, technological, industrial, etc.) should be built. It should be realized that both target characteristics and resource limits can be changed or modified, which requires an immediate (within about a week) change in the multilayer matrix—*change management*, which ensures an uninterrupted development process and represents another crucial feature of the new design paradigm.

The "smart" model is very adequate, mainly thanks to big input and output data; it is quite so close to the real object and demonstrates but $a \pm 5\%$ difference between virtual and full-scale test results (for example, during car crash tests, "smart" models are validated by data coming in from 500 sensors). As a rule, this highly adequate model is referred to as an object's/product's digital twin.

However, as exemplified above by the automobile industry, the model's adequacy is greatly supported by data on manufacturing technologies ("intelligent molding," "intelligent forging," assessment and management of the preceding strain and deformation state, local thin-outs, warping, etc., which follow technological processes). Consequently, this highly adequate smart model, with account for the specifics of a particular factory, is referred to as its digital twin.

It should be noted that successful formation of digital product and factory twins necessitates combining traditional design, product, material, computing engineers, as well as other specialists, into a new type of engineer—a *system engineer.*

Unifying a digital product and factory twin within a single digital model with scores of virtual tests completed as part of a specially organized *digital certification process* leads to the formation of a first-category digital twin.

Later, during the operating stage, which includes, for instance, repairs, the first-category digital twin makes it possible to "generate a smart digital shadow" based on the "smart" model, which adequately describes a real object's/product's behavior in all operating modes (launches, stops, regular work, improper work, emergency situations, etc.) The smart digital shadow is formed through obtaining latest information about the functioning of a particular object/product with the help of Industrial Internet technologies and diagnostics. This additional information received at the operating stage enables further "training" of the digital twin and makes it even "smarter" by boosting its adequacy and using it modeling various unexpected scenarios (for example, assessment of possible damage or residual operation life). Thousands of virtual tests carried out as part of this *digital certification* of newly created digital twins have given us a clear understanding of the location of critical zones, in which various sensors can be placed (accelerometers, tensimeters, temperature-, pressure-, speed-sensors, etc.) This makes it possible to radically reduce the number of sensors and on a regular (daily) basis receive big data, ensure faster big data processing and twin modification, so that it can be transformed into a second-category digital twin.

Notwithstanding the crucial role, which a digital twin plays in Industry 4.0, effective and consistent development by the world's top high-tech companies over the past decade, from total digitization and digital mockup development, it was not before 2017 that a digital twin pioneered in the traditional Gartner maturity model (Fig. 6.12).

Today, there are different approaches to identifying a digital twin. For example, Siemens PLM Software looks at it as four overlapping fields: product design, production planning, structuring industrial facilities, and the real world, with a special emphasis on production. The Dassault Systemes uses the term *virtual twin*, which is development of a system engineering strategy. A virtual twin helps a development team create a product that should combine mechanical, electric, electronic, hydraulic and other systems, then test it and study its behavior in different environments (loads, vibrations, operating software, control systems, etc.) [1].

A digital twin can help optimize a product's prime cost at the design stage, when developers face the challenge of providing a planned prime cost and competitiveness.

6.3 The Process of Tailoring Products to a Fixed Prime Cost and Competitiveness with the Help of Intelligent Automated Systems

Prime cost formation is a well-studied problem, which is described in scientific literature. Many papers focus on economic prime cost analytics, cost calculation, etc. However, the reported solutions are inconsistent and not commonly used. Mostly, these methods apply to a specific activity (production or sales), influence of one or a group of similar factors (reliability parameters, operational readiness, allowances, types of equipment, etc.). The available mathematical methods are very different and, consequently, do not guarantee compatibility or test validity when solving the problem.

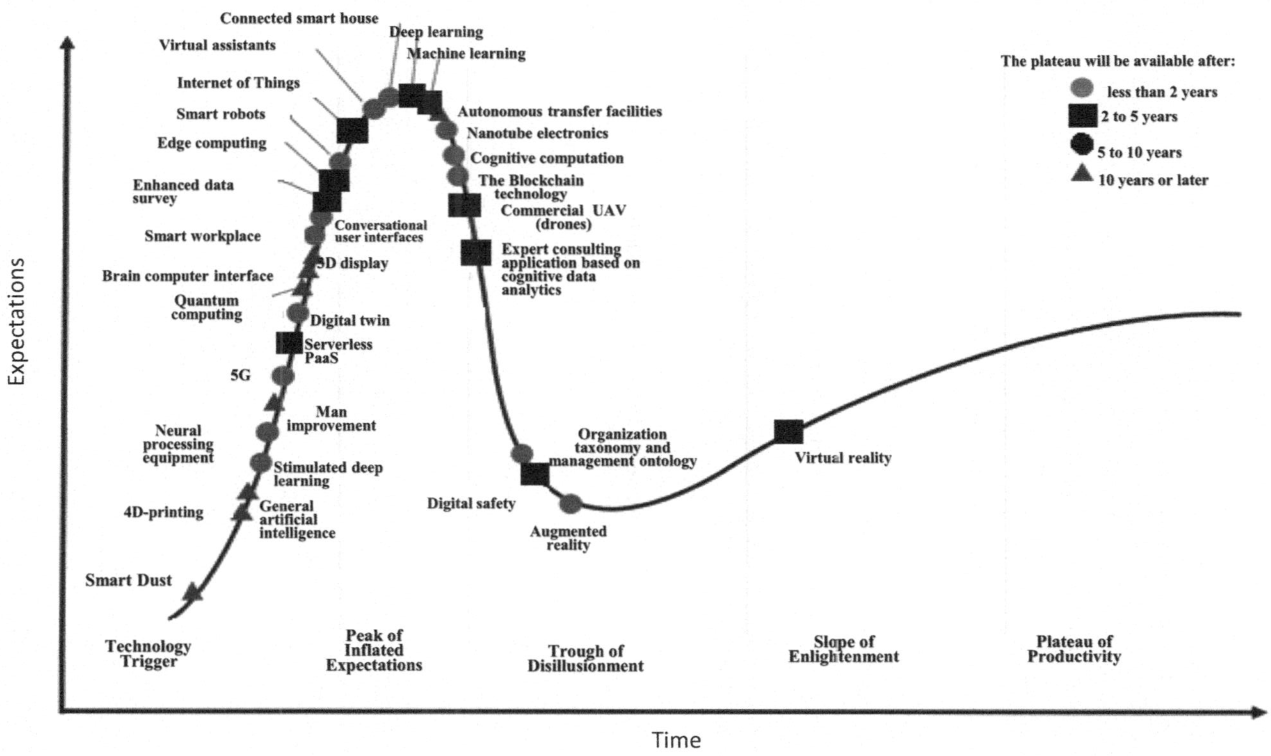

Fig. 6.12 A digital twin in the Gartner's maturity model

A product unit's prime cost is planned in two opposite directions (Fig. 6.13): *upwards*—from the beginning of cost formation during production, and *downwards*—starting from the market sales price.

The former direction forms actual costs per item, in the latter—marginal ones. Modern production ensures a fairly broad range of product and component manufacturing options. Thanks to technological, technical, and scientific progress, this range is growing constantly.

Actual costs reflect planned production expenses, which are calculated based on a specific production's status and culture. Marginal costs reflect expenses, exceeding which will result in failure to achieve the planned profitability level.

When defining a product's prime cost, the calculation starts from actual cost counting and ends in marginal cost counting.

This prime cost counting mechanism uses the following prime cost planning methods:

1. To form actual costs, only cost-plus planning methods are used.
2. To form marginal costs, market prime cost planning methods are used.

Cost-plus and market cost calculation methods use diametrically opposite computing orders.

Cost-plus methods use the *downward* calculation pattern (from material down to general production expenses).

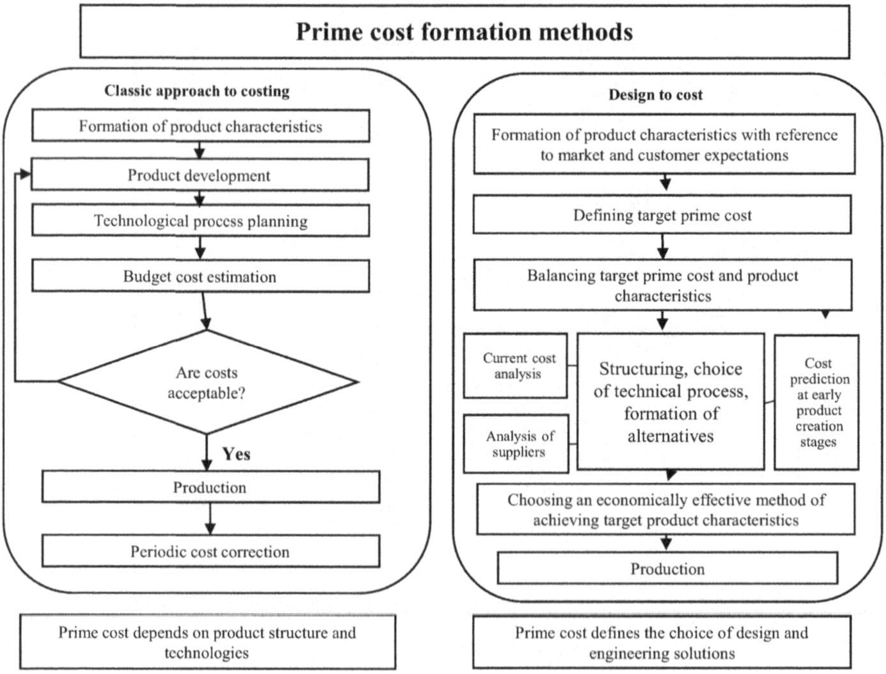

Fig. 6.13 Approaches to forming a product's prime cost

Market prime cost planning methods use an *upward* calculation order—from the production cost value up to material costs.

Different prime cost planning approaches will be discussed herein.

Discussed above are high priority innovative industrial development technologies used to create new materials, develop big data processing systems, artificial intelligence, and machine learning. Such intensive digital technology progress may reveal enterprises' inability to effectively use and implement these technologies in business processes. In order for business to switch to the digital mode, it is necessary to obtain new economic high-tech project management mechanisms. These should include a project resource provision mechanism, which should work at different life cycle stages. It should be based on advanced machine learning and artificial intelligence to automate corporate business processes at all product life cycle stages and to maintain an effective managerial decision-making process at different product life cycle phases through the use of automated expert systems.

Known literature sources do not contain much research data on the influence of early-stage design and technology parameters, automated design, preproduction, and production processes on products' prime cost. Therefore, determining design, manufacturing, and operation costs is one of the greatest challenges. It should be noted that statistical prime cost analytics and calculation methods are quite effective, given the availability of large statistical samples, i.e., with mass serial production, but they do not work for developed one-off products going through manufacture stages. For example, a prime cost calculation method that uses calculation items, given the necessity of defining material consumption according to production drawings and of defining labor intensity according to time standards specified in operation flow charts, is only applicable at the design, prototyping, preproduction, and production launch stages. It is ineffective at early stages due to lack of information about the product. Systemic analysis and mathematical models do not have these drawbacks.

Mathematical prime cost modeling should accompany the entire process of designing a multicomponent product (Fig. 6.14): component design (the prime cost and labor intensity depends on a number of factors, such as geometric parameters, adjustments, allowances, processing quality, materials, technological characteristics of their production), combination of components into a unit, module, and the end product.

At the component design stage, simulation modeling can help specify geometric parameters/adjustments/allowances, which ensure optimal prime cost and labor intensity.

Designing and economic analysis should be connected to each other, and economic analysis results should influence the engineer's choice of adjustments, allowances, and other parameters.

Because it is necessary to possess prime cost and expected price value estimates at the earliest project implementation stages, it is advisable to obtain economic tools through mathematical modeling, as they provide accurate prime cost estimates with reference to types of materials used in production (providing that the product meets target technical parameters) and to admissible runout accuracy, adjustments,

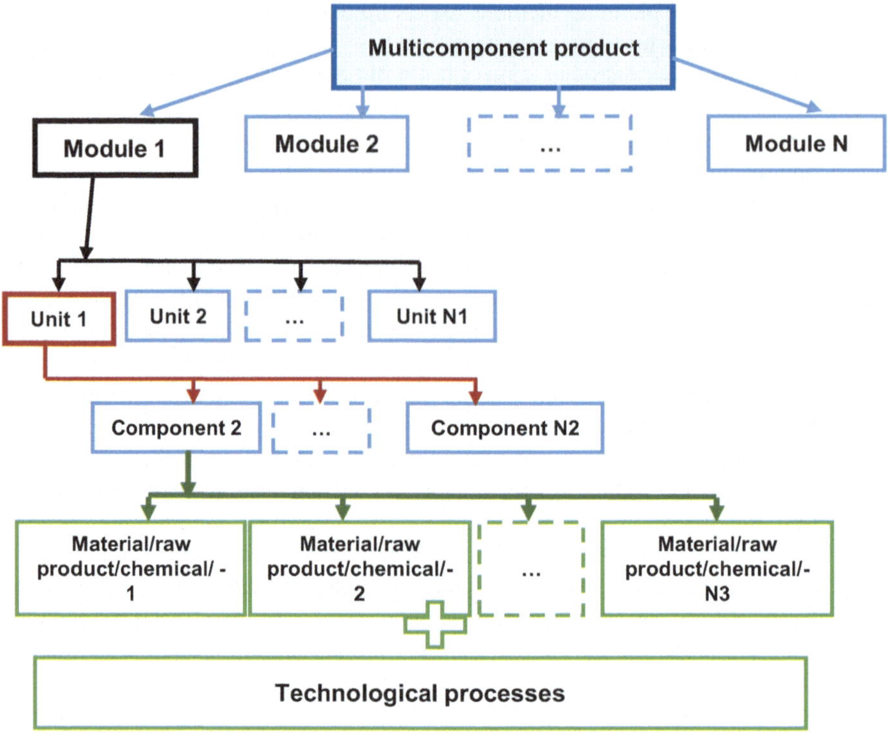

Fig. 6.14 Multicomponent products' prime cost structure

allowances, level of assembly, etc. Also, it is necessary to carry out a simulation modeling of prime cost and labor intensity with varied adjustments, allowances, runout accuracy, materials, etc., given the necessity of meeting planned technical characteristics.

Here is a detailed description of how different factors influence the prime cost:

- Type of material (its machinability, which depends on its mechanical properties, mostly solidity and viscosity)
- Precision standards (dimension, form, and surface position tolerance)
- Finish standards (admissible roughness)

Choice of Material

From the perspective of machinability, the ISO 513 standard highlights six groups. Most machinability charts show specific cutting force values that are characteristic of different metals and alloys, which depend on their state. A higher value means lower machinability. The harder a material is, the harder it is to process and the higher the tool spending is. While turning 60 HRC steel is simple, cutting parts that have this hardness is difficult and not practiced widely. In this case, it is very hard to predict tool spending.

To reduce the cost of a part at the design stage (with optimal workload and operating conditions), it is advisable to use highly free-cutting materials. This is particularly applicable to metal cutting.

Setting Adjustments and Allowances

It is necessary to distinguish between a machine's accuracy characteristics and processing accuracy. For a lathe, it is the apron's repeating accuracy, maximum error not exceeding 0.01 mm. Processing accuracy depends on these parameters and other factors, and it can get close to the lathe's parameters only theoretically. In fact, the apron–workpiece interaction is mediated by the cutting tool, creating the *lathe—cutting tool—workpiece* system.

The first error depends on the cutting tool's size, as it changes during the process. The size can change not only by hundredths and even tenths of a millimeter.

The second error originates in the rotary tool (for example, a milling cutter), as it may have a radial- and face-motion variation. Even long high quality rotary tools may have a beat measuring several hundredths of a millimeter.

The third error results from the deflection of the tool from the set trajectory due to the cutting force. The higher the cutting force and the softer the tool is (for long and thin tools), the greater the deflection is. This can be countered by cutting off a thinner layer of material per pass, but the processing time will increase, and an extra thin and long tool will lose accuracy. Real accuracy depends on the processing conditions, and it may differ greatly from the lathe's accuracy characteristics.

Size tolerance (allowance) should be set very carefully, because the lathe's potential accuracy is 0.01 mm, and cases whereby a ten times lower accuracy is required, complex processing schemes and tool combinations should be used. This can cause the prime cost to grow dramatically. An accuracy level of Grade 12 or lower is not hard to achieve; Grade 11 to Grade 8 is medium high accuracy; Grade 7 or higher characterize high and extra high accuracy. Allowances for different grades and size ranges are shown in the table of adjustment and allowances. For example, making a part with a diameter of 50 mm, a length of 300 mm, and a straightness tolerance of 0.02 mm is not difficult, while making a part with a diameter of 5 mm, a length of 1000 mm and a straightness deviation of 0.1 mm is hardly feasible. Likewise, a nonflatness of 0.01 mm with a length of 200 mm and a thickness of 1 mm is an impossible feat. *Inflated accuracy requirements can lead to a significant increase in the prime cost.*

Surface Roughness Standards

Each type of processing implies a roughness limit. Within the range of attainable values specified for each type of processing, minimum roughness tends to inflict additional labor efforts and therefore a higher cost.

Parts' and elements' workability is defined by a variety of roughness standards, and the most preferred values are specified in tables. Inflated roughness requirements will increase the prime cost instead of reducing it.

Economic accuracy of a processing method implies that costs do not exceed those resulting from the use of another method, which can be applied to the same surface. Economic accuracy is a conditional term, as it characterizes an opportunity to choose

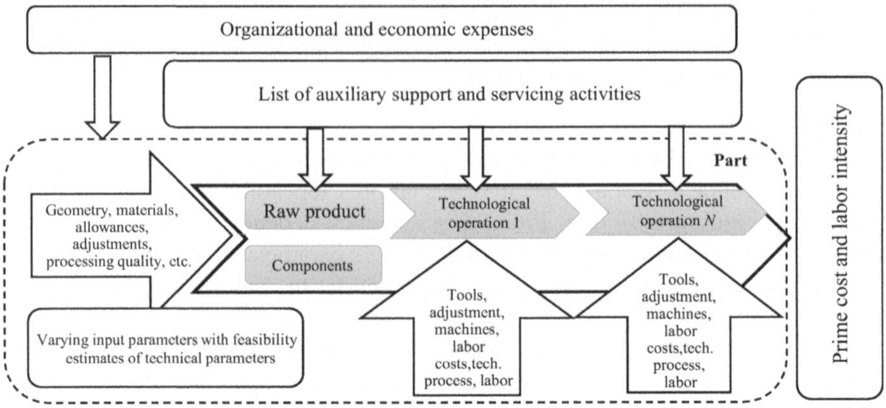

Fig. 6.15 Prime and labor cost optimization

a high-accuracy processing method with minimum time and labor costs. Attainable accuracy is a maximum accuracy, which can be achieved by a highly qualified worker with facilities intended for processing parts at a fixed level of accuracy. A comparison between economic and attainable accuracy characterizes the quality of the technological process.

A mathematical modeling device presents a generalized methodology of building various dependences based on experimental data.

Simulation modeling of the cost and labor intensity of part/unit production relies on available information about factors that influence the prime cost, which includes information about materials used in production and corporate manufacturing facilities. While varying these factors, with set technical characteristics fulfilled in a workpiece, it is possible to determine its optimal prime cost and labor intensity (Fig. 6.15).

The striving to reduce parts' and components' production cost and, consequently, the prime cost, is related to solving optimization problems, which are usually formed based on the choice of quality criteria, limits, and optimized (changeable) parameters (task variables).

There is an optimization algorithm (Fig. 6.16), which can be used to change existing size allowances for parts of mass-produced goods. It is applicable to various types of products. Actually, optimization defines the following parameters:

- Characteristics of the general quality of the system $K = F_1(\Delta A_i) \leq b$; prime cost $S = F_2(\Delta A_i)$, $i = 1, \ldots, m$, where F_1, F_2 are functions relating the quality value of the i system component to the system's general quality and the cost of quality formation; ΔA_i is a characteristic off the quality of the i system component; b denotes limits.
- Linear dependence between values ΔA_i and part size deviations, which form these values: $\Delta A_\text{i} = f_1(\delta_1, \delta_2, \ldots, \delta_n)$.

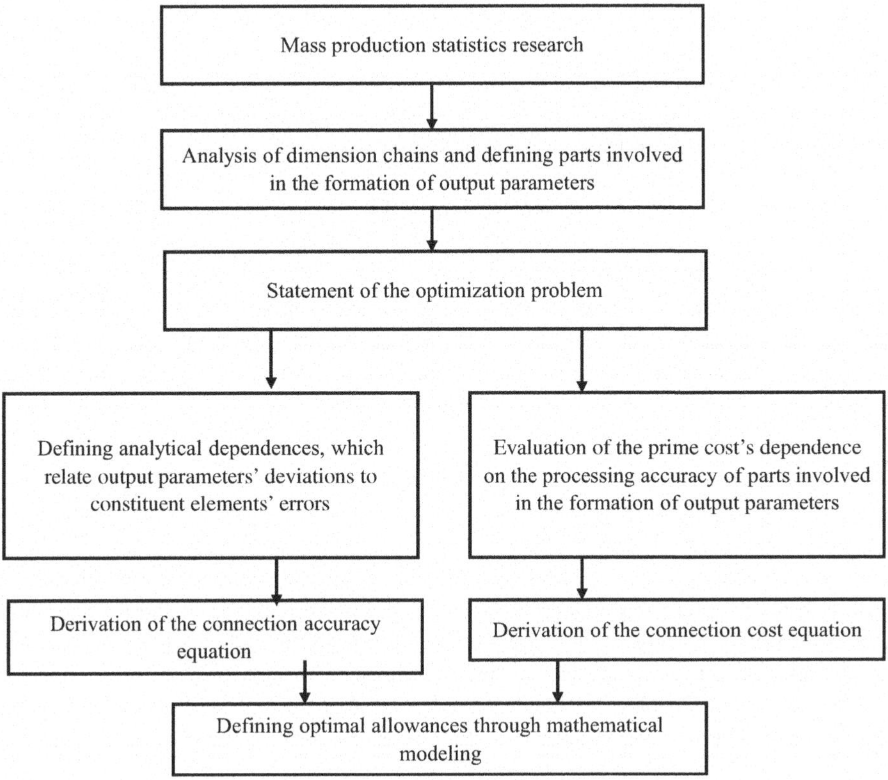

Fig. 6.16 Allowance optimization algorithm used when designing a product

- Linear dependence between accuracy and production cost: $S_{A_i} = F_2[f_i(\delta_1, \ldots, \delta_n)], i = 1, \ldots, m.$
- A complete function expressing production cost $\Delta A_i = f_i(\delta_1, \delta_2, \ldots, \delta_n) \leq b; S_{A_i} = A_0^i + \sum_{i=1}^{n} a_i \delta_i.$

Original data that is used during optimization includes: existing part size tolerance limits and output geometrical deviations; a linear dependence between the cost of forming an output parameter and its constituent elements' dimensional accuracy; a linear dependence between deviations of output geometrical parameters and their constituent elements' size errors.

Assume there are an N number of part sizes of an output parameter's constituents. It is admissible for each size to randomly deviate from the nominal one during manufacture. Meanwhile, to ensure proper assembly and functioning, deviations have limits—size allowances:

$$d_{i1} \leq \delta_i \leq d_{i2}, i = 1, \ldots, N.$$

For each group of random size errors observed in constituent elements, according to the specific linear dependences, there will be an output geometric deviation, which also has limits related to operational reliability:

$$\Delta A_i' < \Delta A_i \leq \Delta A_i'',$$

where $\Delta A_i'; \Delta A_i''$ are the low and high ends of the output geometrical tolerance.

It is known that the cost of processing of each dimension S_i when making components depends on the nominal accuracy. Therefore, the total cost of building an output parameter is a function that characterizes the accuracy of parts' sizes, which are involved in the formation of this parameter.

$$S = \sum_{i=1}^{m} S_{A_i}.$$

Research has shown that initially set allowances can be changed, so that the limits will be fulfilled, and the cost of their formation will be reduced. It should be noted that new optimal allowances d_{i1}', d_{i2}'' cannot exceed the low-end accuracy value, and they cannot drop below high-end accuracy values.

This poses the problem of linear programming of the minimal component production cost with limits applied to the dimension chain allowance variance range:

$$S_{i0} + S_{i1}d_1' + S_{i2}d_2' + \ldots + S_{in}d_n' \rightarrow \min,$$
$$D_{i1} \leq d_i' \leq D_{i2}, d' \geq 0, i = 1, \ldots, n.$$

where D_{i1}, D_{i2} are the low and the high ends of the allowance variance range d'.

This model can help define a relative cost with original allowance values, output parameter dependence coefficients, output parameter limits, as well as results, optimal allowances, respective relative cost values, and output deviations observed during the allowance optimization process.

This algorithm helps optimize all dimensional chains of a product, where any master link can be a limit. Similar algorithms can be used for the choice of a material, setting processing quality standards, etc.

Automation of the prime cost optimization process, both for a separate component and a whole product, is a paramount goal, which can be achieved through the use of digital development and production technologies based on creation of digital twins. With the help of "smart" intelligent mathematical simulation methods, a design-to-cost system can offer the most economically optimal production process configuration. The informational basis for this type of intelligent system is a database that contains information, which must be used when executing design-to-cost projects:

- Materials used in component manufacture: Physical characteristics, raw product cost, processing costs, cost of meeting target allowances/adjustments/processing accuracy, etc.
- Parts' geometric characteristics, component layout.
- Technological operations: Machinery and equipment (technical and marketing characteristics), tools' cost, labor intensity of the adjustment process, etc.
- Production site: Cost of labor, direct production costs, etc.
- Production cost and the length of processing cycle for a component made of a particular material with the use of particular machinery.

The mathematical model of an intelligent design-to-cost and design-to-labor-cost system implies *upward* calculation, i.e., the prime cost and labor cost calculation relies on how costly and labor intensive the production of separate parts is, on their unification into units, the units' unification into modules, and the modules' unification into the end product. The input parameters of the mathematical model are products' geometry, component layout, materials used (processing cost and labor intensity), allowances, adjustments, processing quality, etc., data on technological operations (machines and equipment, tools, adjustment, labor costs, technological process, working hours, etc.) The output parameters include cost and labor-intensity estimates for parts, units, modules, and the end product.

Optimization parameters of this simulation model include the cost and labor intensiveness of the manufacture of parts/units/modules/end products. Simulation modeling is based on parts/units/modules mathematical and digital models. Materials are checked and selected, different geometric, component layout, allowance, adjustment, processing quality, etc. options, as well as alternative technological processes are considered.

Simulation modeling and parametric varying are carried out along with a strict control over a product's meeting of target technical characteristics. For each configuration obtained after simulation modeling, the cost and labor intensity are calculated. A criterion of finding an optimal configuration (for example, a minimum cost or labor intensity) is defined. Optimal parameters are the simulation model's output product, which includes the material used, geometry and component layout, allowances, adjustments, processing quality, technological process.

The intelligent design-to-cost system should meet requirements concerning data on factors, which must be considered when shaping a product's technical and economic image and knowledge bases (large volumes of statistical data). This data is used for making lists of factors involved in the formation of a product's technical and economic image; defining information, which is necessary for automating the choice of optimal materials, allowances, adjustments, etc., based on big data, given the necessity of achieving target technical characteristics; adjusting technological processes to a product's set prime cost and competitiveness while maintaining technical characteristics obtained through big data analysis of equipment's capacity, specifics of technological processes, etc.

The intelligent prime cost optimization system is more effective when used along with a product's digital twin. Indeed, a digital twin is a virtual replica of a real object, and it behaves in the same way. Existing project life cycle management (PLM) tools create a full-scale digital platform, which supports digital twins and faithfully reproduce the end-to-end design and manufacture processes.

For example, the NX package of the Siemens PLM system creates a product replica, which is passed over to the Teamcenter package in a specially designed 3D format. Next, the application builds a virtual version of the product, which replicates the original product. To reveal possible problems, the digital twin uses big data processing tools, design and technology information (PMI) contained in models (allowances, adjustments, links between parts and units) and a basic description of the engineering process. Revealing design flaws at early stages (more feasible with the use of a digital twin) helps save a lot of time and money both when designing and manufacturing a product.

A digital twin can raise the effectiveness of designers and engineers' cooperation, help choose an optimal production site and production technology, as well as provide necessary resources. Product engineers possess information on each added operation ensure that new processing routes meet set productivity indexes. If they do not, they replace or modify technological operations. After that, they resume mathematical modeling and do it until the process route meets all requirements. All designers and engineers have immediate access to the new process to confirm it. Should any problems arise, designers and engineers can join efforts to eliminate them.

Using a digital twin—a real product's replica—helps reveal possible design and engineering issues, speeds up preproduction and reduces the prime cost. Besides, using digital twins gives a guaranteed ability to manufacture a developed product: all design and engineering processes are maintained in an actualized and synchronized state, newly developed technologies prove to be effective, and production is organized in strict accordance with the plan. A digital twin helps test the effectiveness of methods of incorporating advanced technologies into existing production lines. This eliminates risks that arise when purchasing and mounting equipment.

This is a detailed analysis of the digital design-to-cost product engineering process with the use of a digital twin. The following chapters will focus on building a digital replica of the production process, whose interaction with a product's twin poses another economic problem that is optimal preproduction of a new type of goods.

Reference

1. Stackpole, B. *Digital twins land a role iin product design*. Retrieved from http://www.deskeng.com/de/digital-twins-land-a-role-in-product-design/

Chapter 7
Preproduction of Advanced Products with High Technical and Economic Characteristics

7.1 Preproduction of Products with High Technical and Economic Characteristics Based on New Physical Principles

A product's digital twin helps solve quite a number of life cycle cost optimization problems, as described above. Meanwhile, there is a concept of *digital twin of a process*, a virtual prototype of a whole set of production equipment, which can be used for modernizing process implementation options. Primarily, a digital twin of an industry solves cost optimization problems at some life cycle stages, which include preproduction and production itself.

Because products being designed and produced are supposed to win dominating positions in the market, which should be due to, on the one hand, their technical and, on the other hand, economic (cost) characteristics, it is important to use the design-to-cost approach. It ensures setting optimal allowances, adjustments, reliability, and integrity characteristics, etc., which can be achieved with the help of specific types of equipment, tools, accessories, control, and measurement facilities. Such equipment and tools should be chosen during the preproduction phase or while upgrading goods that are already in production, to give them competitive advantages.

This life cycle stage starts when a decision is made to launch production; it includes activities aimed at creating conditions for mass, one-off or single-item production, and it continues until the product obtains technical, technological, and economic characteristics, which make it highly marketable.

From the perspective of life cycle economy, this stage is important, because it encompasses highly intense preproduction work. Optimally organized preproduction predefines effective meeting of the following objectives:

- Providing sustainable economic growth of advanced product manufacturers by reducing the prime cost during product development while using the design-to-cost approach and forming a competitive price

A. Tyulin, A. Chursin, *The New Economy of the Product Life Cycle*,
https://doi.org/10.1007/978-3-030-37814-1_7

- Creating a new highly competitive product that uses new physical principles with account for modern global science, engineering, and technology trends
- Enhancing a company's key assets through creation and introduction of new automated and robotic facilities intended for production of advanced goods
- Winning sales market competition through setting an optimal price (with account for design to cost production), which gives the product a customer appeal that makes it competitive

As it is known, preproduction includes several types of activities, which in a way endow a new product with competitive advantages, as it is produced with the help of new technical tools, methods, and production technologies. For visual clarity, these activities are shown in Table 7.1.

All preproduction activities listed in Table 7.1 require the use of respective types of resources. Practice shows that organizations do not always have all needed resources in possession. Some of them have to be found outside of a company and used in such a way as to ensure maximally effective preproduction. This solves two

Table 7.1 Main preproduction activities

Type of work	Point of work
Organizational work	Project drafting, management of all necessary preproduction R&D activities in all directions; development of preproduction regulations, provisions, and standards; training personnel to qualify for the production of advanced goods; building technological and organizational competences
Project design	Developing project statements; preparing construction documentation; pre-drafting production workshops; complex production building, workshop and section reconstruction; organizational process development; process regulation
Product design	Developing engineering documentation; manufacturability checks; preparing documentation for mass production; preparing equipment operation, repair, and restoration documentation; improving standardization, manufacturability, etc. indicators; improving maintainability while referring to consumer use test results
Production design engineering	Development of technologies: designing new technological processes, improving existing ones, developing typical and batch processes; developing software for CNC equipment, versatile automated processes, processing centers, 3D printing equipment; guidelines, standards; tool preparation: defining jigs, fixtures and tools, standard universal accessories; outlining original instrumentation, specific cutting tools, packaging; development of jig, fixture, and tool production process and production of these; metrological provision: calibration tool nomenclature; original calibration tool design; planning calibration processes; adjustment of calibration tools; regulatory preparation: development of norms for consumption of materials, energy and labor
Process engineering	Development of nonstandard equipment; development of mechanical and automation production tools; mounting and adjustment of equipment and energy supply systems; planned and preventive maintenance; developing safety measures and guidelines

important problems: providing all necessary resources and their effective use during preproduction. Resources required for effective preproduction are shown in Fig. 7.1.

It should be noted that resource costs remain high as long as a company's resource potential and resources provided for implementation of separate advanced projects are limited.

Preproduction itself is a labor-intensive and costly process, as its cost is reflected in the output product's prime cost and, consequently, its price. High preproduction costs increase the output product's price, which results in lower competitiveness, a longer total payback period, and higher risks of ineffective future project implementation in terms of profitability. Today, according to different sources, this life cycle stage's cost accounts for a 10–30% of output products' prime cost, depending on a project's complexity, available tools, etc.

Therefore, it is important to define an optimal resource provision level for preproduction. First, the amount of resources needed for preproduction depends on the complexity and number of parts and units, which a company is planning to manufacture. Also, it is necessary to define products and units, which do not require specific preproduction steps and complex instrumentation, and can be carried out with the help of available corporate tools.

Second, the amount of preproduction resources depends on the degree of automation of the preproduction process. This changes the resource structure: the higher the automation is, the less human resources are needed, while there is a need for a greater amount of information resources (big data) and computational facilities, which are capable of collecting, processing, and analyzing big data and outlining the most effective ad versatile production routes.

Important in resource provision is optimal resource planning based on their availability. For example, some organizations can use their own corporate resources (for example, tool production equipment), while others could adjust a clear resource provision mechanism (this concerns raw products and materials needed for making instruments), or it may take extra time for a company to obtain them (if extra competences need to be developed for preproduction specialists, so they need additional training, which takes extra time and funding). Therefore, it is important to plan timely provision of resources, which are necessary for preproduction, depending on the target product's complexity.

Meanwhile, optimized use of all types of resources implies their use while relying on the concept of lean production, which is based on resource conservation through eliminating ineffective and costly processes (for example, elimination of overlap activities, using simultaneous, not consecutive, electronic document preparation). The scheme shown in Fig. 7.2 presents steps aimed at reducing the cost of each preproduction activity through optimal resource management.

Resource provision process can also be optimized thanks to increasingly automated preproduction. Automation grows and evolves dynamically toward shorter completion periods and lower labor costs (Fig. 7.3).

Preproduction processes' evolution is also explained by the evolution of the end product itself, its units, modules, and devices, which are increasingly compact and functional, when one device or equipment unit (for example, a machine) perform

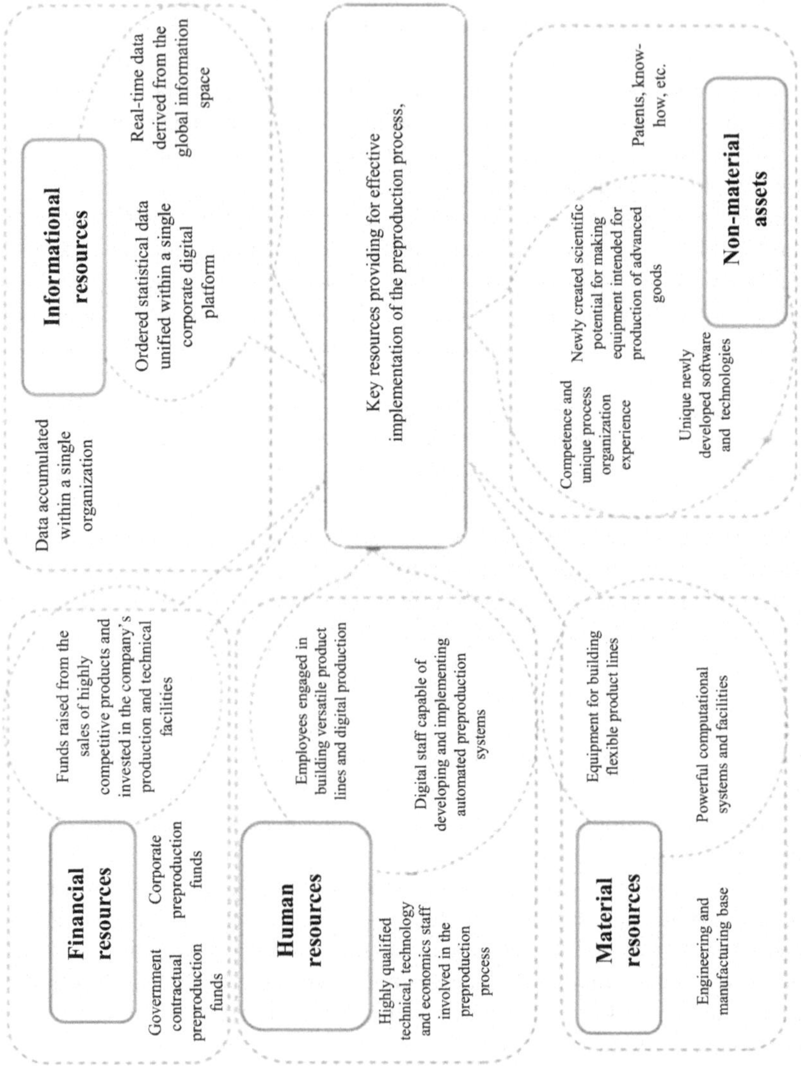

Fig. 7.1 Key resources needed for effective preproduction

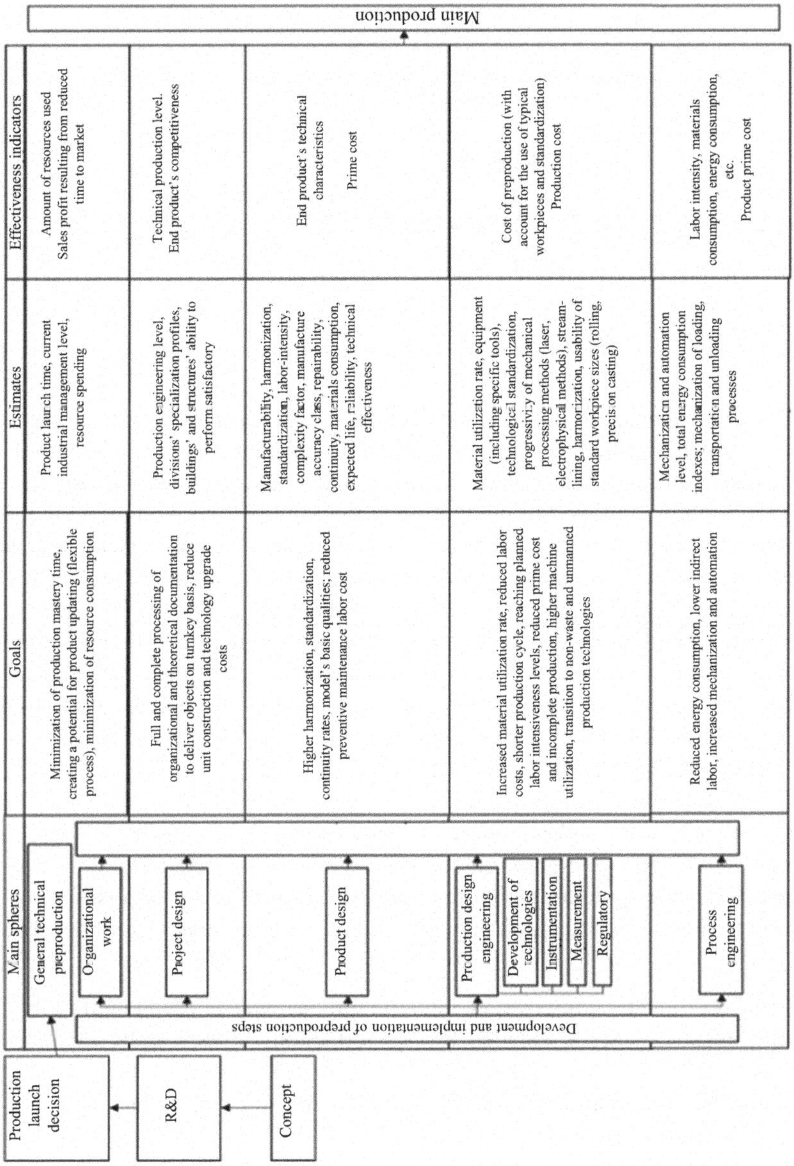

Fig. 7.2 Preproduction at an organization and project cost optimization activities

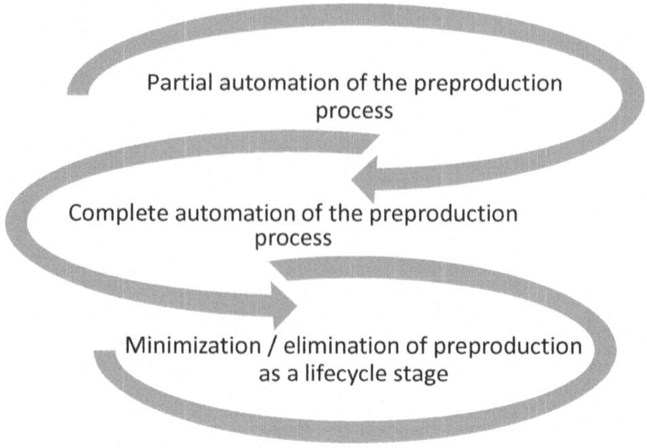

Fig. 7.3 Progressing automation of the preproduction process

functions previously carried out by a group of devices or machines, for example, old computers, which would not even beat a simple calculator in capacity, used to occupy whole buildings. They featured tubes and relays and were enormous in size. Today's laptops, many of which are just the size of a book, are by an order of magnitude more powerful than those obsolete units. The main factor that has contributed to the miniaturization is the invention and use of microchips. This high-tech product has reduced the size of radio-technical elements and boosted their reliability more than 1000-fold. Today, it is quite so common to combine within a single microchip a whole electronic device; receivers, calculators, hearing aid and many other devices operate on just one chip with a minimal number of additional parts. Technologically complex operations and elements, which entail costly and lengthy preproduction, have fallen out of use.

This case demonstrates improvement of a new product's structure, which uses new physical principles and stimulates the development of new techniques and technologies involved in the preproduction process. This triggers a spiraling evolution of preproduction systems, which map the evolution of end products.

Nowadays, some preproduction activities already use automated systems. These include:

- Modeling a product's composition and structure, as well as production tools
- Planning resource spends entailed by preproduction, as well as by running an entire project
- Evaluating the changing labor intensity and materials consumption
- Using CNC equipment
- Preproduction management with the use of computational tools, etc.

The early stage of a preproduction system's evolution based on gradual automation of production and engineering processes is shown in Fig. 7.4.

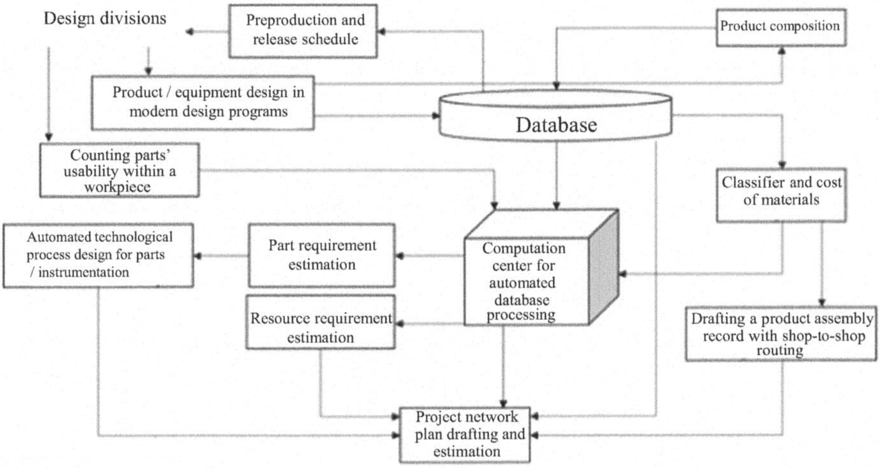

Fig. 7.4 The starting point of a preproduction system

This stage of preproduction systems' evolution is characterized by introduction and implementation of automated fitment and equipping systems along with digital twin building. Once a preproduction and release schedule is developed, product parts' usability is calculated to determine their production volume; instruments are selected automatically, as they must be used for producing these parts according to templates derived from the database and with account for the diversity of materials required for the production of instruments. Also, the demand for resources is formed, which are needed for production, an assembly record and a project network plan are outlined.

To make these processes more effective, it is necessary to develop and carry out organizational and technical work, which mostly includes technical and technological offers relating to quality improvement, prime cost reduction, and release of future products in volumes sufficient for satisfying the market demand. Activities that improve output quality, reduce prime cost, and increase output product volume, include automated production, measurement and control, use of advanced paperless technologies, low-waste or non-waste production technologies.

This case refers to the second level of preproduction automation, which is complete automation, when instruments and equipment are designed and produced with the help of robotic systems and facilities, which use future instruments' (for example, a stamping mold's or molding die's) digital twins and design them without using paper, as the digital twin is fed to a robotic production system.

Complete automation is carried out through building and implementing an automated complex featuring several subsystems that are shown in Fig. 7.5.

Complete automation is closely related to the developing Industry 4.0 concept and technologies, which, when introduced, will lead to in-depth automation of the preproduction process, which will rely on new physical principles. The algorithm of

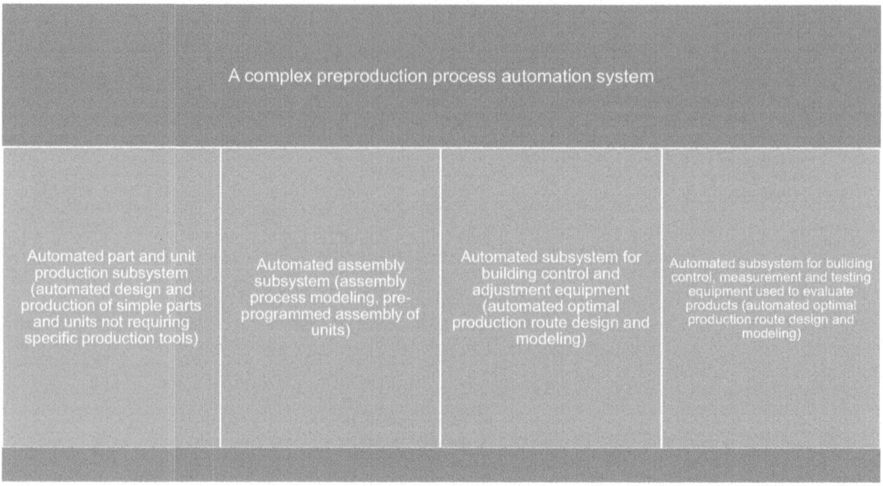

Fig. 7.5 Structure of a complex automated preproduction system

incorporation of the high-tech Industry 4.0 strategy into preproduction processes is shown in Fig. 7.6.

In other words, production of a competitive type of goods should involve all newest scientific and technical achievements, which an organization can financially afford to incorporate, while providing a return period long enough to ensure its financial and economic growth.

The third automation level is the most comprehensive one, as it excludes preproduction from a product's life cycle and includes it in the production phase. A key distinctive trait of comprehensive automation is its responsibility to do the most labor-costly work, which includes adjustment and testing, in an automated mode with the help of intelligent IIOT systems and technologies, which replace (eliminate) human labor.

It is possible to achieve a high automation level in the current dynamic global economic system, as developing and now common virtual communication tools are used in an increasingly digital environment that embraces industrial enterprises activities, become advanced projects and technologies. Industrial organizations integrate into a unified cyber-economic system, digitize business processes and plan them not only with the use of PLM systems, but also through adaptive management of all corporate resources even when these resources are uncertain, but an ultimately new high-tech product is created.

There organizations rely on cloud technologies and a service model, while an industrial system relies on creation of innovative products with the use of processing technologies, big data analytics and artificial intelligence methods, building relationships with the main consumers and suppliers based on communication models and cyber systems. Production processes are transformed through incorporation of versatile production systems with adaptive schedules and resource distribution, robotic complexes, cyber-physical systems, use of modern digital technologies at

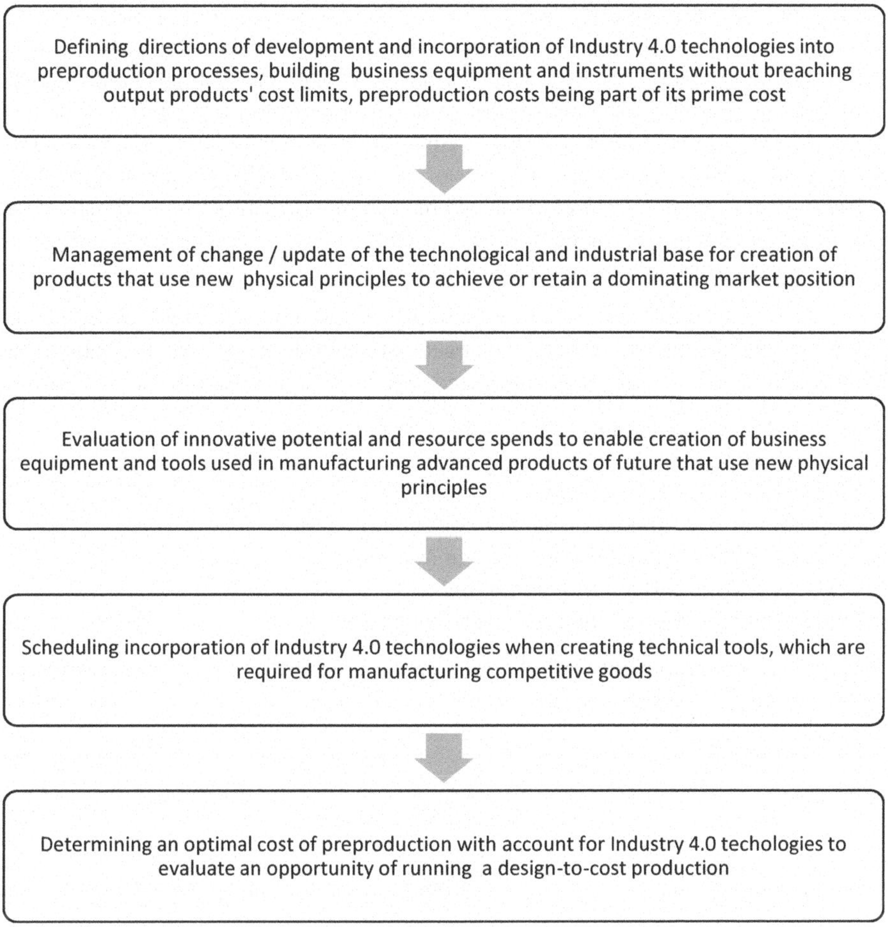

Fig. 7.6 Algorithm of incorporating advanced Industry 4.0 technologies into the preproduction process

all life cycle stages, such as neural networks, Blockchain, the Internet of Things, VR and 3D technologies, etc. The use of these technologies is vital for effective cost optimization at early stages. All products, units, and installations are classified and unified into three groups:

- Standard products not requiring preproduction or ones that can be brought from outside (nuts, bolts, etc.).
- Products not requiring any work preparation thanks to the use of advanced technologies. For example, 3D printing of separate parts and components can be an effective advanced solution when it is technically feasible. In this case, work preparation of such parts and components will be minimal and confined to the programming of printing equipment and printing a template based on the product's digital twin.

- Products, which require preproduction, particularly preparation of technological instrumentation, which is created with the help of robotic and self-learning artificial intelligence systems. That is necessary when each output item or parts of it are structurally and functionally unique and require specific types of instruments, measurement, and calibration equipment. In such cases, implementation of an industrial program entails the use of unique high-tech equipment, individual routing technologies, as well as standard technological processes, which use a substantial amount of manual work. Production of such parts requires using automated instrumentation design systems.

In this case, to implement preproduction that is based on new physical principles and high technical and economic characteristics, it is necessary to implement advanced methods of structuring automated preproduction systems with account for:

- Data and technical documentation management systems
- End-to-end process development systems
- Shop-to-shop routing systems
- Material norm-setting and labor cost calculation systems
- Automated technical equipment design systems
- CNC equipment programming systems
- Process index collection and analytics systems

Each system component should be integrated with the entire corporate communication field for solving various problems within an integrated IT environment (Fig. 7.7).

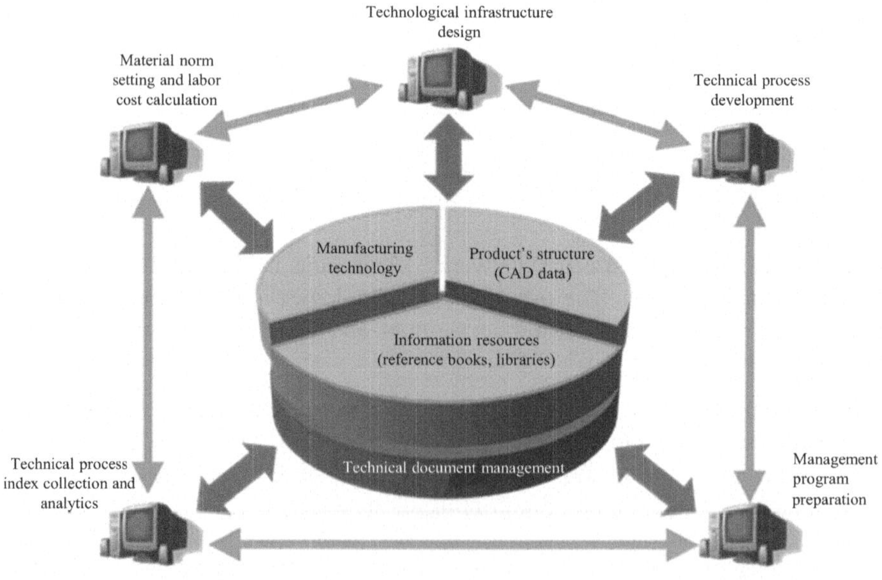

Fig. 7.7 Integrated information preproduction environment using new physical principles

A preproduction process that uses new physical principles begins at product design stages and is performed through the use of *technological masks*—technological limits, with account for CAD-integrated processes' potentials; this can help avoid errors at early stages and enable implementation of optimal design and engineering solutions. This order sets design, manufacture, and operation standards for products' standardized elements.

The equipment, which is being designed and manufactured and is going to determine new physical production principles for the new product, should also use advanced twin-building and optimal twin configuration technologies with the use of intelligent systems. These intelligent systems will be able to automatically model the most advanced equipment versions with reference to globally available data on current technical and technological trends, global academic achievements, scientific discoveries that enable production of goods based on new physical principles.

A universal approach to configuring standardized elements helps adapt technological solutions and sets universal technical requirements to branch competence centers.

Branch competence centers, which specialize in creating specific types of instrumentation or use of advanced equipment, can unify into a *virtual organization*, whose mission is speeding up design and production of goods with competitive end-user performance, which could dominate an existing market and satisfy demand, and goods produced as part of rapid product development programs, whose innovative properties and unique usability could create new markets and stimulate new public needs.

In this case, a virtual organization includes competence centers, which produce one-of-a-kind instrumentation and equipment, and/or resource sharing centers, which possess unique equipment and share it with the original manufacturer to enable manufacture of the end product.

However, when creating this type of virtual organization, which, along with different subdivisions, includes branch competence centers and/or resource sharing centers sharing unique modern equipment, it is necessary to carry out systemic competence evaluation within a branch and related branches, particularly—their conformity with the global scientific and technical progress. This can be done as shown in Fig. 7.8.

The evaluation process comprises four stages.

- Stage 1. Setting an international conformity evaluation standard.
- Stage 2. Filling tables of key properties of a product manufactured on the basis of an evaluated competence and weight rates, including:
  - Outlining requirements to sources and methods of acquisition of information about products made with the help of the evaluated competence and similar ones, derived from available sources
  - Searching for and obtaining information about products (goods and services), which are produced with the help of the evaluated competence and similar ones, from available sources
  - Entering collected data in tables of key properties and their weight indexes.

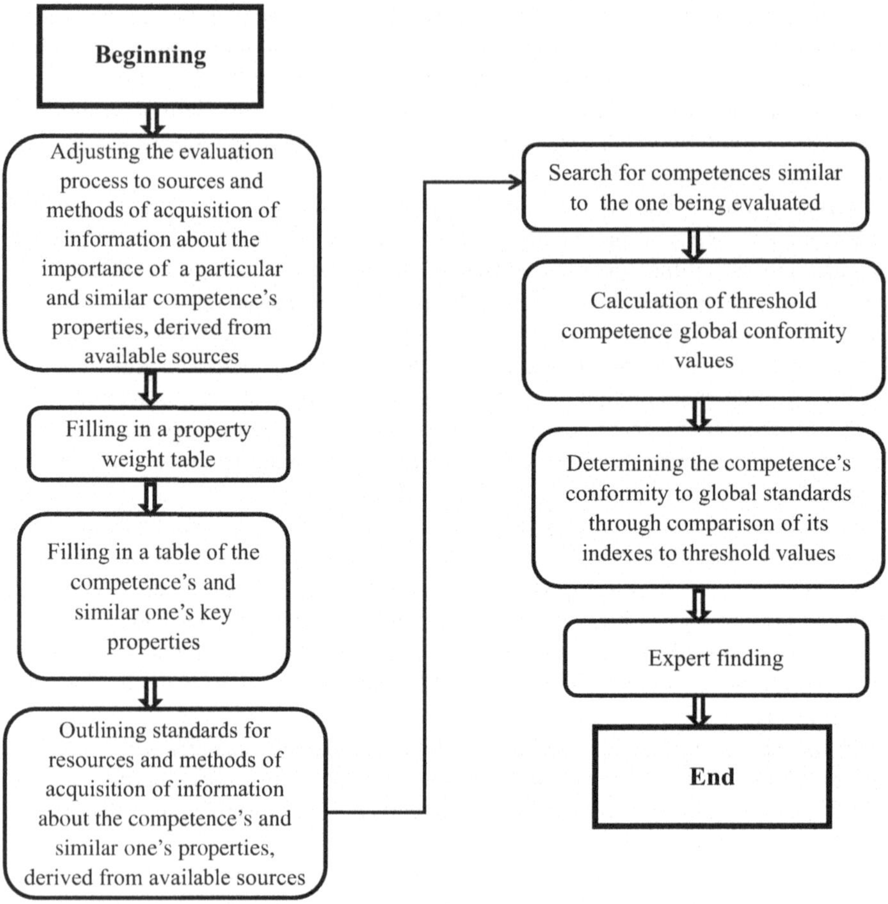

Fig. 7.8 Determining a competence's conformity with international standards

- Stage 3. Filling in tables of property indexes and determining high-end and low-end threshold values characterizing the competence's conformity to international standards:
  - Outlining requirements to sources and methods of acquisition of information about indexes of competence's properties and/or those of the product manufactured with the help of evaluated competences and their equivalents
  - Search for similar technologies and/or products manufactured with the help of similar technologies
  - Search for and obtaining indexes characterizing the properties of products manufactured with the help of similar technologies
  - Entering indexes characterizing the properties of a technology and/or product made with the help of the evaluated technology and similar ones, into property index tables

- Determining threshold and choosing high-end indexes characterizing a competence's conformity with international standards while relying on derived information about the properties and indexes of the evaluated competence and its equivalents.

• Stage 4. Evaluation of a competence's conformity with the international standard and summing up expert findings:

  - Evaluation of the competence's conformity with international standards by comparing its indexes (of those of a technology of goods produces on this competence's basis) to threshold values:

    A. Determining international competence evaluation requirements:

      1. There is an algorithm, according to which requirements to sources of information and methods of deriving it from these sources.
      2. To determine requirements to sources of information and methods of deriving from these sources of information about the evaluated and similar competence's properties, experts, whose competences lie within the same subject field, are invited to participate in the evaluation process.
      3. A list of databases containing information, which can be used for determining the parameters of the evaluated competence and those of similar ones. Each database should contain data of at least one of the following types: Information about fundamental research relating to the competence and similar ones; information describing the use of the competence during the production stage.
      4. Each database should be followed by requirements determining the choice of data search methods and tools. The most common search methods include patent classification index search and semantic search over texts and document abstracts; for this purpose, classifier indexes that are relevant to an object evaluated are determined, as are requirements to the search query shape.
      5. The use of the International Patent Classification (IPC) is mandatory.

    B. Filling in tables of key properties and their weights:

      1. Tables of key properties and their weights are intended for determining the competence's and similar ones' properties, as well as these properties' importance for the goals of evaluation.
      2. The right-hand column should include names of the competences' and their equivalents' properties, which can be qualitatively assessed with the help of measurement tools through calculation or based on values provided by experts.
      3. The remaining columns should contain names of parent groups' properties, which unify several properties named in the next column to the right. Depending on the type of table, the left-hand column should contain the name of a property that characterizes the evaluated

competence's or its equivalents' quality or integral quality. Depending on requirements applied to sources and information acquisition methods, properties can be defined either based on competence classifiers (i.e., based on generalized or individual information) or based on data referring to competences similar to the evaluated one. In case whereby generalized information is used to fill in the table of properties, searching for equivalent competences is not mandatory. In case whereby individual data is used to fill in the table, searching for similar competences is imperative.

4. Prior to searching for equivalents, it is necessary to outline requirements to sources and methods of acquisition of information about the evaluated competences' and its equivalents' values from available sources.
5. Specifics of searching for equivalents are described in Point 10.
6. Determining weights (importance indexes).
7. Filling in tables of property values and counting threshold indexes that characterize the competence's conformity with international standards.
8. To outline requirements to sources and methods of acquisition of information about the evaluated competences' and its equivalents' values, experts are invited to participate in the assessment process, whose competence lies within the same subject field.
9. Search for similar competences.
10. Based on the obtained information and materials, a list of evaluated competences' equivalents is made, which should include their properties and their values. The discovered equivalents of the evaluated competence include: equivalents that, according to one of their properties, can be deemed the most suitable ones for using information about those to determine control properties' values; and equivalents that, according to one of the properties, can be deemed less suitable for using information about those, to determine properties' rejection values.
11. Tables of property values are filled according to each table property based on information derived from available sources (experts, documents and databases). Property values are specified for both the evaluated competence and all discovered equivalents. Threshold quality and/or integral quality values, which characterize a technology's conformity with the global standard, were evaluated with the help of the Analytic Hierarchy Process (AHP) according to the following two formulas:

Formula 1: The standardized value of the competence K_{ij} is calculated as follows:

$$K_{ij} = \frac{Q_{ij} - q_i^{\text{contr}}}{q_i^{\text{rej}} - q_i^{\text{contr}}},$$

where q_i^{contr} is the property's control value derived from the table of competences (and its equivalents') property values; q_i^{rej} is the rejection value of the property obtained from the table of competence's and its analogs' properties; Q_{ij} is the value of a property in the table of competence's and analogs' property values; $i \in \{1, \ldots, m\}$ is a property's number, m is the number of properties; $j \in \{1, \ldots, n\}$ is the number of a competence or its analog in the table of property values, n is the number of evaluated subjects, which includes the evaluated competence and its analogs.

Formula 2: It is used to calculate the quality or integral quality indicator:

$$P_j = \sum K_{ij} G_i,$$

where K_{ij} is the standardized index of the i-property for the j-subject (a competence or its equivalent); G_i is the combined index of the importance of the i-property of the last column (layer) for the j-participant, and each for each group and subgroup the index is defined with the help of the AHP method with the use of expert estimates; $\sum G_i = 1; i \in \{1, \ldots, m\}$ is the number of a property, where m is the quantity of the last column's (layer's) properties in the table of properties; $j \in \{1, \ldots, n\}$ is the number of a competence or its equivalent in the table of property values, where n is the quantity of evaluated subjects, which include the evaluate competence and its equivalents.

Threshold values of indexes characterizing the competence's conformity with the global standard are defined as maximal and minimal quality and/or integral quality indicators P_j, for the world's best equivalents.

A competence's conformity with the global standard is calculated in the following way: if the evaluated competence's quality and/or integral quality indicator P_j is below the low-end threshold, which was determined based on information about the best equivalents, the competence will fail to meet the international standard. If the evaluated competence's quality and/or integral quality indicator P_j is within the interval, which was determined based on information about the best equivalents, the competence will meet the international standard. If the evaluated competence's quality and/or integral quality indicator P_j is above the high-end threshold, which was determined based on information about the best equivalents, the competence will surpass international standard.

Thus, this algorithm helps evaluate competences' and even technologies' conformity with the global scientific and technological process.

The selection of the most promising competences and technologies, which surpass the international standard at the stage of technological preproduction and their further use in production, should help manufacture products that use new physical principles and can create new markets by satisfying new needs and being exceptionally valuable for consumers.

Meanwhile, acquisition and accumulation of competences will lead to an intelligence boom, which will produce a new class of industrial equipment—*ultrasmart machines*, which will fully eliminate human labor. This will radically change the current industrial order, as these ultrasmart machines are going to build equipment of different types and complexity and run it. Thus, a technological singularity will be achieved—an extremely rapid technological progress resulting in full and complete replacement of human labor with machine labor.

Once the third level of automation is achieved thanks to technical and technological development, preproduction will disappear, because all operations will be performed by intelligent systems in an automatic or automated mode.

Thus, the amount and labor intensity of preproduction work is gradually decreasing. This is going to help reduce the preproduction share in the prime cost and the risk of exceeding the planned cost of a future product, which is being created using new physical principles.

To solve this problem, it is necessary to effectively manage optimization and reduce the cost of preproduction with the goal of excluding it from the product life cycle, by way of switching respective business processes, which are managed by intelligent preproduction systems, to the production stage.

Eventually, while discussing effective management of preproduction of a new dominating product, it is necessary to pay special attention to optimizing its planned cost through effective running of all technological preproduction and cost optimization systems, resulting from process automation. A product's prime cost and labor intensiveness are the key factors determining its competitive price and competitiveness.

An important role in the formation of a product's prime cost belongs to process design and engineering stage, which comprises a whole set of interconnected steps that ensure creation of new and improvement of existing products, introduction of advanced technologies, effective labor management, production and management processes. Effective preproduction is an important resource, which helps optimize a product's prime cost and the process's labor intensiveness through the use of new-generation equipment and technologies. Introduction of intelligent production management systems can eliminate some manual and semiautomatic procedures.

7.2 Prime Cost Optimization Through Effective Preproduction

As noted above, effective management of the preproduction process is a powerful tool that helps optimize and reduce a product's prime and labor cost.

According to research carried out by McKinsey & Company, the key preproduction trends within Industry 4.0 are intellectualization of processes (consists in automation of the decision-making process through big data analytics and artificial intelligence, including the use of intelligent machinery), setting up a stable relationship between an industrial system's components (this increases productivity), and ensuring its flexibility (to ensure effective response to rapidly changing factors).

The effectiveness of these steps has been confirmed by a number of manufacturers. For example, new preproduction principles, which leverage intelligent robotic production systems, 3D printing and big data analytics, have helped Procter and

Gamble raise productivity by 160% and reduce production costs by 20%. New technological preproduction principles have helped the company adapt to constantly changing customer preferences and the toughening competition. Preproduction that uses new physical principles has prompted most employees to develop new competences, such as big data analytics, intelligent robotic systems, and additive manufacturing. The new competences have contributed to a flexible digital process setup, effective automatic quality control, and end-to-end synchronization of the supply chain. Effective preproduction has made the production system more flexible and thus enabled the manufacturer to quickly adapt to the changing market demand and new technological achievements.

The transformation of the preproduction process within a company that strives for market domination should rely on latest achievements in the sphere of digital economy and cyber-economy, which offers a system for management of various corporate resources, as it uses advanced machine learning methods. An important goal of the cyber-economy in the sphere of design and engineering preproduction is management of the technological platform and corporate manufacturing system, to produce a dominating type of goods.

While implementing an effective preproduction project, the following factors should be considered, as they directly influence a product's prime cost at the preproduction stage:

- Order of design and engineering preproduction with the use of advanced automated and intelligent tools and methods
- Technical level of production in terms of using new and advanced technologies, raw products and materials and the degree of processing
- Production and labor organization
- Implementation of one-of-a-kind equipment and test rigs

These factors not only influence the prime cost, but also extend the production cycle and sometimes impede the formation of new competences and, consequently, release of new products into the market.

To ensure effective preproduction, it is necessary to use a combined approach to reducing these factors' influence, which includes simulation prime cost modeling with the use of various technological solutions. Simulation modeling comprises industrial processing, assembly and installation of products while relying on effective visualization, which provides data for quick production planning and management in an MES system. Simulation modeling accounts for personnel's skills and competences, as well as manufacture centers' equipment capabilities.

Along with simulation modeling, the following consecutive and concurrent problems can be solved:

- Determining critical points, risks, and possible solutions
- Determining optimal production routes
- Determining sufficient resource availability
- Determining the use of labor, time, and budget resources

While executing effective preproduction of a new type of goods in a digitally transformed environment, it is necessary to take into account interaction between a large number of employees with automated equipment. The preproduction process involves design services (building tools and instruments, preparing measurement tools and test rigs), material and component supply services, CNC adjustment services and processing centers. This whole system can work effectively providing there is an intelligent management system relying on big data analytics. A large-scale intelligent preproduction system is created one time only, and further on it adapts to ever-changing conditions.

Also, technical aspects of organizing such a system should be considered.

As a preproduction system evolves, the priority shifts toward parts, units, and installations, which are made with the help of technologies not requiring preproduction, which include paperless technologies and digital twins that replicate products and production processes.

A traditional company is a unified static system with a rough structure. However, processing standards are changing, and corporate manufacturing systems become as flexible and universal as robotic systems and software tools, which have been in use for a long time. In a traditional company, even a brief production line readjustment is very costly. Contrary to that, a flexible enterprise can easily and quickly reconfigure the production and engineering chain and thus increase productivity, save time and budget resources, and simplify the delivery chain, through the use of big data analytics and artificial intelligence.

A flexible transformer factory is an assembly and production system, which can promptly readjust the production of complex high-tech products without having to develop and produce equipment. This can be achieved through the use of advanced intelligent robots and complete automation of the production process through the use of digital twins. An intelligent platform unifies machinery, conveyors, and other production elements with the help of the Internet of Things.

A flexible enterprise effectively uses its production line's potential, as the line can be reconfigured for production of various product versions and performing a variety of tasks with minimal downtimes and, possibly, greatly simplified supply chains. This flexibility is the result of a new production paradigm that is based on progressing hardware and software and uses big data and artificial intelligence. Particularly, advanced sensors and modern artificial intelligence tools make robotic and automated systems fail-safe, save space, and solve problems previously solved by humans only. Also, availability of 3D printing tools reduces a company's dependence on remote suppliers, and industrial IT solutions enable it to monitor, coordinate, and predict the operation of machinery used in production.

This production system is quite complex and adaptive (Fig. 7.9). According to the basics of cybernetics and cyber-economy, it is impossible to describe this kind of system by way of studying its separate components. Not infrequently, lists of equipment, personnel, and technological process schemes cannot provide a clear picture of how a particular operation is executed. In practice, execution of specific industrial operations depends on decisions made by a large number of persons: from executive workers responsible for carrying out particular operations through engineers, who are to ensure continuous functioning of the manufacturing and technological chain. Each of these persons, who make decisions, receives new information

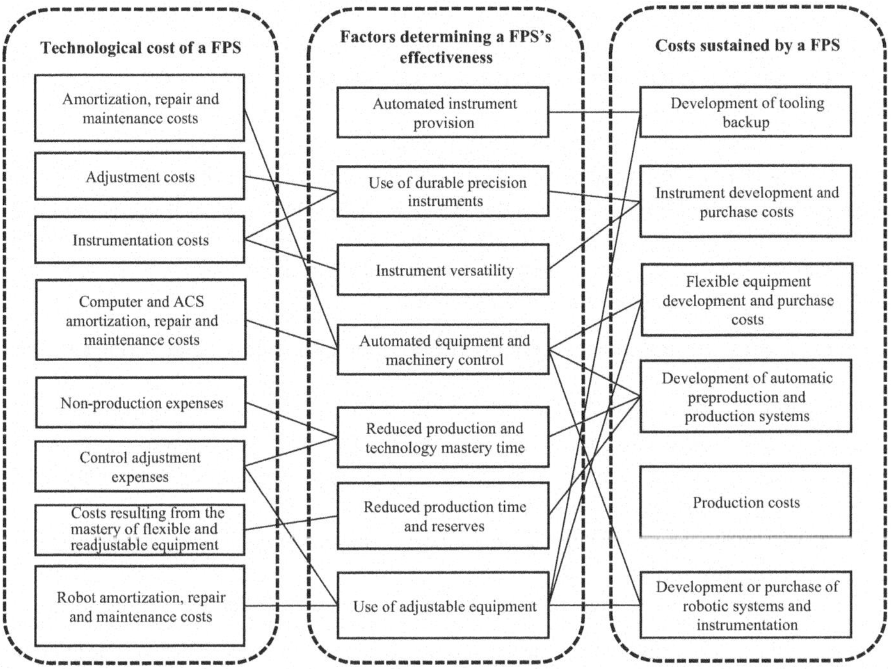

Fig. 7.9 Building an economically effective flexible production system (FPS)

from their circle in a nonstop mode and adjusts his/her behavior in such a way as to reach the highest goals. Within such a complex system, changing operating conditions can exert a nonlinear influence on production, the product's cost, and labor intensity.

An industrial operation can be represented as a succession of activities, each one becoming part of a product's prime cost and labor intensity. This is how the general task of design and engineering preproduction is formed. In order for a product to enter the sales market, it should be able to balance the enterprise's throughput capability against market demand.

The first stage of building an effective preproduction mechanism is development of flexible production elements within an organization to enable production of equipment and instrumentation. This should help the company execute a transition to a new production in a versatile and flexible way and ensure proper, reliable, and effective use of resources throughout the production process. Preparing production tools is part of the preproduction process, which is related to prime and labor cost optimization. This happens because reduction of life cycle and growing diversity of the process results in a continuously increasing share of production tools in a production line's cost. Therefore, the cost management process should include expenses resulting from the purchase of production tools and the choice of an optimal time for investment in production, for example—purchase of cutting and punch tools. These make up a complex technical system consisting of a huge number of one-off parts. However, time study engineers should promptly analyze production tools' cost and outline a cost allocation strategy. Mathematical cost modeling is an effective solution to this problem, as it provides clear and concise indicators generated by parametrical models.

In most branches and industries, precise tool cost calculation is important part of a future project's commercial success; based on such calculation, it is possible to formulate price quotation options and analyze tool purchase prices. In today's highly competitive market environment, a quick, detailed and accurate tool cost evaluation must be executed at the early stages of product development.

Production tool cost evaluation is based on:

- Previously used and checked parametrical models of broad-range tools intended for manufacturing parts
- Algorithms used for achieving various calculation accuracy levels (from quick to detailed evaluation) and modeling different technical and economic scenarios
- Systemic operation of main calculation data and profiles with the help of a centralized database and flexible access system
- Due recording of calculations and change history, options for checking and comparing different versions
- Building clearly structured records and automatic cost breakdown journaling

The design and quality of production tools determine not only its cost, but also the cost of a produced item. Therefore, general cost evaluation implies evaluation of both products' and technological tools' cost. Combined use of cost calculation models for products and production tools (they can work independent of each other) within a fully integrated cost evaluation solution can reveal relationships between a product and production tools' cost, as well as output volume and production tool design. Besides, this integrated solution can collect data on tools' cost while referring to the entire design specification and projects (product portfolios). Changes in production tools' cost result in automatic product or project data updates.

The most effective way of achieving positive effects in preproduction should be the previously mentioned digital modeling, which, along with digital product twin building, can help execute preproduction in a more rational way (Fig. 7.10). Using digital models of products and production processes while building economic cost optimization models for manufactured products can help reduce risks, raise continuity and accuracy of technological solutions at early stages of the life cycle. The main hindrances result from the complexity of preproduction task formulation and a necessity to meet a huge number of different requirements, criteria, limits, growing labor costs resulting from having to provide optimal solutions, etc.

Preproduction in an organization that is going through digital transformation has several ultimate goals (Fig. 7.11), which are related to improving product quality or creating/selecting instruments, production tools, etc., as well as providing effective control of the quality of industrial and manufacturing operations with the help of intelligent diagnostics systems; also, they are related to reduction of prime and labor cost of industrial and management operations through effective selection and

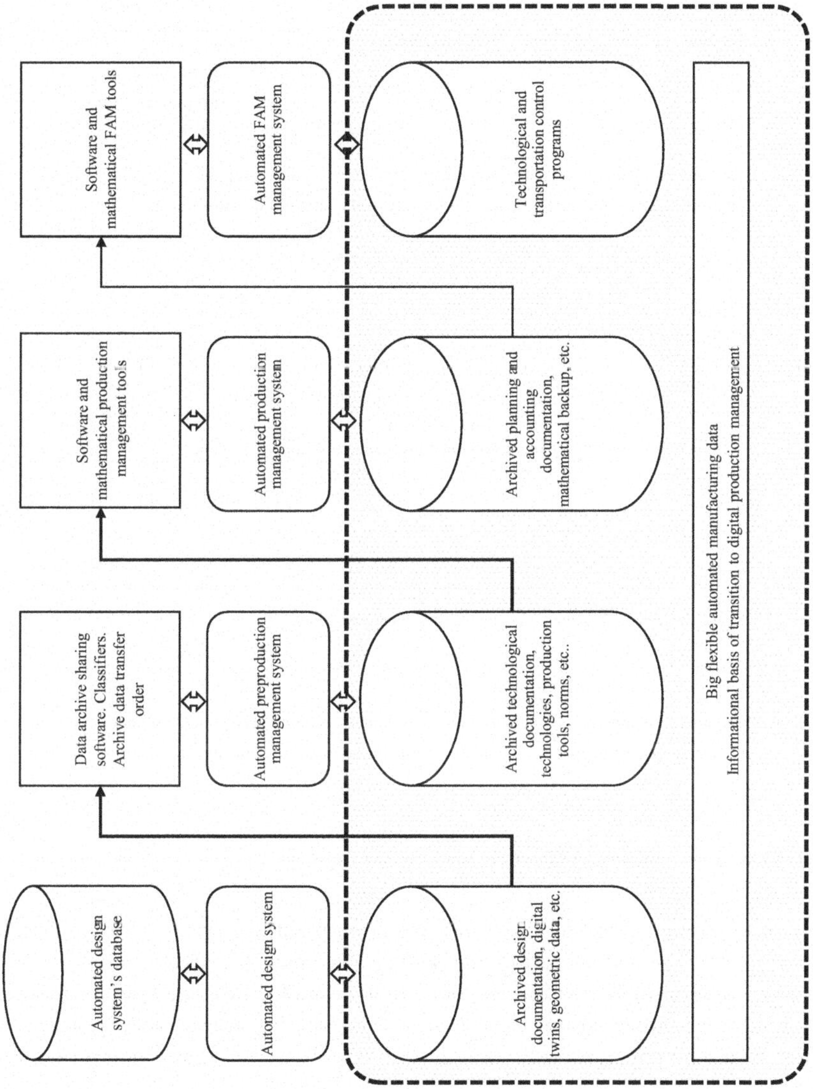

Fig. 7.10 Structure of an integrated preproduction and production management system

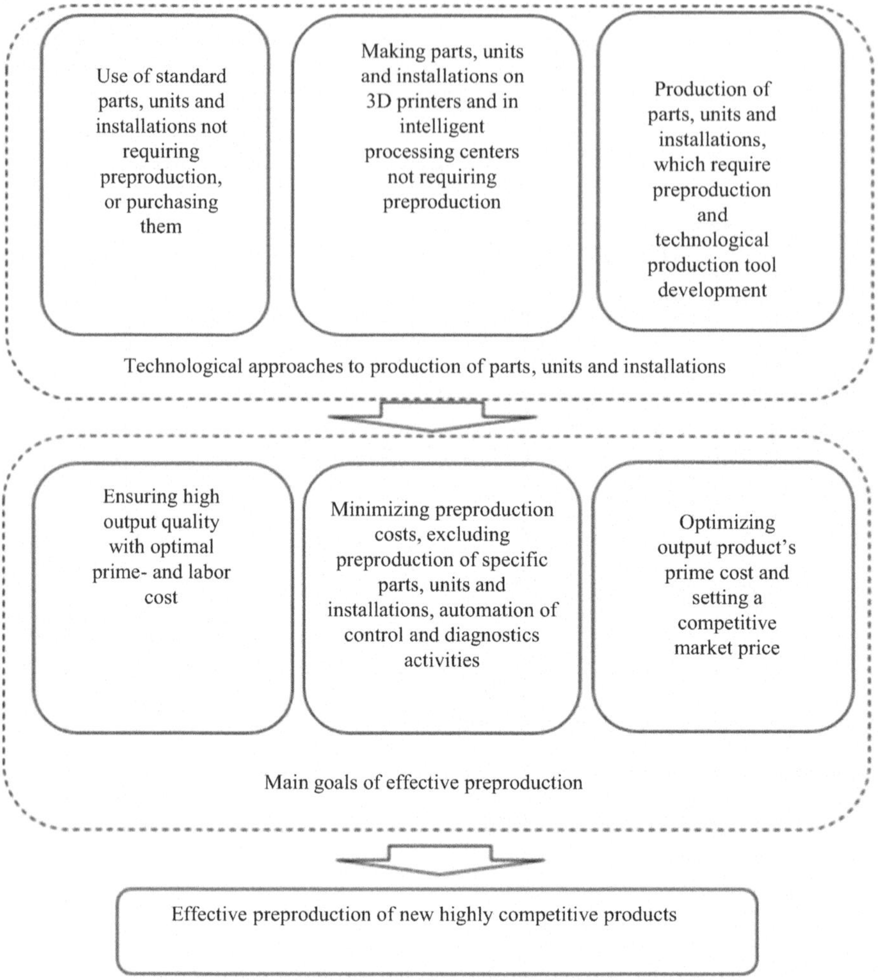

Fig. 7.11 Goals of organizing an effective preproduction system

building of instruments, production tools and equipment; finally, they are related to an output product cost optimization and setting a competitive market price.

There are several mathematical methods that simulate automated preproduction, which should be aimed at improving products' quality along with prime cost reduction, based on paperless technologies that widely use products' and processes' digital twins.

The starting point of an automated preproduction project is described in Sect. 7.1 (Fig. 7.4). This project can be executed in several stages, each implying a product's or technology's evolution through: minimum use of labor-costly parts and units, extensive use of advanced technologies (paperless, additive, intelligent, etc.), use of highly flexible production systems through wide use of advanced systems and methods of supporting design and engineering solutions based on big data analytics, artificial intelligence, automated tooling design, flexible equipment adjustment, preparing for automated assembly, preparing automated output product adjustment systems, testing and diagnostic systems.

The following economic and mathematical model should help select an organizational preproduction pattern with account for a variety of factors and calculate the cost and labor intensity of the preproduction process. The whole product manufactured by the organization will be divided into three groups depending on the preproduction methods used:

- Standard, not requiring any special preproduction or purchased in the form of components.
- Not requiring special preproduction and using paperless digital-twin-based technologies (3D printing, intelligent processing centers).
- Requiring preproduction and development of specific technological production tools. Within a digital organization, these goods can be produced by robotic systems that use digital twins and paperless technologies.

The cost of a preproduction project can be calculated in the following mathematical model.

To evaluate the cost of premanufacture of a new type of goods, it is necessary to consider the general preproduction performance function, which expresses an integrated effectiveness (performance) of preproduction execution through the content and volume of resources used in preproduction and in further production of goods (or purchase of components):

$$Y = F(X_1, X_2, \ldots, X_n),$$

where Y is an integrated effectiveness (performance) indicator (depending on the specifics of an economic problem, it can be nondimensional or dimensional; for example, it can express labor intensity of the preproduction process); X_1, X_2, …, X_n are volumes of n (number) of types of various resources necessary for execution of the production plan, which includes preproduction.

Assume that a particular type of functional dependence is based on an approach, which is used in building a production function. The economic theory has a great number of examples of building adequate production function models, which describe production processes within an organization or even within particular links; therefore, it can be assumed that the model describes a product creation process taking place within a company. Assume that the expected performance of the advanced product development project is Y^0, and it requires respective volumes of resources: $X_1^0, X_2^0, \ldots, X_n^0$, i.e.,

$$Y^0 = F\left(X_1^0, X_2^0, \ldots, X_n^0\right).$$

One of the most renowned and widely used in the economic theory production functions is the Cobb–Douglas production function:

$$Y = AX_1^{\alpha 1} X_2^{\alpha 2} AX_3^{\alpha 3},$$

where $A > 0$ is the technological factor, $\alpha_1, \alpha_2, \ldots, \alpha_n$ are positive values, $\alpha_1 + \alpha_2 + \cdots + \alpha_n = 1$.

Because the model considers three groups of products, the main functional dependence model for the performance of preproduction will consist of three variables: X_1, X_2, X_3, each denoting a particular group. Therefore,

$$Y = AX_1^{\alpha_1} AX_2^{\alpha_2} AX_3^{\alpha_3},$$

where $\alpha_1 + \alpha_2 + \alpha_3 = 1$.

Generally, execution of the technological process has a risk of getting a less pronounced effect from the introduction of a new technology than expected (for example, higher than expected design costs or longer development period due to a necessity to carry out additional work). This risk can be denoted by a random value:

$$0 \le R(I_i) \le 1.$$

Assume that the resource backup of the advanced product development product has a resource-specific degree of internal economic risk. Therefore, values X_1, X_2, X_3 can be treated as random ones. Expected volumes of resources invested in the development of a new product with account for the risk are:

$$M\ (X_j) = X_j^0 (1 - R_j), j = 1, 2, \ldots, n, \tag{7.1}$$

where R_j is the degree of risk related to the backup of production process with an X_j resource.

In this case, the probable value of provision with the resource of the group X_j relative to the planned value of provision with thie resource of group X_j^0, will be a quantitative estimate of the degree of risk in terms of resource provision:

$$R_j = M\left(\left(X_j^0 - X_j\right)/X_j^0\right), \tag{7.2}$$

where X_j^0 is the planned resource provision level; X_j is the probable resource provision value with account for risk factors. In other words, with complete provision of the resource X_j, the degree of risk resulting from the use of this resource will zero.

It is essential to suppose that the degree of risk resulting from, for example, attracting resources from outside, is higher than the risk related to using domestic resources. With these assumptions, the integrated performance indicator of the project Y will also be a random value. In case of pairwise independence between X_1, $X_2, \ldots, X_n$, the will be an equation:

$$M(Y) = A(M(X_1))^{\alpha_1}(M(X_2))^{\alpha_2}(M(X_3))^{\alpha_3}. \tag{7.3}$$

Then, from Eqs. (7.1) and (7.2) it follows:

$$M(Y) = Y^0(1 - R_1)^{\alpha_1}(1 - R_2)^{\alpha_2}(1 - R_3)^{\alpha_3}.$$

The following value will be assumed as one denoting the degree of risk R_Y of failure to fulfill the planned performance indicator Y:

$$R_Y = \max\left(R_j\right). \tag{7.4}$$

With the values X_1, X_2, X_3 and X_{10}, X_{20}, X_{30}, the formula (7.4) can be expressed in the following way:

$$R_Y = 1 - M\left(\frac{Y}{Y_0}\right) = 1 - M\left(\min\left\{X_1/X_1^0, X_2/X_2^0, \ldots, X_n/X_n^0\right\}\right). \tag{7.5}$$

In practice, statistical or expert methods are used to evaluate risks. The more complex a technological process is, the greater number of elements it involves and the higher the risk of an unplanned increase in resource consumption is. Figure 7.12 demonstrates a rough example of a process, where the product requires development of specific production tools and accessories. Each component shown in the example can be a source of risk of high preproduction and production costs. 3D printing can move the product to the non-preproduction group and thus eliminate the sources of risks and make the technological process more sustainable.

Assume that a company is planning to implement three types of manufacturing processes (standard ones, not requiring preproduction and requiring production of specific tools), which ensure achievement of high quality characteristics:

$$I_1, I_2, \ldots, I_N.$$

It can be assumed that the choice of a manufacturing process determines a K number of various technical and economic product characteristics. Each process is characterized by an expected preproduction cost:

$$V = V(I_i).$$

Time required for execution of production operations within the technological process:

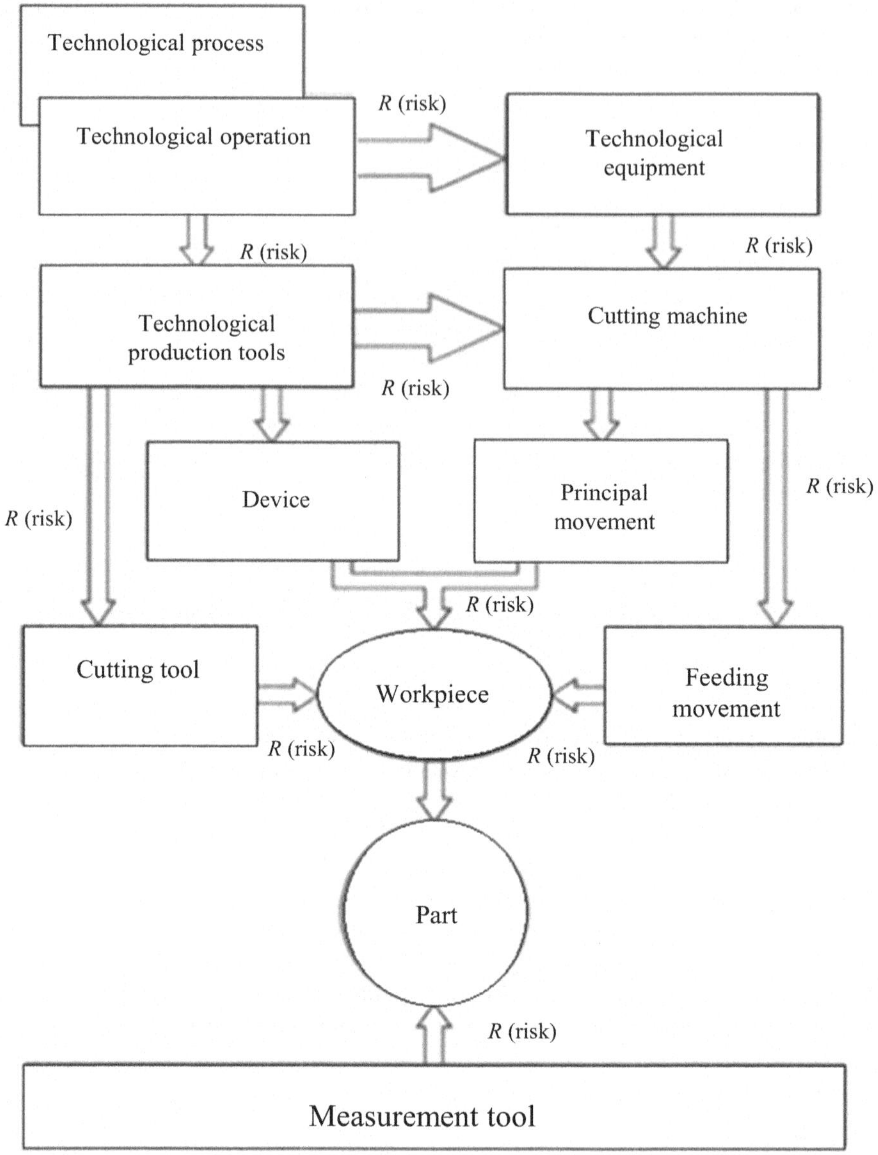

Fig. 7.12 A rough scheme of production with a cutting machine featuring specific production tools and accessories

$$T = T(I_i).$$

Thus, the process can be described by the following vector:

$$I_i = \begin{pmatrix} V_i \\ T_i \\ R_i \end{pmatrix}.$$

This vector is used in the economical and mathematical preproduction optimization model, which ensures optimal cost of manufactured goods.

An economical and mathematical preproduction optimization model, which should give the product expected quality characteristics, can be formed as follows:

$$M = \min_{I_{i_1}, I_{i_2}, \ldots, I_{i_L}} \|V, T, R\|. \tag{7.6}$$

In this mode, M is the performance of the preproduction process, which expresses a maximum effectiveness of product cost optimization and quality preservation technologies applied.

Next, the concept of integrated product quality indicator should be defined, which is described by a vector of product quality indicators.

The quality ratio K_i according to a chosen technical and economic characteristic can be calculated according to the following formula:

$$K_i = \frac{q_i}{q_i^0},$$

where q_i is the value of the individual i quality indicator of the product, derived from a multitude of individual indicators (component of vector Q); q_i^0 is the value of the individual i quality indicator that can be achieved through the use of technologies mastered by organizations, derived from a multitude of individual indicators.

Next, it is necessary to compare private product quality indicators to those reflecting scientific and technological achievements concerning a particular type of product. The ratio can be calculated according to the following formula:

$$K_i' = \frac{q_i}{q_i'},$$

where q_i' is the value of the individual i quality indicator derived from a multitude of those reflecting scientific and technological progress.

The integrated product quality indicator is calculated as follows:

Table 7.2 A reference table reflecting a possible choice of technologies

	Standard part/ purchased as components		Does not require preproduction (3D printing, automatic processing centers)		Requires preproduction and development of specific production tools	
Part/unit/installation	Value	PRisk	Value	RRisk	Value	PRisk
D_i	V_i^1	R_i^1	V_i^2	R_i^2	V_i^3	R_i^3

$$M_i = \sum_{j=1}^{m} \alpha_j \left(K_j + K_j' \right),$$

where m is the number of recorded individual indicators; α_i stands for weight coefficients reflecting each indicator's significance, $\sum_{i=1}^{m} \alpha_j = 1$.

This integrated quality indicator algorithm can help select a cluster, in which technologies are going to play an even more important role in determining a product's quality. The final choice of a technology, which ensures sufficient competitiveness, will be made with account for the degree of risk.

The final choice of a technology and respective cluster will be made based on the calculated M indicator, which demonstrates a maximum effectiveness of the technological process aimed at product cost optimization and quality preservation. The M indicator can be calculated according to the following formula:

$$M = \max_{I_1, I_2, \dots, I_S} \left(M_i \cdot V \cdot (1 - R_i) \right),$$

where $i = 1, \dots, S$.

Here is a reference table reflecting a possible choice of technologies needed for execution of a company's production plan, which can help achieve target product quality indicators (Table 7.2).

In practice, organizations manufacture a broad portfolio of products, which can include hundreds or thousands types of items. Therefore, combinations of various technological approaches come as big data, which can be effectively analyzed with the help of artificial intelligence and machine learning methods. The total production performance Y, in this case depending on the type of the preproduction process (based on the data shown in the reference table) can be calculated in the following way:

$$Y = \left(\sum_{i=1}^{l} V_i^1 \right)^{\alpha_1} \cdot \left(\sum_{j=1}^{m} V_j^2 \right)^{\alpha_2} \cdot \left(\sum_{k=1}^{n} V_k^3 \right)^{\alpha_3} \cdot \left(1 - \max_{l} R_l^1 \right) \cdot \left(1 - \max_{m} R_m^1 \right) \cdot \left(1 - \max_{n} R_n^1 \right);$$

$$\alpha_1 = \frac{\sum_{i=1}^{l} V_i^1}{\sum_{i=1}^{l} V_i^1 + \sum_{j=1}^{m} V_j^2 + \sum_{k=1}^{n} V_k^3},$$

$$\alpha_2 = \frac{\sum_{j=1}^{m} V_j^2}{\sum_{i=1}^{l} V_i^1 + \sum_{j=1}^{m} V_j^2 + \sum_{k=1}^{n} V_k^3},$$

$$\alpha_3 = \frac{\sum_{k=1}^{n} V_k^3}{\sum_{i=1}^{l} V_i^1 + \sum_{j=1}^{m} V_j^2 + \sum_{k=1}^{n} V_k^3}.$$

The choice of an optimal preproduction project execution pattern is determined by its maximum performance indicator with a total sum (monetary) of all resources that are necessary for preproduction and production; the sum should not exceed the set value V, which is the high-end prime cost threshold. Thus, the optimal solution meets the preproduction pattern, which provides for the expected prime cost value with set quality parameters achieved:

$$Y = \left(\sum_{i=1}^{l} V_i^1\right)^{\alpha_1} \cdot \left(\sum_{j=1}^{m} V_j^2\right)^{\alpha_2} \cdot \left(\sum_{k=1}^{n} V_k^3\right)^{\alpha_3} \cdot \left(1 - \max_{l} R_l^1\right) \cdot \left(1 - \max_{m} R_m^1\right) \cdot \left(1 - \max_{n} R_n^1\right) \to \max;$$

$$\sum_{i=1}^{l} V_i^1 + \sum_{j=1}^{m} V_j^2 + \sum_{k=1}^{n} V_k^3 \le V;$$

$$\alpha_1 = \frac{\sum_{i=1}^{l} V_i^1}{\sum_{i=1}^{l} V_i^1 + \sum_{j=1}^{m} V_j^2 + \sum_{k=1}^{n} V_k^3},$$

$$\alpha_2 = \frac{\sum_{j=1}^{m} V_j^2}{\sum_{i=1}^{l} V_i^1 + \sum_{j=1}^{m} V_j^2 + \sum_{k=1}^{n} V_k^3},$$

$$\alpha_3 = \frac{\sum_{k=1}^{n} V_k^3}{\sum_{i=1}^{l} V_i^1 + \sum_{j=1}^{m} V_j^2 + \sum_{k=1}^{n} V_k^3}.$$

The value $W = \sum_{i=1}^{l} V_i^1 + \sum_{j=1}^{m} V_j^2 + \sum_{k=1}^{n} V_k^3$ is the total cost of the preproduction (also in case whereby creation of production tools is necessary) and production process.

Therefore, product cost optimization with effective preproduction systems used can be achieved through:

- Switching to automated technologies, which remove the necessity to create production tools and accessories (through the use of 3D printing and intelligent processing centers)
- Digitization of product processing (through the use of intelligent robots, intelligent manufacture execution systems, etc.)
- Designing products and, if necessary, tooling the process to a set cost
- Reducing production costs through the use of flexible production systems

Execution of each of these stages of preproduction evolution results in lower preproduction costs that therefore decreases output prime cost and helps establish a more competitive market price.

In a digital company, this can be achieved through intertwining production and product design technologies, including technical process engineering, simulation modeling and production management within a unified PLM platform. This platform automates the preproduction process and reduces its duration by a parallel execution of the design and engineering process, and it reduces production costs by, for example, running automated tooling preparation. New approaches to preproduction, which use advanced software, allow production engineers to promptly find solutions and optimize labor and process execution costs. Designing products, preproduction, and production are interconnected (Fig. 7.13).

Improvement of a product's structure, which implies exclusion of labor-costly parts or units, requires new approaches to preproduction and production, as they use additive technologies, 3D printing, intelligent systems, etc. This results in gradual and evolutionary decrease in the amount of preproduction work, and it moves to a specific production process. Therefore, retaining high quality of the production process reduces costs, which influence its bottom-line price. Effective execution of preproduction and production, which necessitates wide use of intelligent big data processing methods and artificial intelligence, is crucial to designing new products to their specified cost and competitiveness.

During the transition to Industry 4.0, manufacturers who are able to manage preproduction of a complex technical system using cost-effective technologies in the best possible way, will have a sustainable competitive advantage over those practicing traditional production management and preproduction patterns. Modern preproduction patterns rely on extensive use of intelligent systems, which play the role of an automated system with a cloud architecture, in which decisions are made on the basis of big data analytics with the help of artificial intelligence and machine learning. Therefore, the use of big data relating to execution of production and manufacturing operations within a company, which is processed with the help of

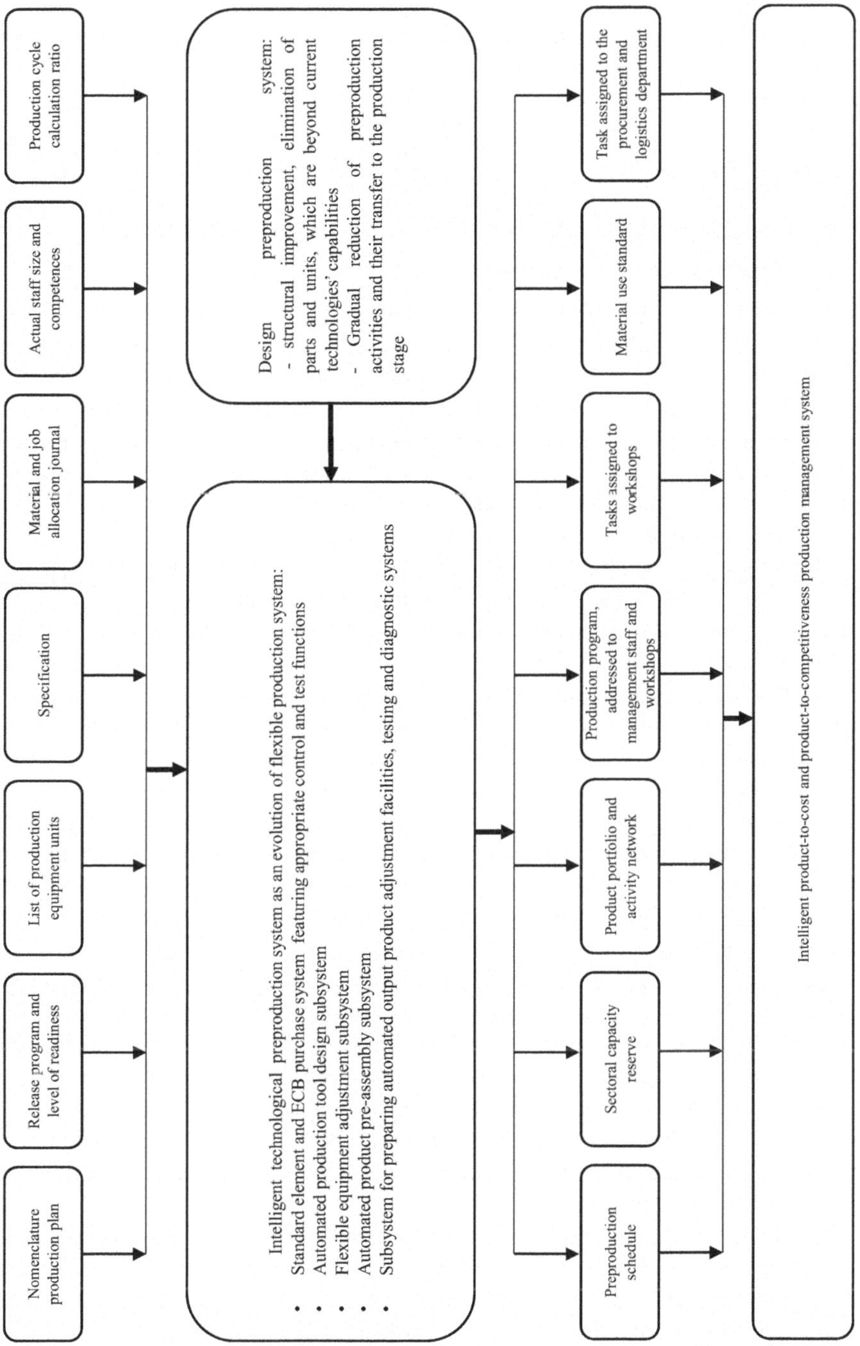

Fig. 7.13 Automated preproduction and production management system

artificial intelligence, helps evaluate a production system's economic state and solve the task of its adaptation to a new type of production, without any use of manpower. This helps reduce preproduction's and production's labor cost and output products' prime cost. More detailed reviews of intelligent preproduction and production of new types of goods will follow.

7.3 Building an Intelligent Automated Preproduction Management Systems

As previously concluded, creation of an advanced preproduction system should rely on intelligent management systems, which broadly use big data analytics, as well as artificial intelligence and machine learning to enable automatic decision-making processes in the sphere of preproduction and production of new types of goods. As shown in Fig. 7.3, a large nomenclature of manufactured products, changes in technical documentation, procurement, result in having to do a lot of calculation and evaluation when planning and during preproduction.

Intelligent preproduction systems can automate most preproduction activities. Also, an effectively organized preproduction system should not undergo serious adaptation, should it be necessary to launch production of a new product; however, it should be able to adapt to the specific of the production process. The need for a highly flexible preproduction system is predicated by increasing personification of demand, when customers wish to obtain a product boasting individual characteristics that are not related to any physical principles of its functioning or software (for example, the color of the body or a signature). The preproduction process should not undergo any transformation. An intelligent preproduction system should automatically adjust the technological process and properly tune the execution of manufacturing operations. The economic effectiveness of this approach to preproduction lies in not having to hire personnel or carry out labor-costly activities. This excludes preproduction costs.

Another case of preproduction system's adaptation consists in endowing a product with a new competitive property. In this case, solving the preproduction cost minimization issue is considered in interaction between preproduction and product design. Part of the preproduction of a product having a new competitive feature remains unchanged with respect to the old-generation product. Changes, which are directly related to implementation of the new competitive feature, are occasional and usually do not require labor-costly operation, which the intelligent system cannot perform in an automated mode.

Finally, development of ultimately new product that uses new physical principles may require development of new technological processes; their successful implementation is closely related to development of new competences within an organization. Costs resulting from the advancement of the preproduction system will contribute to its evolution and more effective incorporation of new competitive qualities into output products.

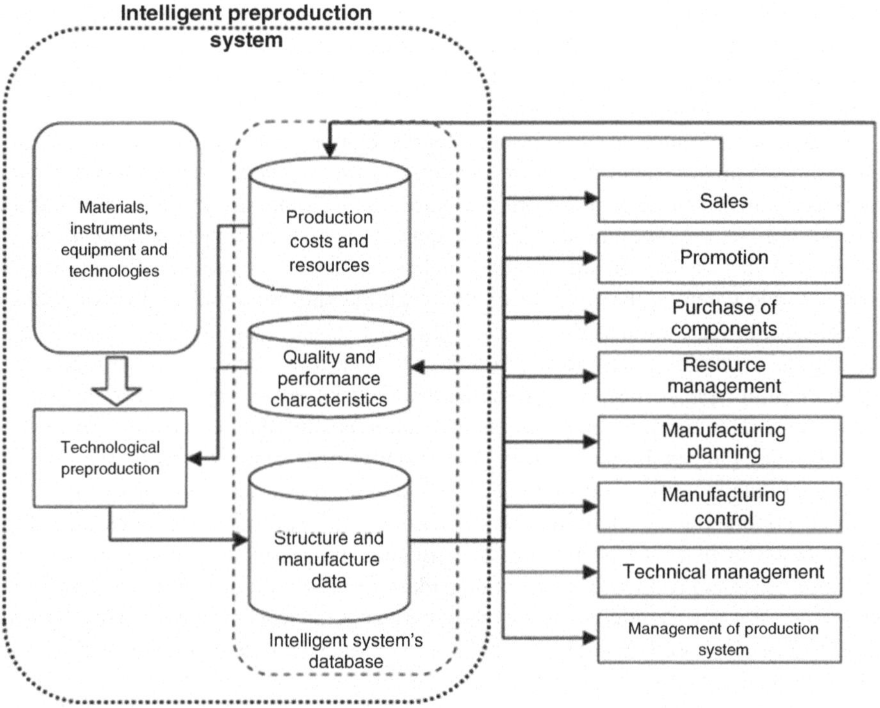

Fig. 7.14 An intelligent preproduction system as a corporate source of big data

The informational base of an intelligent system of preproduction of a new product consists of production data derived from corporate communication systems, sensors, and devices featured in the equipment, as well as from the global information space. Variable data, which depends on the effectiveness of the preproduction process, influence the effectiveness of most divisions (Fig. 7.14). An intelligent preproduction management system is an extremely important source of big data, which can be used by various corporate functional services.

In traditional production systems, production is presented as a relatively static system. Process engineers, who manage preproduction in such systems, determine various types of input data for the system and achieve the most effective production possible. For example, when building an assembly line, the most important thing is even distribution of resources over a certain segment of the value chain. The assembly line allows for a maximum throughput capability, as it minimizes the likelihood of a human error by limiting human labor. This environment helps each employee to do more work, do it quickly and better, without needing a lot of training. The concept of assembly line production becomes effective when input data and system requirements are static and clear (i.e., they change over time).

However, nonstop emergence of new technologies and their constant evolution extends design and engineering horizons, making them available for manufacturers.

Every new technology introduced into a production system results in an exponential growth of the number of possible combinations, which should be taken into consideration when doing a design and technological preproduction to ensure an optimal engineering and manufacturing process.

Today, raw product and component supply chains have no geographical limits and are distributed around the globe. Separate components can be produced on the spot or obtained from several suppliers at once. In this context, simple understanding of a system of engineering and manufacturing operations and interdependences within it is impossible. Besides, static understanding of an industrial operation is only possible when a relatively stable high-volume and low-confounding operation is analyzed. In fact, industrial operations should withstand any disturbance within a system and changing customers' preferences. Therefore, there is no optimal static preproduction option, because interrelated industrial subsystems never go into a static state. The modern preproduction system is one that constantly adapts in response to changes in the surrounding environment resulting from the use of intelligent data analytics and making effective managerial decisions based on it.

Such understanding of industrial operations and preproduction steps aimed at their execution makes a foundation for effective preproduction management. For example, the lean production concept considers a production operation as a complex adaptive system, as it reflects the fact that there is no universally optimal implementation option. It should be noted that an industrial operation is regarded as a dynamic set of embedded subsystems, which never achieve an optimal state (according to theoretical lean production assumptions); however, the intelligent search for it and, consequently, preproduction correction, goes on without interruption. For this reason, the lean production system includes feedback loops, which ensure nonstop monitoring of the manufacturing and engineering process (these feedback loops are largely definitive of a process's cost and labor intensiveness); these include throughput capability, cycle length, improvement and reject level. Smart monitoring and constant improvement of these indicators enables a dynamic and uninterrupted preproduction, and the manufacturing and engineering system evolves toward a fairly effective state.

Despite the huge number of technological innovations introduced in production, instruments used by leading enterprises, which they use to reveal, monitor, and constantly update manufacturing and engineering operations, remain unchanged. Their optimization and switching to paperless modes offers a substantial resource for product cost optimization through more effective preproduction. Creation of an information and software infrastructure for collection and distribution of important business data is not a novelty, and is known as *manufacturing execution system* (MRS).

Manufacturers often rely on MES to ensure effective processes and better control over them. The biggest downside of this approach is that MES systems execute preproduction while striving to encompass an entire production operation when designing a product. Effective complexity is based on clay–clay data models, which require that a software engineer to have a deep understanding. Once such systems are implemented in a workshop and begin to face the business reality, changes are inevitable. These changes should be managed by the software engineer, who may not have business expertise. Because processes constantly adapt to the ever-changing environment, changes in a MES system should be nonstop too; this requires using a versatile toolkit, which enables them to accept these changes and manage the preproduction process in a decentralized mode.

This toolkit, which contains flexible instruments used in the execution of the preproduction process, presents a new generation of software solutions (for example, Tulip, which uses goods produced by top companies, such as Siemens, Toyota, Merck, etc.) Instead of developing a universal preproduction solution, which uses a prescriptive pattern of managing a manufacturer's routine operations, the flexible platform uses a set of digital tools to pass the changes over into the process and for real-time monitoring of the changes' consequences. Changes can be generated by the process engineer or by an intelligent system, which makes managerial decisions in an automated mode. Neither the process engineer nor the system has to use a specific programming code, because the artificial intelligence produces effective algorithms, which are formed from data coming in from computers, sensors, and server systems installed in factory buildings.

This enables process engineers to promptly incorporate improvements into engineering and manufacturing processes, monitor these changes' influence on their cost and labor intensity and make decisions based on original data. Applications can be organized into a unified complex or deployed separately by scores of engineers working on different problems. Further improvement of these systems results from the automation of decision-making procedures in the preproduction sphere, which are governed by an intelligent preproduction system.

Analysis of literature reports shows that introduction of an advanced preproduction system implies running a complex project, which requires development and transformation of many employees' competences. Introduction and use of an effective preproduction system is extremely important given the diverse nomenclature of manufactured products, which use a large number of structural options and production technologies of different complexity. For example, the design-to-cost approach requires a continuous preproduction and entails production costs. Ideally, engineers (designers and technologists) should accept responsibility for products' prime cost. Because not all engineers have expertise in the economy of new goods production, cost control and optimization should be performed by automated intelligent systems, which have knowledge and, depending on circumstances, can draw logical conclusions that form a basis for effective decisions.

The main procedures carried out as part of production engineering refer to effective designing of a product and manufacturing technology. Figure 7.15 shows general preproduction and production steps executed by an intelligent system that is part of a universal new production launch system.

It manages subprocesses, such as design of a product, manufacturing technology, prime cost calculation with account for preproduction costs, running production records, and backup. Within this intelligent system, a decision to manufacture a new type of goods that has a certain structure and manufactured according to a certain type of technology, is made with reference to acceptability of production costs, as well as the prime cost and the product's qualitative characteristics.

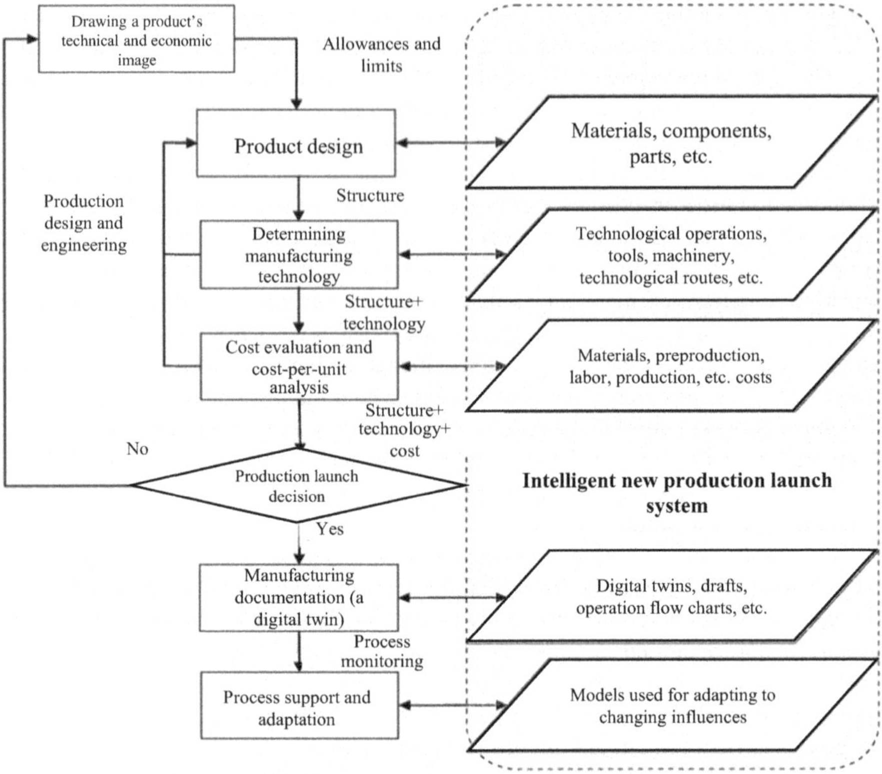

Fig. 7.15 Intelligent production launch system

An intelligent automated preproduction system should reduce the influence of cost factors through analyzing such parameters as overproduction, idle time, irrational transportation and distribution, unnecessary processing and reject rates. These sources of losses exert direct and indirect influence on general expenses and affect a product's market price.

The intelligent system influences the following key indicators:

- Increased productivity
- Reduced life cycle
- Reduced reject rate
- Increased product quality

Various data on materials, manufacturing operations, tools, process routes, etc., comes into a production launch system. The same type of data is used in digital twins of products and processes. Integration between the intelligent production launch system and products' digital twins (Fig. 7.16) implies formation of data not only for an effective real-time launch of a new production process. Also, it implies accumulation of information in data pools, so that they can be used later when launching future projects.

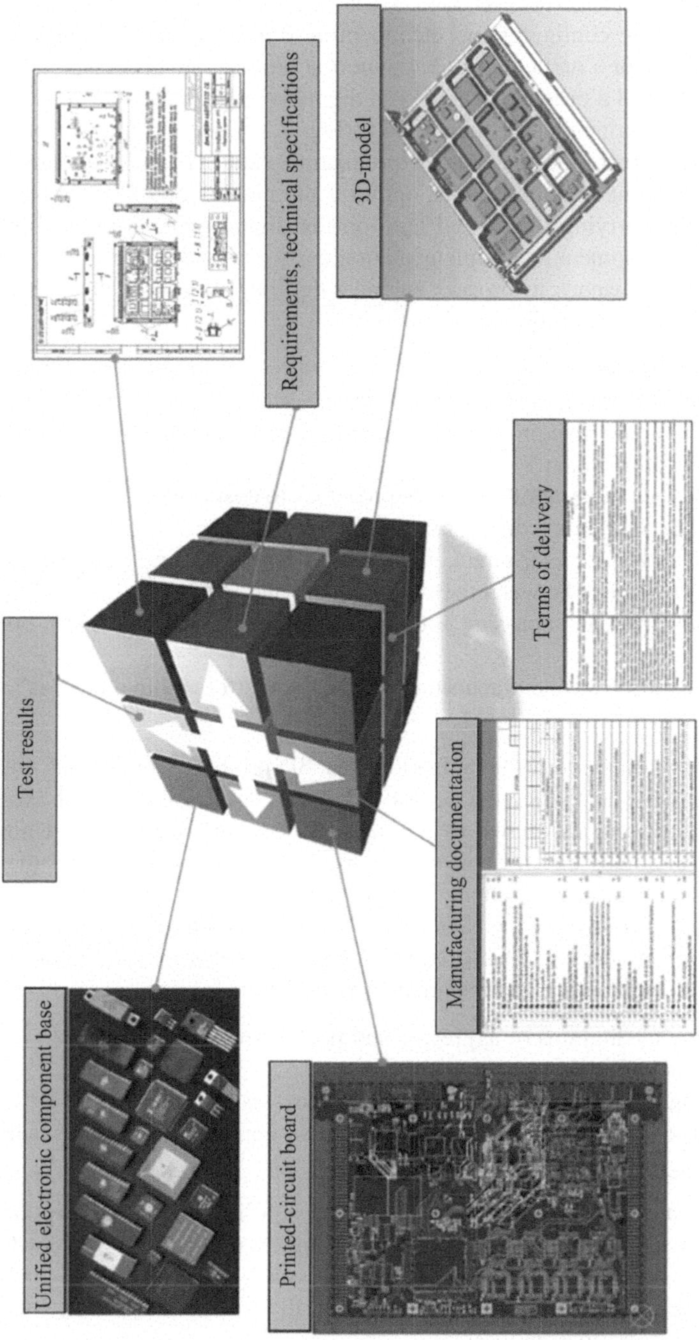

Fig. 7.16 An example of a digital twin's architecture

A digital twin of a product carries out real-time telemetry collection and stores product history and state records, data on various process indicators collected from sensors, operations, configurations, etc., mentioned in Fig. 7.16. This helps create a data architecture for a particular digital twin of a product and process.

A digital twin of a product, as long as a digital twin of a process is implemented with account for principles of Industry 4.0 (Fig. 7.17) and the Internet of Things, helps execute a real-time analysis of production data and, consequently, reveal bottlenecks, prevent structural faults, optimize the number and labor intensiveness of real tests by carrying out virtual tests on the digital twin. Practice shows that digital twin development and implementation costs pay back quite soon.

Advanced approaches to process engineering with the use of digital twins are going to develop in the following directions:

- Broader presentation and integration of all aspects of production of mechanical and electrical units, control systems, software, etc.
- Ensuring more flexible process of twin modeling of manufacturing processes (for example, heat treatment)
- Better all-chain integration of supply processes to ensure concerted creation (and updating) of a process's digital twin
- Broader opportunities for direct connection to manufacturing equipment to minimize the difference between the virtual model and reality
- The digital twin encompassing equipment's life cycle (preproduction, operation and maintenance)
- Better integration of manufacturing and quality control within a process's digital twin

Progressing computer, artificial intelligence, 3D printing technologies, introduction of affordable and effective sensors, and big data analytics tools will help integrate within a process digital twin of all processes described above. Unification of these technologies can help nearly all machinery businesses create their own digital twins and automate production engineering processes (Fig. 7.17).

Centralized software solutions are worth using in preproduction when there is a need to design processes, which require interaction between various objects and subjects of an engineering and manufacturing process:

- Developing cyclograms of aggregate and final assembly
- Assigning shop-to-shop routes
- Developing aggregate and final assembly processes
- Verifying technological processes for correct assignment of components from the engineering or modular bill of material

An aggregate and final assembly cyclogram is a hierarchical description of an assembly process succession with assigned components. Each process within the cyclogram, apart from unit and installation assembly subprocesses, includes a manufacturing plan, which is a storage unit for all process information concerning a manufacturing or repair process.

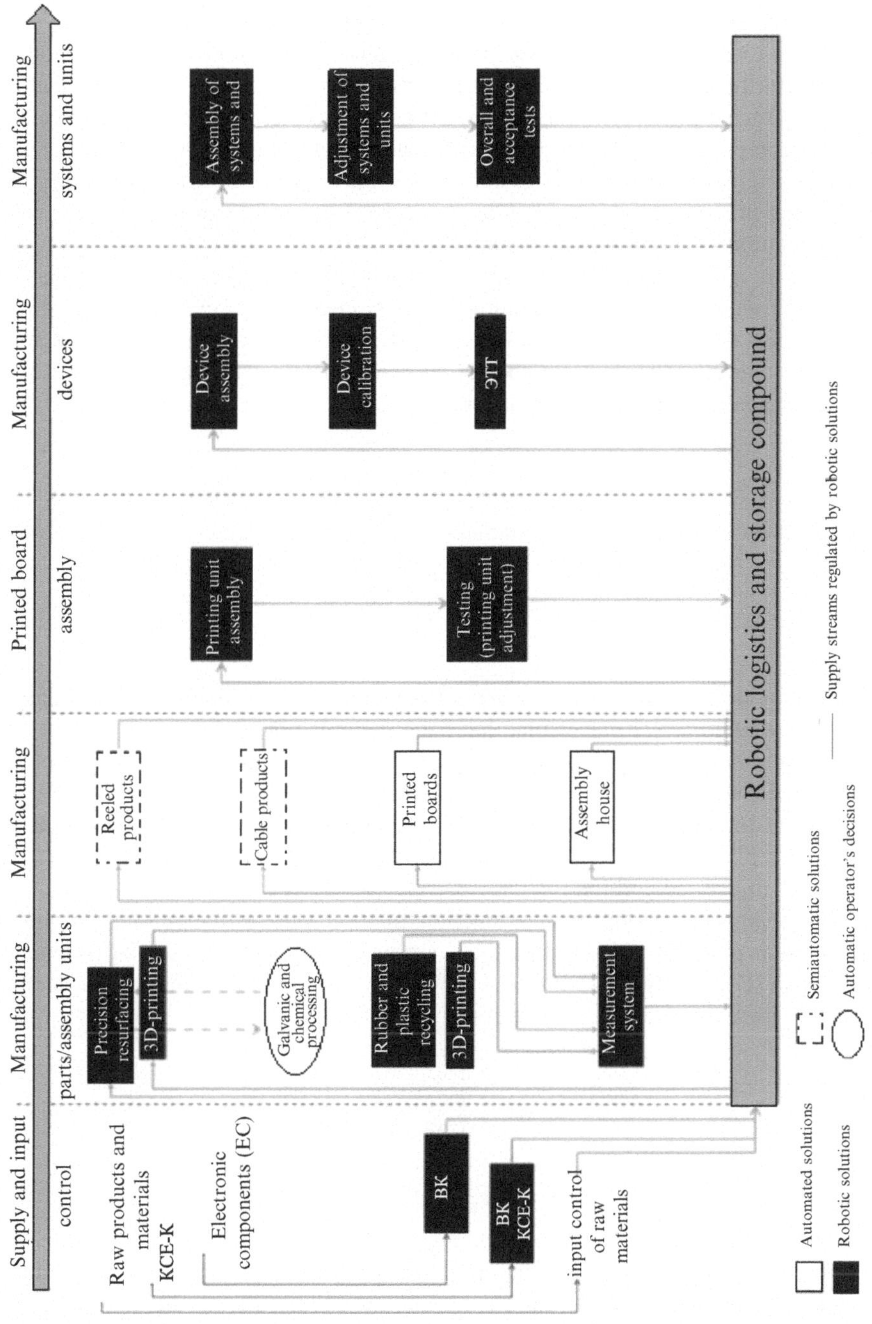

Fig. 7.17 Automated preproduction of electronic devices

To solve the problem of production engineering of a new product, it is necessary to use mathematical manufacturing method definition models, on which the automated decision-making process within an intelligent system is based.

There is a mathematical algorithm of automated choice of a method of manufacturing new products, which the intelligent system uses to launch production.

Based on a digital version of the project documentation, as well as objective data on the company's engineering and manufacturing potential and competences, a process engineering standard for the new product can be formulated. Next, an intelligent procedure, which relies on a Hopfield network, can help find the most suitable automated production method.

There is a formal model for selection of technology at the process engineering stage. For each project stage, required or recommended parameters can be developed, which describe the result of using a particular production method (costs, time, etc.). Assume there is an N number of such parameters. Then an N vector can be introduced to describe these parameters:

$$x = (x_1, x_2, \ldots, x_N),$$

where each component can take on values:

$$x_i = \begin{cases} +1 \\ -1 \end{cases}.$$

Assume that the value +1 means that the i requirement is applied to an engineering process, and $a - 1$ means that the requirement is not applied.

For automatic selection of a production method, a Hopfield network should be used. This network is a single-layer network of neurons, where each neuron connected to all the rest ones. The Hopfield network principle means that the network replicates the associative memory mechanism, which can be used effectively for image recognition.

The search for an optimal production method can be regarded as an image recognition process. Assume that there are an M number of potential production methods (3D printing, referring to a robotic processing center, etc.). Each method caters to its own vector of parameters:

$$\begin{aligned} x^1 &= \left(x_1^1, x_2^1, \ldots, x_N^1\right); \\ x^2 &= \left(x_1^2, x_2^2, \ldots, x_N^2\right); \\ &\ldots \\ x^M &= \left(x_1^M, x_2^M, \ldots, x_N^M\right). \end{aligned}$$

Next comes a description of requirements applied to the output product, which is expressed in a vector:

$$y = (y_1, y_2, \ldots, y_N).$$

The image recognition task is formulated in the search (recognition) of such a vector x^k, which is in a way close to the vector y. To solve the task in an intelligent space system, it is recommended to use a Hopfield network.

A Hopfield network is used in two stages:

1. Teaching the network
2. Production method recognition

These stages can be described mathematically. Teaching consists in building a matrix of neural connectivity weighs:

$$W = \begin{pmatrix} w_{11} & \dots & w_{1N} \\ \dots & \dots & \dots \\ w_{N1} & \dots & w_{NN} \end{pmatrix}.$$

The teaching procedure includes several steps:

Step 1: Assume that all weights are zero.

$$w_{ij} = 0.$$

Step 2: Perform a weight teaching procedure for each performer vector:

$$w_{ij} = w_{ij} + x_i^m x_j^m, \quad m = 1, 2, \dots, N.$$

Step 3: Assume

$$w_{ii} = 0, \quad i = 1, 2, \dots, N.$$

Step 4: Set norms:

$$w_{ij} = \frac{w_{ij}}{N}.$$

The recognition (selection) of an optimal production method lies in performing the following procedure. Assume that the project task is expressed in a vector:

$$y = (y_1, y_2, \dots, y_N),$$

Then the following algorithm should be used:

Step 1: For each j from 1 to N:
Assume
$d = 0$.
Step 2: For each i from 1 to N:
Assume
$d = d + w_{ij} y_i$.

Step 3: End of cycle (i).
Step 4: If
$d > 0$,
it should be assumed that
$z_j = 1$,
otherwise,
$z_j = -1$.
Step 5: End of cycle (j).
The vector is built:
$z = (z_1, z_2, \ldots, z_N)$.

If the vector z coincides with a potential production method, this method is found with the help of the algorithm.

Step 6: Assume y = z
Step 7: See Step 1.

Thus, the intellectual algorithm of a Hopfield network's work helps determine the most optimal way of making a new product.

A theoretical basis for a Hopfield network is a recurrence equation:

$$z_{k+1} = F(W_{z_k}), \quad k = 0, 1, \ldots,$$

which reveals an optimal variant.

While using a Hopfield network, the procedure may reveal several production methods, providing they have the same signature mark for required parameters. In this case, it is necessary to perform additional assessments of the effectiveness of production, possible with the use of new criteria.

Existing production management and process engineering systems do not provide users with any clear economic recommendations, which should automatically adapt to rapidly changing internal and external influences. However, the current level of technical and technological development is sufficient for creating ultimately new types of products—intelligent preproduction systems, which, through complex use of a company's entire communication infrastructure, deeper and more thorough processing of industrial data coming in from sensors and devices featured in production equipment using intelligent mathematical algorithms, create a new image of the digital production launch process. Also, exact economic problems are solved automatically with a high integrity. Effective organization of the process can be achieved through collection and integration of the best competences (including those developed in branch competence development centers), innovative technologies and technical solutions, the latest scientific and technical achievements generated in universities and the Academy of Sciences.

Therefore, preproduction may imply selection of the best competences and technologies both at a corporate and divisional level, as divisions can exert the greatest influence on a product's competitiveness because, on the one hand, it is going to be unique and one-of-a-kind and, on the other hand, boast optimal cost values. Together, these factors will make the product competitive on a global scale. An optimal scheme of an intelligent preproduction system is shown in Fig. 7.18.

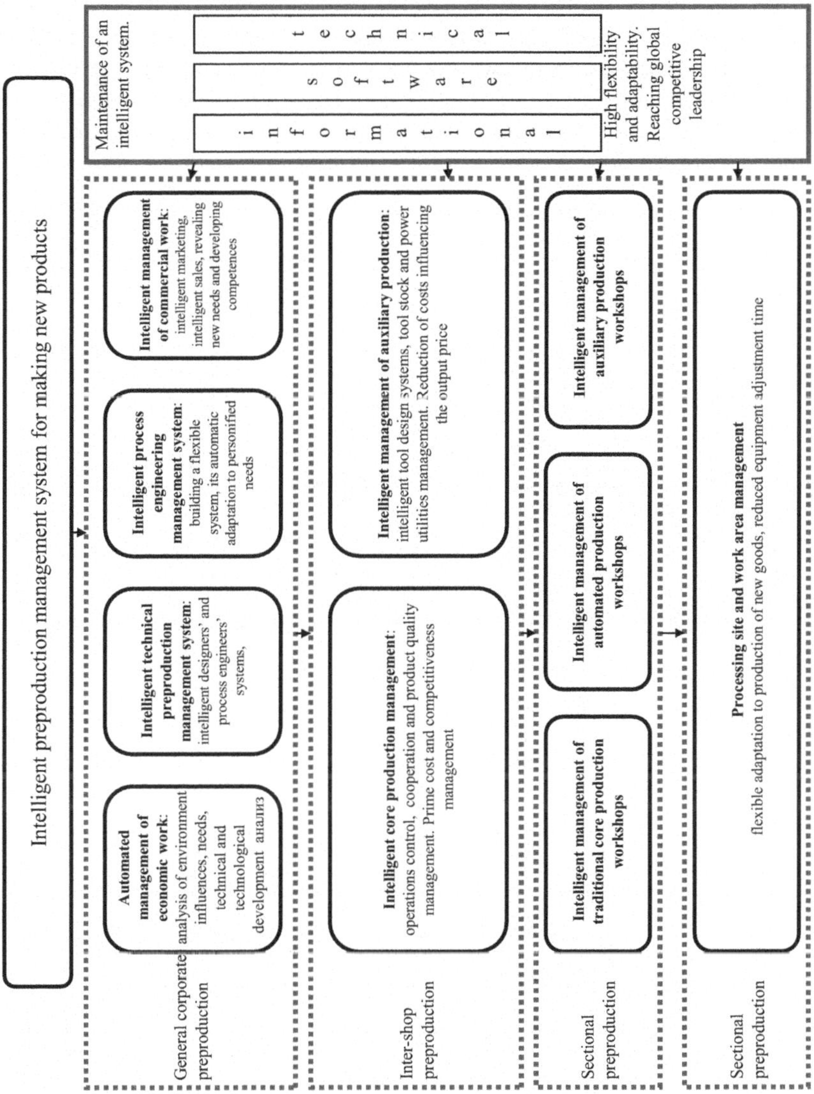

Fig. 7.18 Intelligent preproduction management system for making an advanced product

The potential of intelligent industrial data processing can significantly raise the economic effectiveness of using corporate resources. Intelligent preproduction management systems, combined with intelligent process engineering and control systems (will be reviewed later) make up a whole system of design-to-cost production system. The economic basis of this system's effectiveness is intelligent processing of design, production and other data, which characterize the internal and external environment; this helps optimize the product's labor costs and other expenses, which influence the market price. This multiplicative effect of digital transformation of an organization increases its competitiveness and lays the groundwork for global competitive leadership.

Chapter 8
Modern Manufacturing Process Management Methods

8.1 Setting Up Flexible Automated Manufacture Processes Relying on Digital Technologies' Advantages

Given the in-depth digitization and automation of various processes executed at different life cycle stages (automated product design, digital twin building, enabling non-preproduction cycles through incorporation of automated artificial-intelligence-driven preproduction tools, etc.), influenced by rapidly evolving high-tech product market, with increasingly rapid updates, organizations and tasks, which are solved during production, are undergoing a transformation (Fig. 8.1).

Transformation of an organizational framework, which also influences the flexibility of a manufacturing process (Fig. 8.1), is caused by a group of circumstances. A company built on a functional basis, boasting high flexibility and high integration of the know-how into the manufacturing process, can promptly adapt to changes in the production line and output volume. However, this takes a lot of time, during which the product may lose its competitive advantages that were embedded in its technical and economic image.

Given the increasing customer orientation and a need to speed up the release of a new product, businesses tend to choose the divisional pattern, which shows a higher flexibility, quick response to changes in the outer environment, and a closer relationship between the business and customers.

Today, a matrix type of organization is evolving rapidly, within which factory area networks, business cooperation groups, and competence centers originate, which focus on the production of unique and popular goods. However, this pattern still fails to ensure full and complete flexibility, although it does help reduce labor intensiveness and the output price.

Next-generation businesses and industries should be formed as adaptive, highly transformable and highly productive ones, as they use the benefits of digital technologies, artificial intelligence, big data analytics and self-learning systems, which enable to produce advanced goods on a design-to-cost basis with minimal use of

A. Tyulin, A. Chursin, *The New Economy of the Product Life Cycle*,
https://doi.org/10.1007/978-3-030-37814-1_8

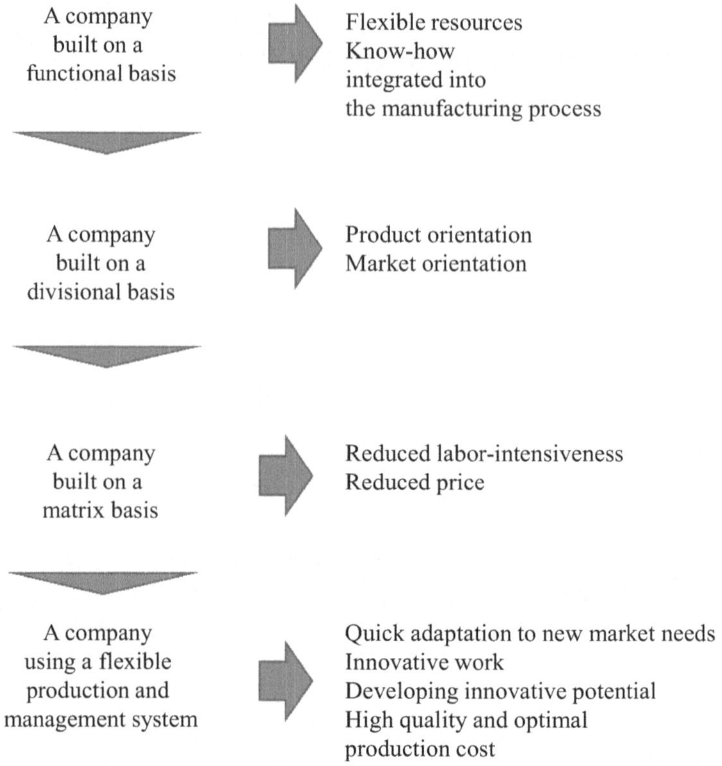

Fig. 8.1 Organizational framework transformation

manpower. This type of design and manufacturing process reduces the period of advanced product's manufacture by shortening each of its life cycle stages. Preproduction stage will no longer be a separate stage, and it will be managed by intelligent systems and assimilated into the production stage.

Thus, the preproduction subsystem is becoming part of a whole production system, in which products will be designed to a specific cost and competitiveness level. In this case, man has to surrender adjustment, measurement, control, testing, and other preproduction functions, to intelligent robotic systems, which should perform them in an automated or semiautomated mode.

For this reason, one of the biggest priorities is arranging a flexible business that should rely on ultimately new manufacturing process engineering methods and new technical tools.

Flexible manufacturing implies simultaneous processing of a number of components, a quick transition to production of advanced goods, as well as a potential for running these processes without stopping or readjustment of equipment, because these are carried out as long as the system operates.

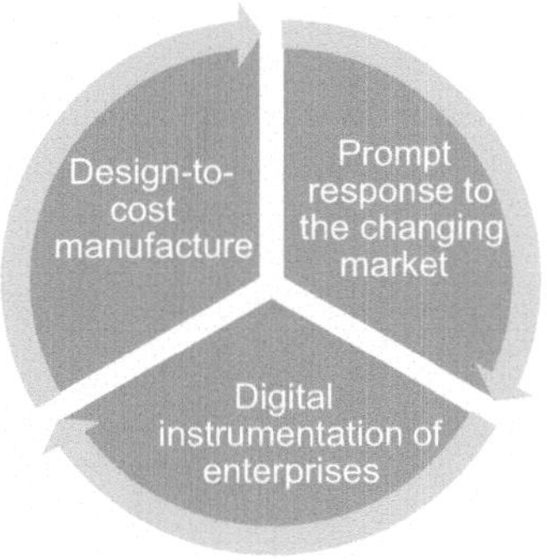

Fig. 8.2 Requirements to a company manufacturing advanced products

Creation and use of flexible manufacturing systems exerts a great influence on an organization's work in different spheres. Using them significantly increases productivity, reduces the prime cost and manufacturing time through reduction of some life cycle stages (for instance, preproduction, prototype testing, etc.), exclusion of a huge amount of obsolete and ineffective equipment, which is replaced by new systems that use artificial intelligence and can integrate. This should result in the formation of a digital organization.

Using flexible manufacturing systems is extremely important and advisable for companies that produce broad ranges of goods that have a similar size, similar forms, require similar types of processing, etc. Effective use of flexible processes is closely related to evolution of automated product design (building a digital twin) and preproduction management; this can be considered an end-to-end automation encompassing the entire *design–technology–process* cycle. The product built with the help of advanced software tools is put into production and is manufactured by digital commands with but minimal participation of man. This predicates the following advanced process setup standards that ensure production of marketable goods (Fig. 8.2).

- Manufacturing systems' adaptability to changing customer needs and values and ability to switch to design-to-cost manufacturing of new high-tech products
- Use of advanced equipment and technologies, including cutting-edge digital tools that are based on new physical principles
- Maintaining economic effectiveness of projects aimed at design-to-cost manufacturing of advanced products

As follows from the above, to set up an effective manufacturing process, companies should use flexible automated systems that meet the requirements mentioned above.

Setting up a flexible automated enterprise is a great engineering challenge, and its main goal is developing a highly economic project relying on latest scientific achievements and benefits of digital technologies. This project should demonstrate high productivity, product quality, and optimal prime cost. Using flexible automated industries embraces two principles: self-regulation and self-organization (Fig. 8.3).

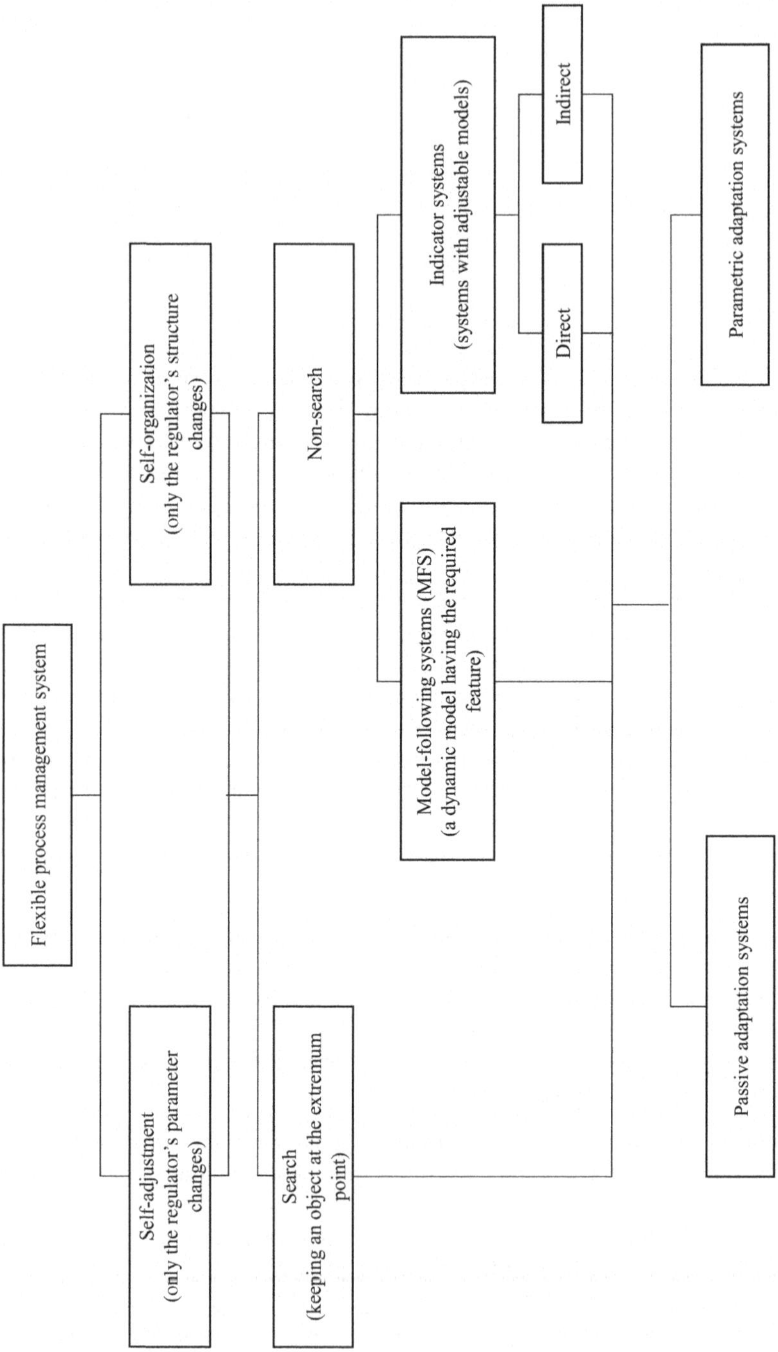

Fig. 8.3 Types of flexible industries

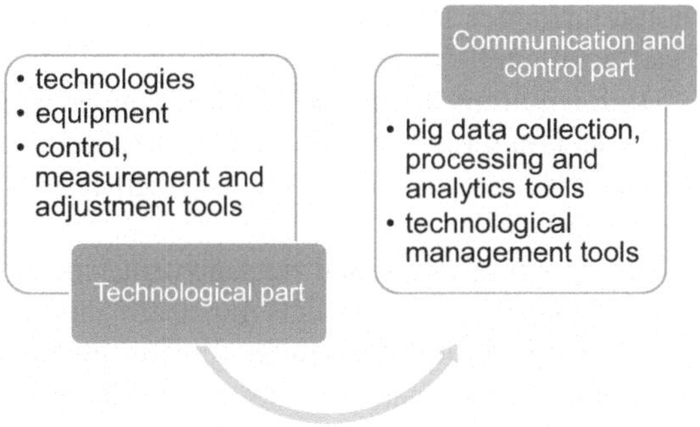

Fig. 8.4 Parts of a flexible manufacturing system

The goal of building flexible and automated manufacturing processes is reaching a maximum effectiveness and cost-efficiency, usually through launching full-scale production, and maintaining flexibility, which is mandatory for small- and medium-batch products.

The most important trait of a flexible automated manufacturing system is its adaptability to changes that occur during the manufacturing process. Indeed, it is common for a flexible manufacturing system (FMS) to operate in uncertain economic situations, which are manifested in frequent and mostly undetermined task changes, which concern product nomenclature and output volume. In this situation, an FMS should demonstrate an adequate response to changes and generate proper solutions to adapt to these changes. Today, smart adaptation is an important property, as it relies on advantages of rapidly evolving digital technologies.

Smart adaptation to possible changes in the surrounding environment makes flexible businesses adaptable to outer conditions, so they do not have to resort to any structural, hardware or software modifications.

Elements of a flexible automated manufacturing system can be divided into two groups, as these elements make two parts of the system respectively (Fig. 8.4).

The structure of a product manufactured at flexible enterprises should be maximally high-tech in terms of its potential for automation. For example, automatic assembly units should consist of a minimal number of parts, and products should have specific structural signs and properties enabling their automatic recognition, orientation, and positioning; they should demonstrate a high family resemblance, which should lay the ground for mass production, which is a promising option given the frequent nomenclature changes. This can be achieved through standardization of product and component production technologies, as well as of parts' and components' structure.

In turn, all parts of the manufacturing process, including production tools, instrumentation, devices, equipment, and technological route, should work toward flexibility. Particularly, the manufacturing process should be built according to grouped technology laws. Production tools, instruments, and devices must be standardized and suited for automation, installation, readjustment, and replacement. There should be robotic, programmable, readjustable technical equipment, which should guarantee implementation of various engineering processes through the use of artificial intelligence.

High requirement s to technical equipment can only be met in case of switching from traditional (manual) to automated manufacturing methods, which use computational devices, robotic systems, self-learning systems driven by artificial intelligence.

The communication and management part of an automated FMS is facing the increasingly strict requirements too, and it should be built in such a way as to ensure digital compatibility of all subsystems to guarantee their optimal interaction when performing target functions. This requires incorporation into the system of intelligent algorithms, strict regulation of input and output module parameters at all levels of input/output signals and management methods.

Today, with digital technologies and communication systems evolving rapidly, software appears to be the most expensive element of FMSs. Given the continuous growth of prices for software with the lead over those for technical tools increasing, it is important to ensure intra- and interlayer program software compatibility of the entire equipment.

Limited flexibility and insufficient reliability of the central data exchange and processing system on the one hand, and difficulty coordinating separate management and processing activities on the other, pose the necessity of using a distributed and centralized method of management of information, diverse big data, and their processing within FMSs. Characteristic of this method are: flexible data management and processing achieved thanks to software adaptation of respective technical tools to changing functionality standards; high operation speed thanks to an opportunity of temporal and spatial parallelization of management and processing activities, high reliability and survivability of respective systems. Flexible management with the use of digital technologies is schematically shown in Fig. 8.5.

Practical implementation of distributed centralized data management and processing systems became possible thanks to extensive development and use of advanced technologies at all management layers, which unify all equipment units into an integrated modular system.

Use of modular structuring of system in different technical spheres has demonstrated its ultimate effectiveness. In complex systems, as automated MFSs are, the modular structuring principle is naturally implemented in combination with its hierarchical structure. This refers to both the technological and communication and management part of an automated FMS. In the technological part, the modularity is observed in products' structure and equipment packaging. The mathematical backup of the information and management part also uses the modular principle. This provides for high standardization and typification of the system's elements, which can facilitate and speed up their incorporation into the manufacturing process. Also, it creates an opportunity for step-by-step introduction of flexible automated systems, which contributes to their extension through total automation and assimilation of new elements without modifying already implemented stages.

The experience of creation of flexible automated businesses shows that in order to get them to work effectively, it is important to define a nomenclature of parts, which

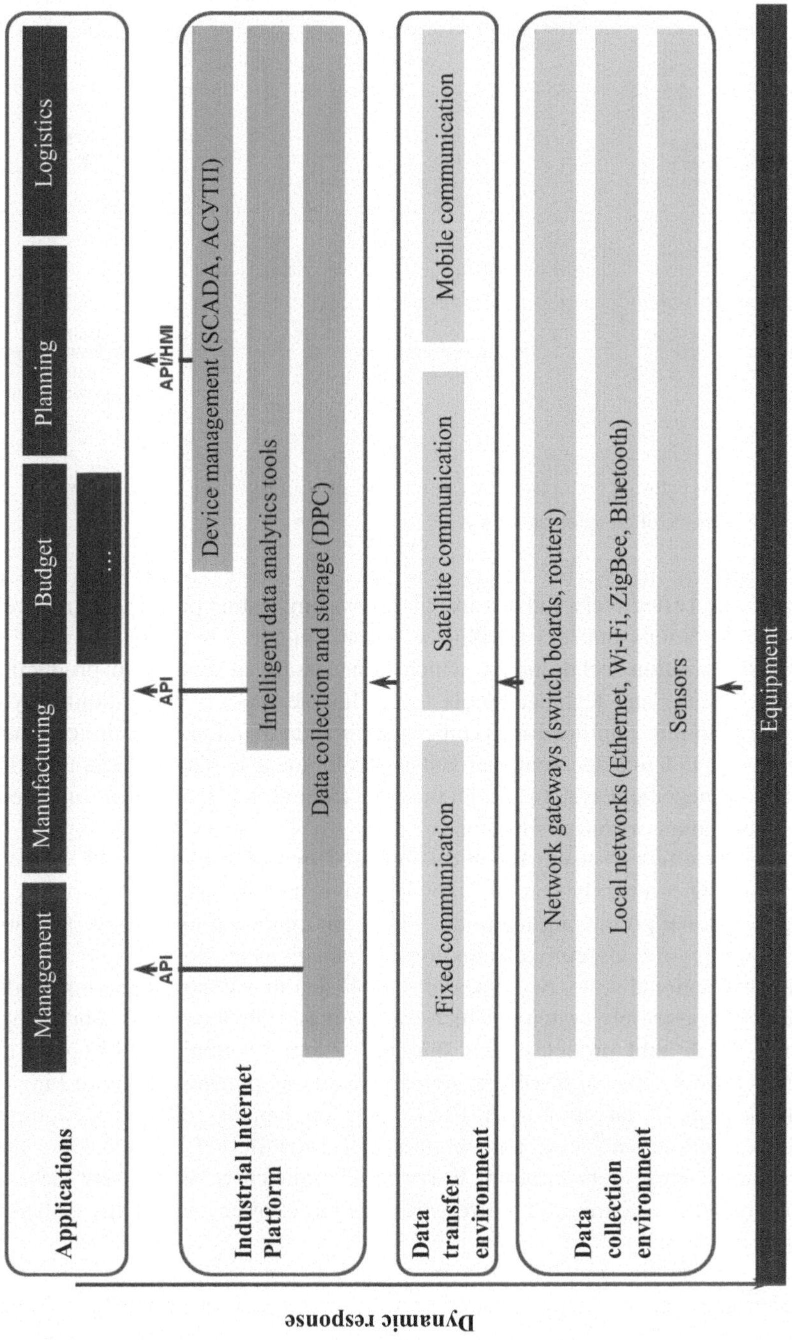

Fig. 8.5 Ensuring flexible management by relying on digital technologies' advantages

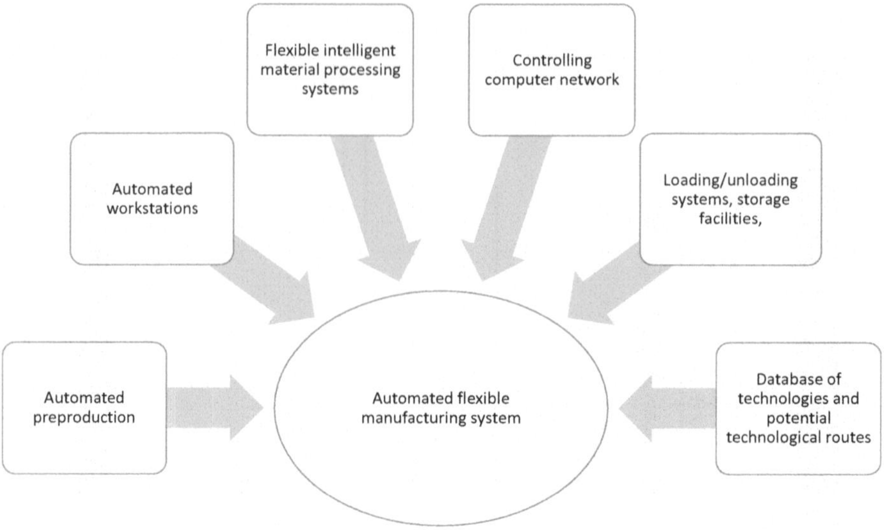

Fig. 8.6 Components of an automated FMS

can be processed effectively with the use of this system; testing parts' performance in processing conditions that exist within a system; in-depth reconfiguration of the routing and operation technologies, which is necessary in automated production; choosing elements and technical tools for a flexible process and rational layout arrangement options; unification, typification, standardization of technical tools, organizational and management elements; development of an intelligent flexible production management system, which includes technological, organizational, technical, and economic management loops.

The biggest effect of using flexible automated manufacturing can be achieved when processing complexly shaped, labor-intensive, and operation-rich parts in case of using traditional processing methods, with a minimum number of reinstallations and high number of consecutive technological transitions.

On an international scale, two types of flexible automated manufacturing models are considered: assembly systems, which carry out assemblage of parts and components into whole end products, and shaping systems, which produce parts and components. Both types of flexible systems include components shown in Fig. 8.6.

Each element of an automated FMS relies on benefits of advanced digital technologies and thus achieves an economic effect manifested in lower labor cost, materials and energy consumption, lower production costs, which will help the product achieve the technical characteristics and target price and occupy a share in the market.

A set of workstations includes machines, which do not require lengthy adjustment or switching between consecutive missions. In most cases, these machines do the milling, boring, drilling, thread cutting, honing, etc. The readjustment process is automated and is driven by self-learning systems.

Table 8.1 Flexibility categories for automated manufacturing

No.	Category	Manifestation
1	Flexible products	A system can produce various goods using the same equipment
2	Flexible operations	An opportunity to produce several types of goods on different equipment providing that operations are organized in different orders
3	Flexible production pattern	Changing output volumes of different goods with account for the changing demand and future needs, without losing profitability
4	Flexible material processing	An opportunity to modify operation routes between equipment units
5	Flexible work	Number of manufacturing routes
6	Flexible control	An opportunity to carry out uninterrupted manufacturing with the help of intelligent machinery and system control software
7	Flexible manufacturing	Number of all types of products, which can be manufactured without adding large equipment units

The intelligent material processing system is also automated and flexible, as it allows to switch missions between the processing workstations.

The controlling computer and microchip network perform some or all of the following tasks:

- Routing
- Monitoring of statuses of all current missions to define units, to which the next mission should be assigned after completion of the current one
- Transfers guidelines on each operation to be performed by each workstation and provides appropriate tools for work
- Ensures tracking of correct performance of operations and signals problems that need to be solved

Loading and unloading systems and storage units are also governed with the help of intelligent systems, which ensure uninterrupted manufacturing and even use of production capacities.

Missions, which are to be processed by the system, are entered on the central computer, which loads processing, production, and assembly programs.

Flexible automated production should rely on flexibility of specific categories described in Table 8.1.

Because launching automated flexible projects requires large capital investments, their future effectiveness depends on decisions made at the stage of FMS drafting. To achieve a maximum effectiveness, the some basic optimal design principles should be followed.

All system elements should conform with the planned product nomenclature. The probability of different system states should fully match the probability of appearance in the nomenclature of production surfaces generated by automated FMSs in these specific states. The higher the ratio of states to the number of elements is, the more effective the system is.

Increasing power of feedback channels in fabrication management with increased flexibility (diversity) and growing volume of available design information specified by the production plan. This requirement is explained by the fact that increasing volumes of information passing through design elements (for example, guides) that are stored in the fabrication module's memory, causes a dramatic increase in the number of these elements and the entire flexible manufacture.

To achieve optimal results from the use of automated FMSs, it is necessary to carry out a preliminary statistical analysis of the planned product nomenclature and nomenclature trends, which define quantitative flexibility and throughput capacity criteria. These indicators form a basis for rationality, power, and content of an automated flexible manufacturing business.

It should be noted that the speed of introduction of automated flexible businesses will depend on the quality of software.

In FMSs, software improvement should focus on the following goals:

- Creation of factory, branch, and interbranch management software databases
- Automated management software preparation
- Equipment diagnostics, including self-resets after failures
- Fault prevention through suspension of processing in case of approaching tolerance limits
- Collection, processing, and registration of information about process parameters
- Quick response to unplanned changes in manufacture processes

In general, the algorithm of setting up flexible automated businesses is implemented as shown in Fig. 8.7.

Automated flexible businesses reduce or eliminate wastes and reduce production costs through automation of activities, such as loading/unloading, fixing and fitment of parts, calibration, measurement, replacement of tools and maintenance outside of work cycle, and thus ensure higher productivity. Automated monitoring and adaptive feedback and failure correction provide for nonstop work without using manpower and longer use of expensive equipment. The structure of a flexible manufacturing system is shown in Fig. 8.8.

Compared to traditional manufacturing systems, flexible systems have the following advantages:

- Extensive use of the machine (up to 90%)
- Smaller number of machines used
- Smaller production area
- Quick response to changes
- Less strict requirements to production tools
- Removal of the preproduction stage
- Reduced amount of incomplete production
- Shorter production cycle
- Reduced direct labor cost, manpower, and increased productivity

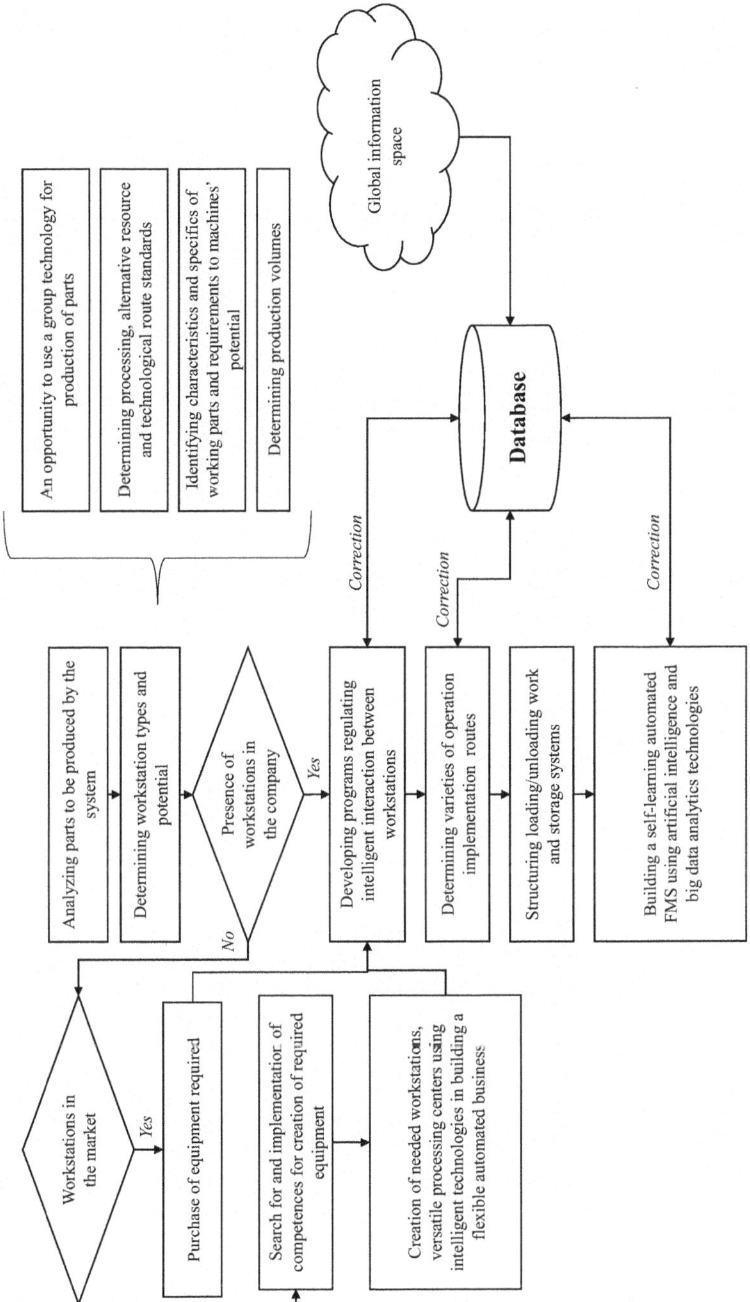

Fig. 8.7 The algorithm of setting up automated flexible projects

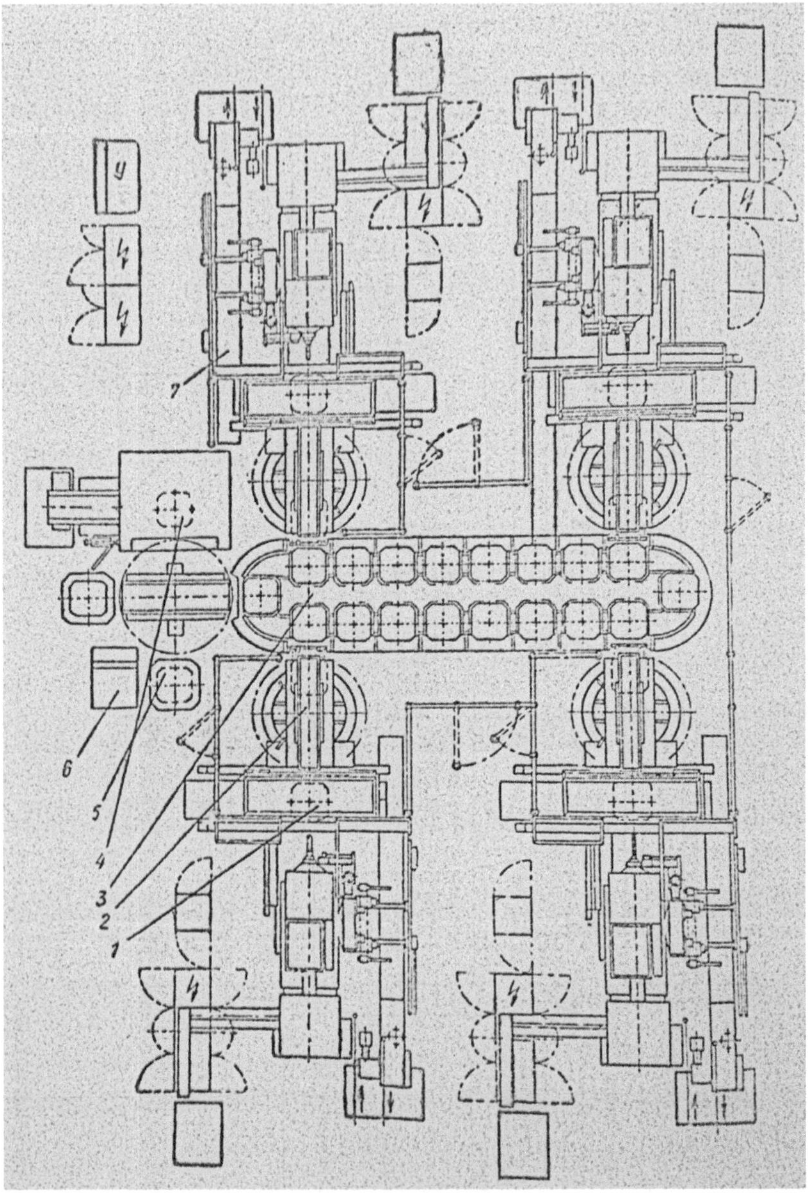

Fig. 8.8 A flexible manufacturing system (FMS). 1—processing center; 2—automatic tool replacement facility; 3—storage element; 4—part washer; 5—clap; 6—system control unit; 7—waste removal unit

Many of these benefits are achieved through the use of robotic systems in automated projects, as they help eliminate the preproduction phase from the life cycle and fulfill functions, which used to be carried out manually.

Therefore, automation of a business should come a long way from automation of separate manufacturing processes through general automation and down to complex automation with the use of automatic self-learning systems operated by artificial intelligence. These systems can optimize workload and operation modes, minimize various operations' (including transportation, loading and unloading) technological cycle, and provide full and comprehensive information about process states at any time.

Based on the above, it can be concluded that using automated flexible solutions increase the effectiveness of the production process.

Increased effectiveness and reduced cost (prime cost) in automated flexible production is achieved through:

- Reduced labor costs
- Release of production sites and equipment
- Quicker mastery of new equipment
- Stable energy spends
- Saving on personnel's salaries
- Reduced production cycle and incomplete production
- Reduced technical control cost
- A large share of run time in the floor-to-floor time
- Reduced downtime due to traumas and occupational diseases

To evaluate automated FMSs' effectiveness, the following indicators are used:

- Annual economic effect
- Economic effect from reduced preproduction
- Reduced production costs (current costs)
- Capital (additional) investment
- Reduced personnel
- Profit resulting from replacement of manpower with flexible manufacturing systems
- Payback period

As a criterion of effectiveness of flexible manufacturing systems, annual economic effect is used; it is based on the difference between reduced expenditures with reference to profit and release of additional products resulting from introduction of flexible manufacturing, as well as an additional effect resulting from reduced preproduction.

The annual economic effect of using an automated FMS, regardless of level (module, production line, division, workshop, etc.), is calculated as follows:

$$Ec_{\mathrm{y}} = [(C_1 + E_{\mathrm{n}} K_1) - (C_2 + E_{\mathrm{n}} \mathrm{K}_2) + \Delta \Pi] + Ec_{\mathrm{add}},$$

where C_1, C_2 is the total cost of the yearly product (work) volume in a baseline scenario and with flexible manufacture technologies used respectively (in rubles); E_{n} is the norm coefficient of capital investments' economic effectiveness, which is .14;

K_1 is the baseline capital investment referring to the annual production volume and received with the use of flexible manufacturing facilities (in rubles); K_2 is the capital investment that ensures production of the annual volume with the use of flexible production technologies (in rubles); $\Delta\Pi$ is additional profit from a volume produced by a laid-off worker over the target year (in rubles); Ec_{add} is additional yearly economic effect achieved thanks to reduced preproduction period (in rubles).

Additional profit from production in the target year by laid-off workers is calculated with the help of a simple equation:

$$\Delta P = P_y W_{layoff},$$

where P_y is average yearly profit per worker (in rubles); W_{layoff} is the number of laid-off (due to introduction of automated FMSs) workers, who create additional profit (persons).

Calculations should note the extra economic effect of the shortening of design, engineering and technological preproduction periods, which can be calculated as follows:

$$Ec_{add} = E_n[(C_1 + nK_1) - (C_2 + E_nK_2) + \Delta P](T_1 - T_2),$$

where T_1 and T_2 are the length of technological preproduction periods in a baseline and FMS scenarios respectively (years).

The most pronounced economic effect is achieved through building flexible businesses with a modular structure.

The modular structure, broad nomenclature of standardized prefabricated units and blocks, along with the benefits of digital technologies and artificial intelligence used in design and engineering, speed up the design/engineering and introduction of flexible modules and units four- to fivefold. It is possible to promptly build a machine or installation with required characteristics of standardized units. If necessary, these flexible facilities can be disassembled into units and blocks, and new equipment can be created with different technical parameters and with the use of automated assembly and readjustment systems.

Steps to improve the structure of flexible automated industries, combining them into a complex facility, pose a serious challenge that is interaction between technical, organizational, and economic factors, which depend on the speed of the automation process and changes in the structure of the entire national economy.

Production of advanced goods with the use of the same production facilities, basic technologies and competences (with addition of new ones when specific products require that) for different market segments and consumer groups, will ensure economic stability through successful sale of price- and quality-competitive goods.

Meanwhile, when building an industrial system in a company, it is necessary to structure the production process in an optimal way to ensure that introduction of automated flexible facilities will produce a maximum effect, with account for their specifics. Therefore, it is necessary to determine optimal structures of production areas, list of parts and components, which make up a potential for an automated flexible manufacture, and works that are going to be assigned to cooperating enterprises, for production of specific high-quality goods over short periods with account for their competences and technical potential. Cooperating organizations should arrange for low production costs in design-to-cost projects.

8.2 Allocation of Production Costs Between Parent and Cooperating Enterprises

A must for reaching global competitive leadership is sufficient resource provision, which is determined by an organization's ability to accumulate financial resources gained from their activity and invest them in creation of new competences and technologies, which eventually lead to development and production of advanced goods. There are two directions, in which this problem should be solved:

1. Production of large volumes of goods using corporate facilities with strict production cost control and maintaining appropriate prime cost, as well as cooperative material and component supply price management, as the materials and components are shipped to the parent manufacturer
2. Extension of the product portfolio to satisfy the needs of different branches of economy and customer groups (if one segment shows a decline, sales, and profit loss can be compensated by an increase in another one)

Production of a complex product implies the use of a large category of competences, for which many manufacturers address cooperative companies or outsource part of the work; they do not employ their own innovative potential or competences, which can be sufficient for solving specific industrial purposes. Therefore, it is necessary to be careful about today's growing cooperation trends in product design, engineering, production, and operation.

Industrial outsourcing is explained by the following factors, which are determined by the ongoing globalization:

- Increasing cost and amount of R&D to be carried out within limited periods of time to maintain product competitiveness give the shortening product life cycle.
- Increasing mobility of the human capital assets. An organization that makes substantial investments in R&D and personnel training is at risk of losing the "legacy" and employees, who may either be hired by competitors or launch their own projects while relying on competences obtained while working at the company.
- The growing popularity of venture investments, which stimulate the development of small investment organizations providing competitive goods and in a way threatening large companies.
- Increasingly tough and strict corporate requirements to the concept generation process, which stimulates intensive geographic distribution of R&D laboratories

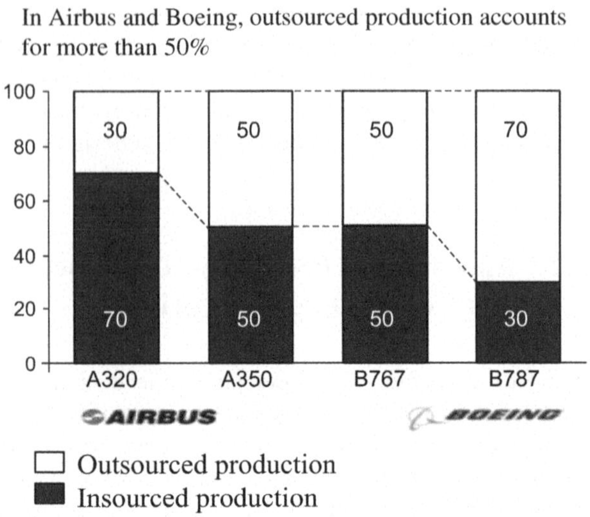

Fig. 8.9 Outsourcing and industrial cooperation in Airbus and Boeing

and their integration into corporate R&D networks; there is a growing number of distributed international design teams, which are not localized in a single center (virtual competence centers).

- A continuing increase in the number of international design teams and third-party organizations' intensive involvement in the research process.

Figures 8.9 and 8.10 demonstrate cooperation between the world's top aircraft producers—Airbus and Boeing. Evidently, cooperation is an increasingly important factor of modern global companies' development. Analysis shows that intensive cooperation is a desperate measure, and it is aimed at increasing a company's competitiveness. Largely, this happens because development and manufacturing of a complete high-tech product requires a lot of components produced by companies employed in other branches and industries. The design and engineering of these requires different competences and equipment, which help produce high-quality goods in respective fields.

Outsourcing is relevant in fields, where massive competition is observed, and hiring an outside company is less expensive than hiring new staff. For example, outsourcing is commonly practiced in the IT sphere. Apple, an IT technology whale, hires a number of teams to complete a single mission. This 10-to-3-to-1 approach allows ten groups to simultaneously and independently work on a specific task. Once the task is finished, an independent expert team reviews the results; first, it leaves three projects in order to eventually accept one. Although this approach is labor costly, it helps an organization endow its products with new competitive advantages for a long time.

Another goal of using outside company's work in high-tech industries is having to look for and obtain data and offer concerning conceptual developments and new technologies, which are necessary for giving future products new competitive high-

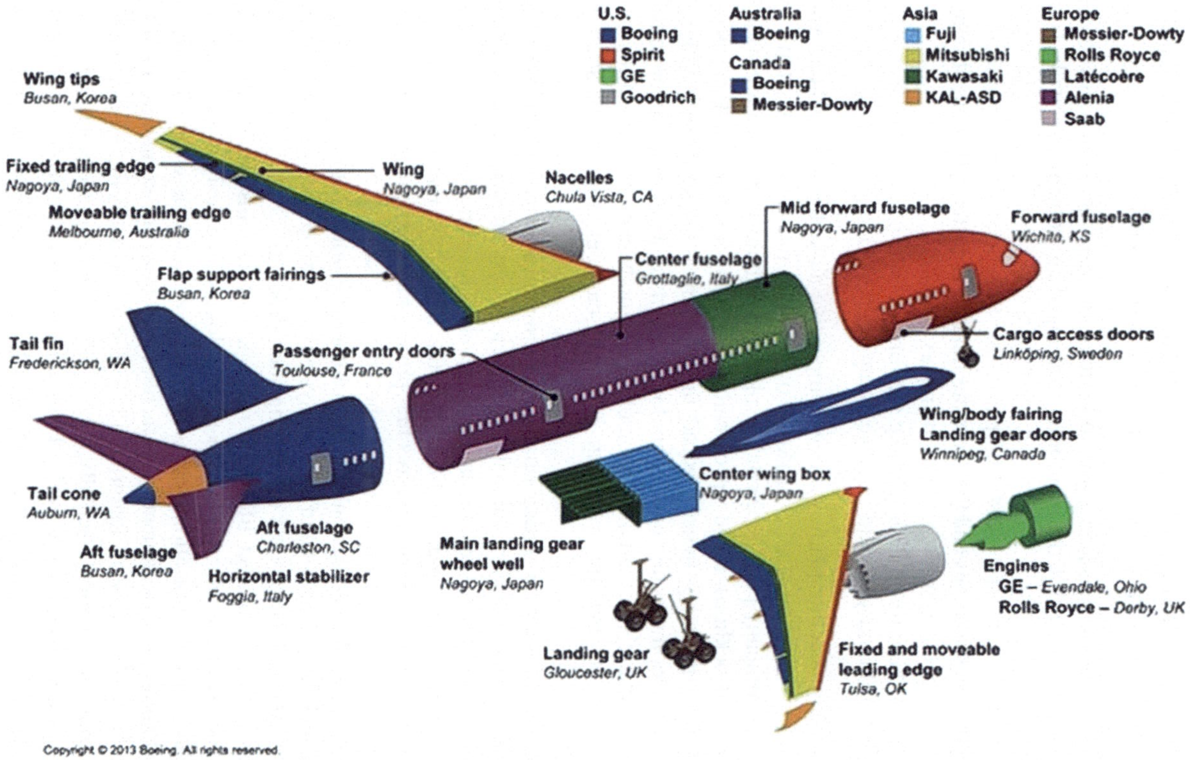

Fig. 8.10 International cooperation in production of Boeing 737 (G. Tchesbro. Open innovations. Creation of profit-making technologies. M., 2007)

tech features. One of the ways of finding this data is winning the startup market, which can provide innovative competences and technologies. A good example of attracting outside organizations is Lockheed Martin's LM XXX satellite platform experience. In cooperation with other companies, Lockheed Martin has created more than 280 standard components, which work on al, satellite platforms run by the company. Their structurally developed platform series contains:

- The LM 50 series, which is going to be used in nano-satellite production. The family includes 10–100 kg platforms developed in cooperation with Terrain Orbital.
- The LM 400 family of 140–800 kg platforms. Specifically, these platforms use 3D printing technologies and therefore are much less costly. Their target segment is low-orbit, geostationary, and interplanetary missions.
- The LM 1000 family of 275–2200 kg platforms.
- The LM 2100 family of platforms weighing over 2300 kg.

To give these satellite platforms new competitive advantages, Lockheed Martin has decided to enter the startup market. To make the most important step in this direction, the company published satellite platform technical specifications (for LM 2100/400/50). Lockheed noted specifically that it is not interested in ideas or products that are privately owned; instead, the company is open for offers addressing conceptual developments and new technologies. Lockheed described an opportunity to use the company's potential for promoting its goods and services on the public market, noting the benefits a startup company can get from cooperation with Lockheed.

Section 8.1 focuses on setting up a flexible manufacturing system based on a matrix-type project management pattern, which necessitates integration between participants on the process and management levels. This paragraph describes issues related to hiring outside organizations with the use of the matrix approach. It is noteworthy that this approach requires that project management processes and procedures, within a separate organization, should be the same as respective processes and procedures set by the design bureau. Another important aspect of forming this type of virtual design and manufacturing system is using communication technologies, which manage the process of creating such products.

Modern high-tech project life cycle management systems (PLM) use options, which are not available when using paper technologies. The design-to-cost approach or reasonable cost minimization clearly implies cooperation between all design and unit, installation and component manufacturers. In terms of communication technologies, this problem means integration between a product being designed and the manufacturing environment. Actually, it is about parallel engineering throughout the entire design and manufacture cycle. The main criterion of effectiveness for these processes is minimization of the cost of the end product's life cycle. Based on the above, it is possible to outline a standard for an advanced intelligent system of a virtual geographically distributed design and production structure, which unified the parent manufacturer's resources and opportunities, as well as those of potential outside contractors; also, it is possible to name the types of systems, which ensure fulfillment of these requirements. This virtual geographically distributed design and manufacturing structure, which comprises the parent organization (corporation) and cooperating ones, should be created under the parent company's supervision for solving its strategic tasks:

- Designing a product to a set cost of its life cycle
- Labor effectiveness growing throughout the product's life cycle
- More effective use of material and nonmaterial assets
- Higher capitalization of the company

The virtual geographically distributed design and manufacturing structure comprising the parent company and cooperating ones, has the following common characteristics:

- Relying on a specific project
- Mobile and flexible configuring of resources involved in the project with the incorporation of flexible project management procedures
- A single information space unified at least by interfaces that ensure data exchange between the parent company's divisions and outside contractors
- A unified and available for all participants (with account for access rights) system of project databases (those of technological modules, norms and regulations, materials and purchased integrated parts, technological equipment, etc.)
- A uniform for all parties model of project documentation and data (including digital ones)
- A uniform for all processes and procedures project management model
- A uniform system of reference documentation, which regulates the interaction between all parties within a virtual structure

Building a virtual design and manufacturing structure should begin with the development of the new product's concept, the core parts of which are:

- A market concept intended for specifying all external requirements to the project in question; i.e., one evaluating the market and expected batch size, as well as potential customers' expectations concerning post-sale servicing
- A technical and economic concept reflecting the future product's technical and economic image, possible versions (options), planned design and engineering innovations, requirements to the production environment (including investment in its modernization), planned development, certification and quality management plans, list of activities aimed at fulfilling the design-to-cost approach
- A production concept with a preproduction plan and production cost estimates
- A supplier management concept, which defines the component purchase strategy/plan, as well as a supplier management plan

It should be noted that all four elements of the new production concept are inalienable, and they are directly involved in fulfillment of the company's goals and therefore must be coordinated by a single center. It can be a design bureau, as it ensures end-to-end planning of work and shipment, budget and resource management and control over all aspects of the project. When organizing a design bureau, it is necessary to take into account the main ways of management technology modernization:

- Developing a product-based approach: all processes and the organizational structure of the management and design/manufacturing systems should be calibrated to creation and maintenance of a specific type of product
- Developing a process approach: the organization should work as a network of end-to-end business processes directly related to the project's and the company's goals and objectives
- Developing a design approach: execution of processes in the form of a project with functional divisions and outside partners delegate (appoint) workers for a period stated by the project manager (a matrix structure)

The biggest part of the design bureau's functions can be delegated to the intelligent new *product creation management system*, which can distribute tasks between the parent company and outside partners, as well as monitor and adjust the process of solving these missions. The intelligent management system is a large organization and economic product creation project management system, which relies on the virtual geographically distributed design and manufacturing structure that is being tailored to this project. This intelligent system belongs to the PLM category and includes subsystems executing the following functions:

- Informational backup of modernized management processes—process management systems (PMS)
- Informational backup of parallel engineering processes—portal decision-making (PDM) systems
- Informational backup of organizational and technical preproduction—systems that include MPM, CAD/CAM, PDM
- Informational backup of management of data (documents and records) on a ship/product throughout the life cycle—project data management (PDM) systems
- Informational backup of production resource and post-sale support management processes—enterprise resource planning (ERP) systems
- Informational backup of development and actualization of interactive electronic technical documentation—interactive electronic technical manuals (IETM) and PDM systems

This approach can provide a complex life cycle project management solution. Also, it can help unify into a single information system mathematical models of the product, its parts, units, installations, with the parent company's, its geographically distributed divisions' and outside organizations' manufacturing systems, which can help create mathematical models of processes and optimize them, as well as develop relevant steps to ensure reequipment and creation of new industrial facilities (or allying with new companies).

Outsourcing the production of components can be efficient at the beginning; however, prices will grow inevitably and necessitate specific agreements with the contractor. This will be much costlier than developing and incorporating advanced

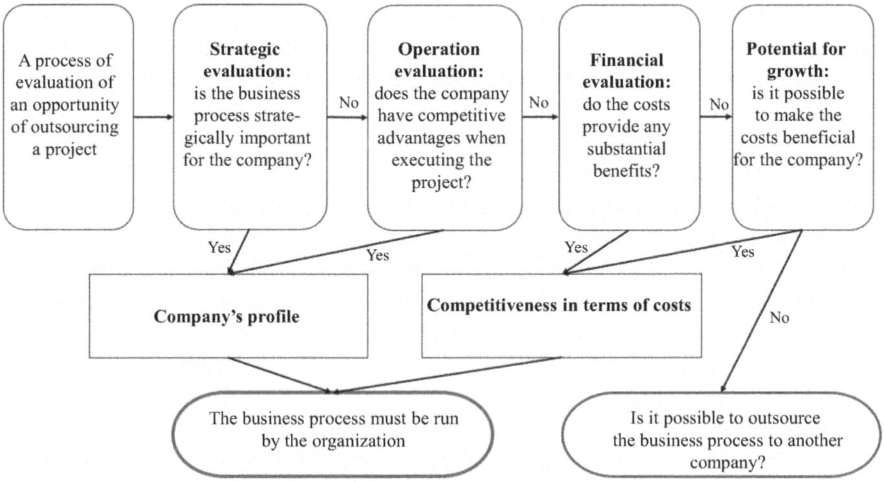

Fig. 8.11 The algorithm of making a decision to transfer the business process (operation) to outside companies

innovative solutions, which help produce highly competitive high-tech goods, on an insource basis.

To determine an optimal ratio between corporate and outsourced work, an intelligent system of a virtual design and manufacturing structure is endowed with a production cost distribution mechanism, which allocates production costs between the parent manufacturer and outside companies. This mechanism is based on an outsourcing algorithm (Fig. 8.11).

Making a decision to outsource a project is a complex task, which affects the parent company's interests, which ensure its sustainable economic development. Actually, it is a matter of choice between in-house production and purchasing third-party products and services, and it relates to the analysis of both the current conjuncture within the organization and possible risks. The central point of choosing between in-house production of using third parties' services is analysis of in-house production management processes and product sales based on strategic perspectives.

With the help of mathematical modeling, it is possible to describe the distribution of work between the parent and cooperating companies, as the process reflects an optimal balance between corporate cost mitigation (achieved through outsourcing part off the work) and making the parent company economically stable (achieved through retaining of a substantial share of in-house production). According to the world's top industrial enterprises, it is necessary to retain at least 30–40% of in-house production of key categories of goods.

Substantial in-house production is one of the methods of making it economically stable. The economic effect of retaining a large share of corporate production, first, consists in a greater profit, uninterrupted, and breakeven work and, second, in reduction of possible losses per unit of operating profit, i.e., minimization of average production risk values. In most cases, outsourcing a separate stage of production, for example, purchasing semifinished goods from a third party instead of producing them, results in higher variable costs. This can be explained: the purchase price for a semifinished product tends to be higher than variable costs of its production (cost of materials, energy, labor costs, etc.) However, the main economic effect of partial outsourcing of production lies in reduction of fixed (or semifixed) costs. The transition's influence on production processes will be visible in production volumes and, consequently, the variable cost increase/fixed cost decrease ratio.

Also, when evaluating changes in fixed costs resulting from outsourcing part of the project, it is necessary to take into consideration not only current expenses, but also the likelihood of facing additional ones; these may include salaries paid to new employees hired to interact with outsource service providers, as well as expenses resulting from communication with the outsourcer. These expenses are often not included in the accounting process.

It is possible to find an optimal balance between the in-house potential and outsourcing with the help of economic indicators. As a quantitative indicator of this balance's effectiveness, the EBIT indicator can be used, which reflects a company's operational profit (%).

It is calculated in the following way:

$$\mathrm{EBIT}_{\%} = \frac{\sum_{i=1}^{N} (p_i - v_i)Q_i - \mathrm{FC}}{\sum_{i=1}^{N} p_i Q_i},$$

where p_i is the price of an item of type i; v_i stands for variable production costs per item of product type i; FC is fixed costs; Q_i is the output volume for product of type i; N is the number of product types.

Fixed costs do not change much when the output volume increases. Variable costs do not change by items, but they do change by the combined output in proportion to the output volume. The analysis will consider ready products' components instead of different types of products, which can be produced by the parent company and, partially, by an outsourcer.

Here is an example of a company manufacturing a type of product that consists of ten components. The task is to evaluate the change in the $\mathrm{EBIT}_{\%}$ indicator resulting from outsourcing part of the project. According to current assumptions, fixed costs decrease when the work is partially outsourced. It should be noted that outsourcing slows down the speed of fixed costs' decrease. Variable costs increase in proportion to the amount of the outsourced work.

Assume that with fully in-house production, fixed costs are: FC = 3700. The output volume of each type of product, price per item, fixed and variable per item and calculated $\mathrm{EBIT}_{\%}$ values are shown in Table 8.2. According to the example, 10–80% is outsourced. Changes in the $\mathrm{EBIT}_{\%}$ indicator are shown in Fig. 8.12.

The operational profit reaches a maximum value with 40% of in-house production, which is quite in keeping with the world's top manufacturers' practice. Thus, an

Table 8.2 Operation profit in a diversified business

Type of work I	Price per item p_i	Variable in-house \outsource costs v_i	Fixed costs resulting from the outsourcing of the i amount of work	Required amount of work of type i Q_i	Operational profit $EBIT_{\%}$, %
1	20	10\12	3500	50	26.4
2	30	10\12	3320	45	26.9
3	27	12\15	3150	30	27.6
4	26	10\13	2990	55	28.3
5	35	13\14	2840	45	28.2
6	30	9\11	2700	40	29.1
7	30	15\17	2570	50	29.6
8	25	10\15	2410	45	29.3
9	20	11\12	2480	30	28.7
10	35	15\16	2330	45	27.5

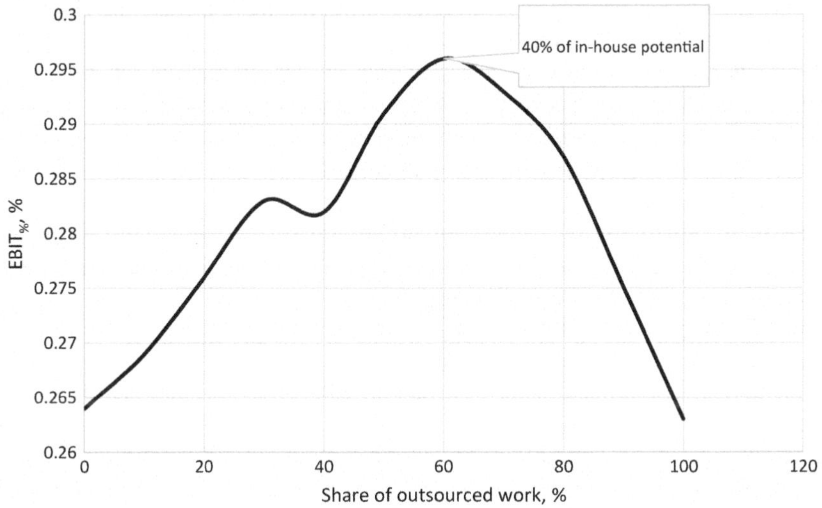

Fig. 8.12 Change in the operation profit $EBIT_{\%}$ resulting from a change in the amount of outsourced work

organization can maximize its profit and obtains a resource, which can be used for advanced development.

It is possible to judge a company's economic strength by the influence of the leverage DOL, which is calculated as follows:

Table 8.3 Calculating the DOL

Types of work	1–2	1–4	1–6	1–8	1–10
DOL	5.3	3.3	2.62	3.45	4.07

$$\mathrm{DOL} = \frac{\sum_{i=1}^{N} (p_i - v_i)Q_i}{\sum_{i=1}^{N} (p_i - v_i)Q_i - \mathrm{FC}}.$$

Therefore, the higher a share of fixed costs is laid in a product's prime cost structure, the greater an influence the leverage exerts and, consequently, the higher the risk will be.

The previous example demonstrates changes in the leverage's influence that result from partial outsourcing. Based on the data shown in Table 8.2, it is possible to calculate the leverage's influence depending on the amount of outsourced work and when including new types of products in the production schedule (Fig. 7.10). The results are shown in Table 8.3.

The DOL indicator is directly related to the risk of operational work: higher level of influence results in a higher risk. A decrease in the leverage signals an increase in a company's economic strength, while an increase signals a lowering. As follows from the calculation, outsourcing more than 60% of the work makes a company economically weaker.

This is an illustration of an increase in the effectiveness of a company's functioning and, consequently, an increase in its economic strength through retaining of a substantial share of in-house production.

After defining the list of business processes (technological operations), which can be handed over to outsourcers, the intelligent system of a virtual geographically distributed design and manufacturing structure will carry out a simulation modeling of work distribution between the parent company's divisions and outsourcers; within this structure, total costs and use of the parent company's and its divisions' production capacities are evaluated; the use of production capacities should not drop below the level that ensures its optimal functioning. Within this intelligent system, the simulation modeling of work distribution is carried out by a subsystem, which provides information support for production resource management processes. There is a mathematical model, based on which the distribution of costs between the parent and outsourcing organizations can be simulated.

The production and engineering potential of a virtual geographic distribution design and manufacturing structure, which comprises the parent company, its divisions and outsourcers, is reflected in the maximum volume of work completed as part of a specific industrial mission with account for optimal use of a company's resources during the period in question.

It is essential to assume that the production potential is influenced by a number of factors, which cause it to increase or decline. They can be augmented thanks to technical reequipping or diminish due to lack of qualified personnel.

Ways of assessing a company's industrial capabilities or its specific areas may be dictated by lists of equipment units, personnel, available competences, products' labor intensity (for example, reference samples) and further simulation modeling of the production cycle with an opportunity to review different structures of work performed by the geographically distributed structure.

Based on the information about available equipment and personnel, usable work time fund limits are set. Next, a model is built, which calculates a maximally feasible production volume with set usable work time limits for a given period of time. This model's target function can be minimization of project implementation costs. Therefore, the optimal production resource utilization model relies on the following data:

- The size of funds of useful time for specialists trained to operate specific types of equipment
- Labor intensity of each type of equipment (with a dissection into all required industrial operations)

A useful time fund limit estimate for equipment S_i in the given modeling period can be determined by the following formula:

$$S_i = d \cdot t \cdot k \cdot n,$$

where d is the number of workdays in the evaluated; t is the working time pattern (the duration of a shift with account for the number of shifts); k is the allowance for equipment adjustment and servicing; n is the number of equipment units.

The staff should match the types of equipment used. The planned useful work time fund equals the sum of shift lengths over the given period. The planned useful work time fund is compensated by the correction ratio, which accounts for time losses resulting from a worker's activity. This makes the value T_i, which expresses the planned work time fund with account for losses.

Based on equipment and personnel limits, the useful time value for a production operation is calculated by the formula:

$$b_i = \min(S_i, T_i).$$

Values a_{ij} (i—equipment type; j—product type) expressing the duration of each production operation are fixed as well. Values a_{ij} make up the matrix A, the number of lines in which matches the number of equipment types, and the number of columns matches the number of types of released products.

Assume that a company's production potential is evaluated based on calculations of maximum workloads; the workload set is presented by operations (business processes): $Q_1, Q_2, \ldots, Q_N$, where Q_i is the cost required for fulfillment of the operation. To each product type, the known value is its labor intensity with the known work allocation structure inside the geographically distributed system.

The target function is minimization of costs, i.e., minimization of resources spent on operations and business processes:

$$a'Q \to \min,$$

$$\begin{cases} AQ = b, \\ Q \in \mathbb{Z}_+, \end{cases}$$

where Q is the column vector reflecting the unknown business process implementation costs sustained by parent and outsourcing companies; a is the column vector reflecting the combined labor intensity of works; b is the column vector reflecting the useful time fund for each business process (operation).

In the optimal production facility utilization model, it is convenient to apply interval estimates of labor intensity values, as well as those of useful time funds. Thus, these values can take on any value from a given interval. The density of value distribution inside these intervals is unknown, because these factors manifest themselves randomly. For this reason, the use of averaging procedures that are commonly used in stochastic programming and a transition to deterministic models is difficult. In this case, a solution to the interval triggered task can be found to be an acceptable deterministic solution to the entire group of linear programming tasks generated by different combinations of the model's parameters derived from respective intervals.

Thus, the interval linear programming task for the production facility optimization model is:

$$a'Q \to \min,$$

$$\begin{cases} AQ = b, \\ |A - A_0| \leq \Delta A, \\ |b - b_0| \leq \Delta b, \\ |a - a_0| \leq \Delta a, \\ Q \in \mathbb{Z}_+. \end{cases}$$

Values A_0, ΔA, b_0, Δb, a_0, Δa are fixed, and they define intervals, which may contain the model's parameters.

The existence of an optimal solution to the interval task can be explained.

Assume that ε is a nonnegative residue vector. The solution Q to the interval problem is its ε-plan, if $|AQ - b| \leq \varepsilon$ for all possible labor intensiveness and useful time values. The vector Q is the ε-plan of the optimal solution providing that it matches the following conditions:

$$\begin{cases} -\underline{A}Q + \bar{b} \le \varepsilon, \\ \underline{A}Q - \bar{b} \le \varepsilon, \\ Q \in \mathbb{Z}_{+}. \end{cases}$$

where $\overline{A}, \overline{b}, \underline{A}, \underline{b}$ are the high-end and low-end estimates of respective parameters.

The linear programming interval task takes on the following shape:

$$a'Q \to \min,$$

$$\begin{cases} -\underline{A}Q + \bar{b} \le \varepsilon, \\ \underline{A}Q - \bar{b} \le \varepsilon, \\ Q \in \mathbb{Z}_{+}. \end{cases}$$

The residue of the ε problem is an unknown value, and its rough assessment can take away the solution's optimality or result in incompatibility between limiting inequations. To avoid this situation, an auxiliary linear programming task can be formulated, which can help define the minimal residue. The minimal residue (the norm) is found from conditions: $\varepsilon_{\min} \to \min$, $e'\varepsilon - \varepsilon_{\min} \le 0$, where e' is a single factor. Once the minimal residue is found $\widetilde{\varepsilon}_{\min}$, it is possible to find an optimal solution to the main optimization problem, which take on the following shape:

$$a'Q \to \min,$$

$$\begin{cases} -\underline{A}Q + \bar{b} \le \varepsilon, \\ \underline{A}Q - \bar{b} \le \varepsilon, \\ e'\varepsilon - \varepsilon_{\min} \le 0, \\ \varepsilon \ge 0, \\ Q \in \mathbb{Z}_{+}. \end{cases}$$

The limitedness of the multitude of optimal solutions follows from its statement (given the limitedness of useful time funds for each operation). To find a solution, the standard *M* method can be used in solving linear programming task.

The result of the calculation based on the offered model will be the lowest possible total cost $Q_1, Q_2, \ldots, Q_N$ of business process implementation with all possible job allocation options between the virtual geographically distributed parent company's structure and outsourcers. With the help of simulation modeling, the intelligent system of management of new product creation process helps define optimal cost distribution between the parent company and outsourcers. The offered methods can help the intelligent design and manufacturing management system can automatically decide to outsource an operation or business process.

In the next section, other subsystems of the intelligent product development and production will be described in detail, as will be the main methods of building large organizational systems driven by the artificial intelligence, neural networks, and basic approaches to their creation with descriptions of recommendations concerning the formation of the basis of the large production management organization.

8.3 Building Advanced Production Management Systems

Advanced production management systems rely on creation and implementation of automated flexible projects and formation of flexible approaches to management of all business processes in a company through the use of digital technologies' advantages, which help reduce expenses at each life cycle stage of a highly competitive product of future designed to a fixed cost.

This production management system lays the ground for a "smart plant," the structure of which is described in Sect. 8.1. This case describes a production management and business process combining system, which includes:

- Creation of digital intelligent platforms, specific cutting-edge ecosystems. Based on the predictive big data analytics, the platform approach helps unify geographically distributed parties involved in designing and production of a future product (creating so-called "virtual companies" to unify the best technologies and competences to design a high-quality product to a fixed cost), raise flexibility and customization while referring to customers' requirements.
- Development of a system of digital models of products and processes being designed. Digital models should be highly adequate to real objects and processes (convergence of the material and digital environments, which generate synergistic effects).
- Digitization of the entire product life cycle (from concept through production, operation, servicing, and disposal). The later changes are introduced, the higher the cost is; therefore, there is a shift toward design processes, which lay the ground for global competitiveness or high customer requirements with account for a competitive price.

Creation of a smart plant based on a system of advanced product management, stimulates the formation and development of new key competences, which boost a manufacturer's innovative potential (Fig. 8.13).

These competences determine:

- A quick response to market and individual requirements
- Use of systemic approaches (systemic engineering), when it is necessary to always keep a whole system in sight with all interacting components
- Formation of a multilayer matrix of target indicators and limits as a basis for a new type of design with significantly reduced risks, field tests and prototyping work through the use of future products' digital twins

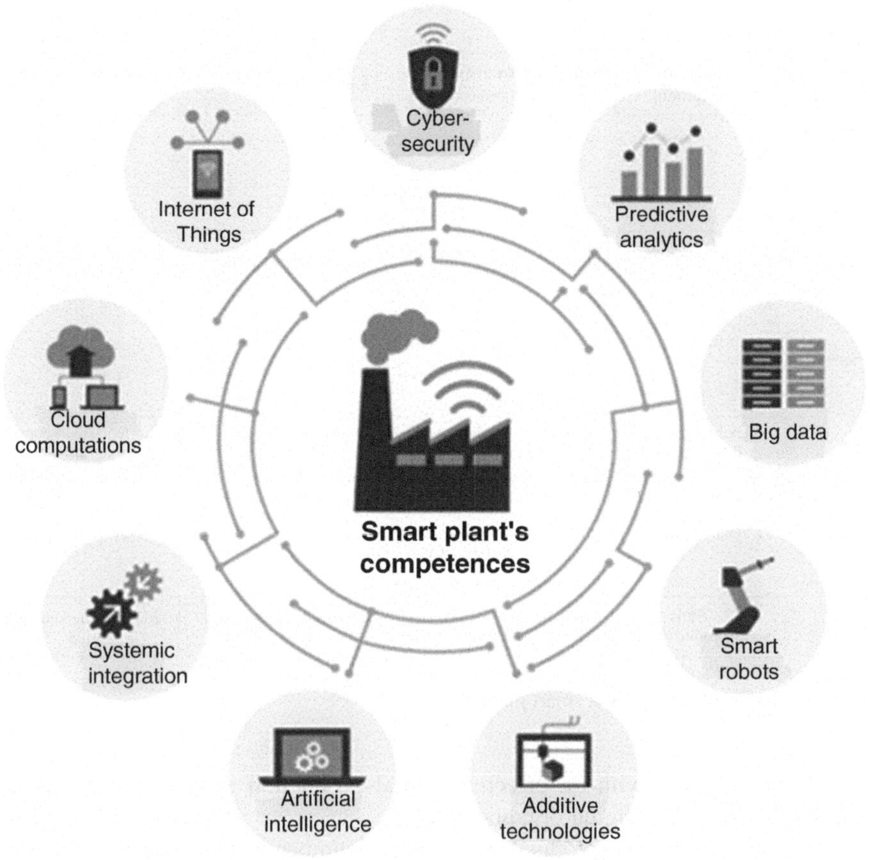

Fig. 8.13 A smart plant's competences

- Life cycle-long change management
- "Digital certification" following thousands of virtual tests of separate components and the whole system

Building a smart plant comprises several stages as shown in Fig. 8.14. The principal goal is launching the "smart work" process.

As shown in Fig. 8.14, setting up a smart plant comprises several consecutive stages, each one focusing on modernization of the technical system, which is governed by an organizational/economic one, which is followed by establishing an effective business that uses virtual companies (Fig. 8.15), which can substantially boost the agility inside the company, as well as in working with suppliers and customers.

Based on automated FMSs and smart production and with account for the benefits of digital technologies, which enable production of a broad and diversified product

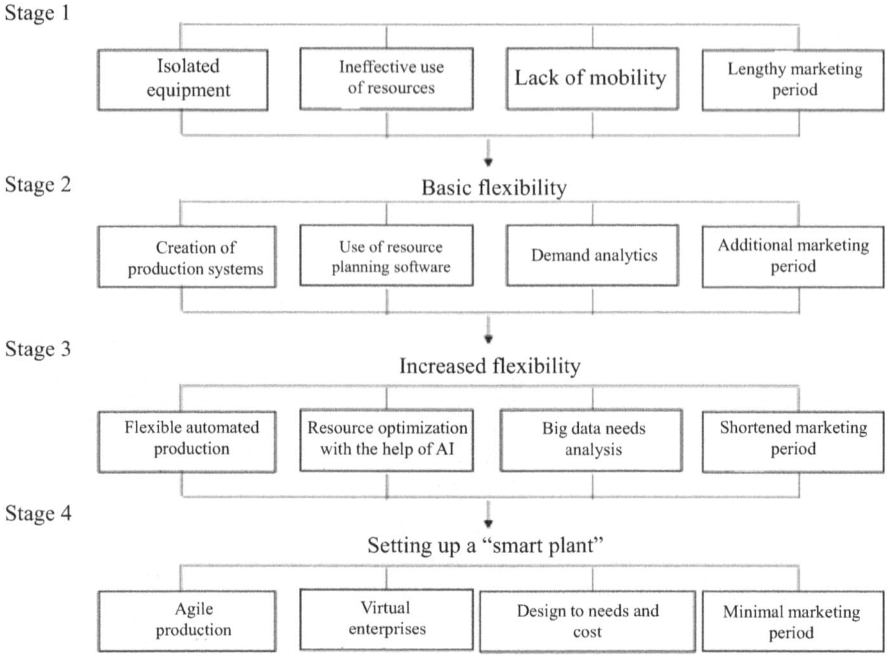

Fig. 8.14 Stages of setting up a smart plant

range, a smart plant with an effective digital production system can be set up (Fig. 8.16), which is a digital organization.

To create an effective production system, a number of approaches have been developed, which differ in product customization level, use of resources and intellectual potential:

- Lean Manufacturing is aimed at continuous removal of manpower, reduction of equipment use and production time with minimization of production area size, and increasing quality and competitiveness. A postulate of the Lean Manufacturing concept states that less continuous readjustment and higher flexibility help produce goods in small portions and successfully compete with large businesses running a full-scale production of similar goods.
- The Quick Response Manufacturing (QRM) strategy consists in lead time reduction and applies to an entire company. QRM focuses on lead time reduction through optimization of all operations, both inside and outside the organization.
- Agile Manufacturing (AM) can quickly adjust to the changing situation and operate in unstable markets.

The main differences between these approaches are described in Table 8.4.

To make the technical system function more effectively, it should be tailored to external trends through adding LP, QRM, or Agile mechanisms (increasing the

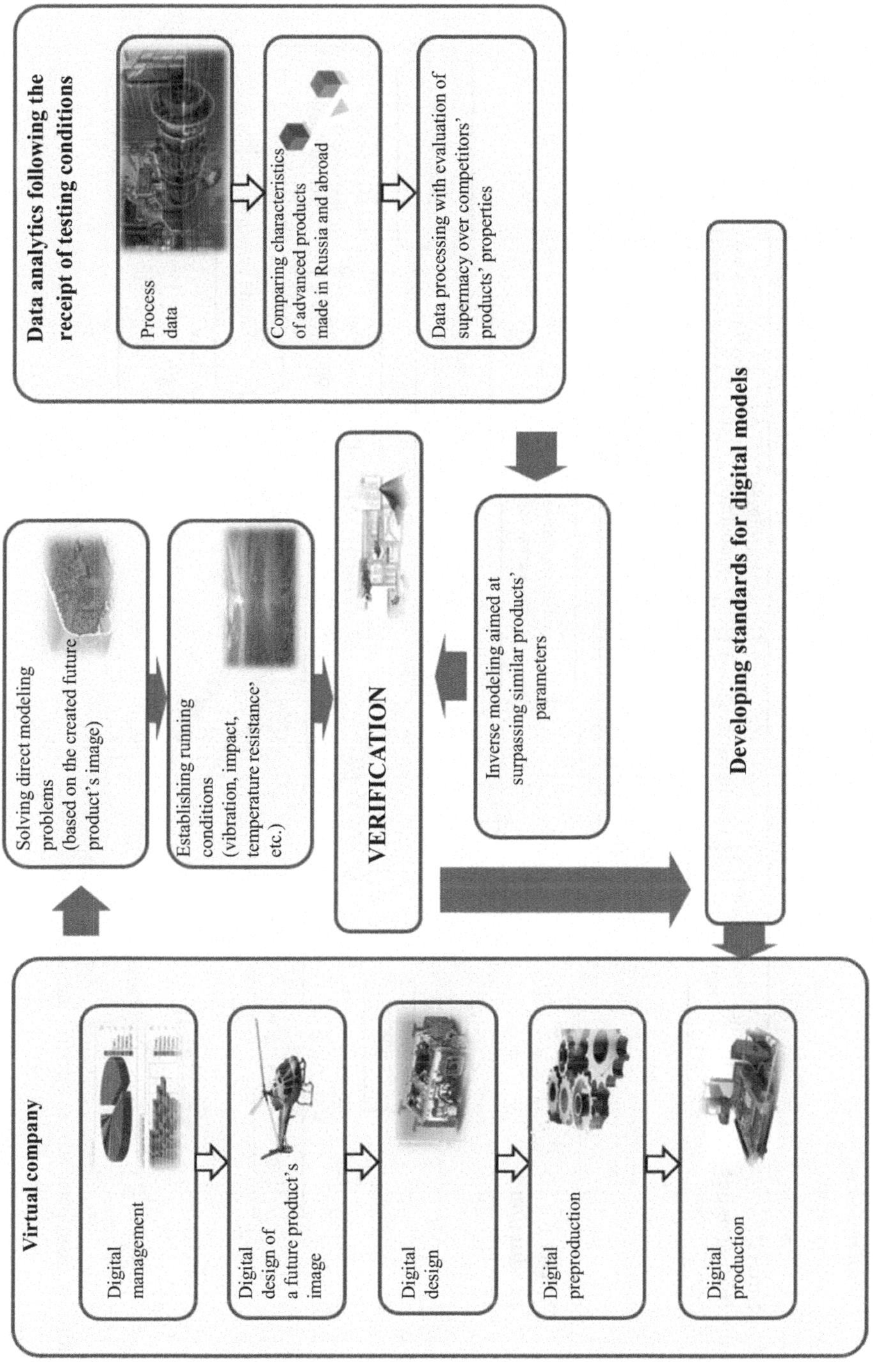

Fig. 8.15 Transition to a virtual digital company

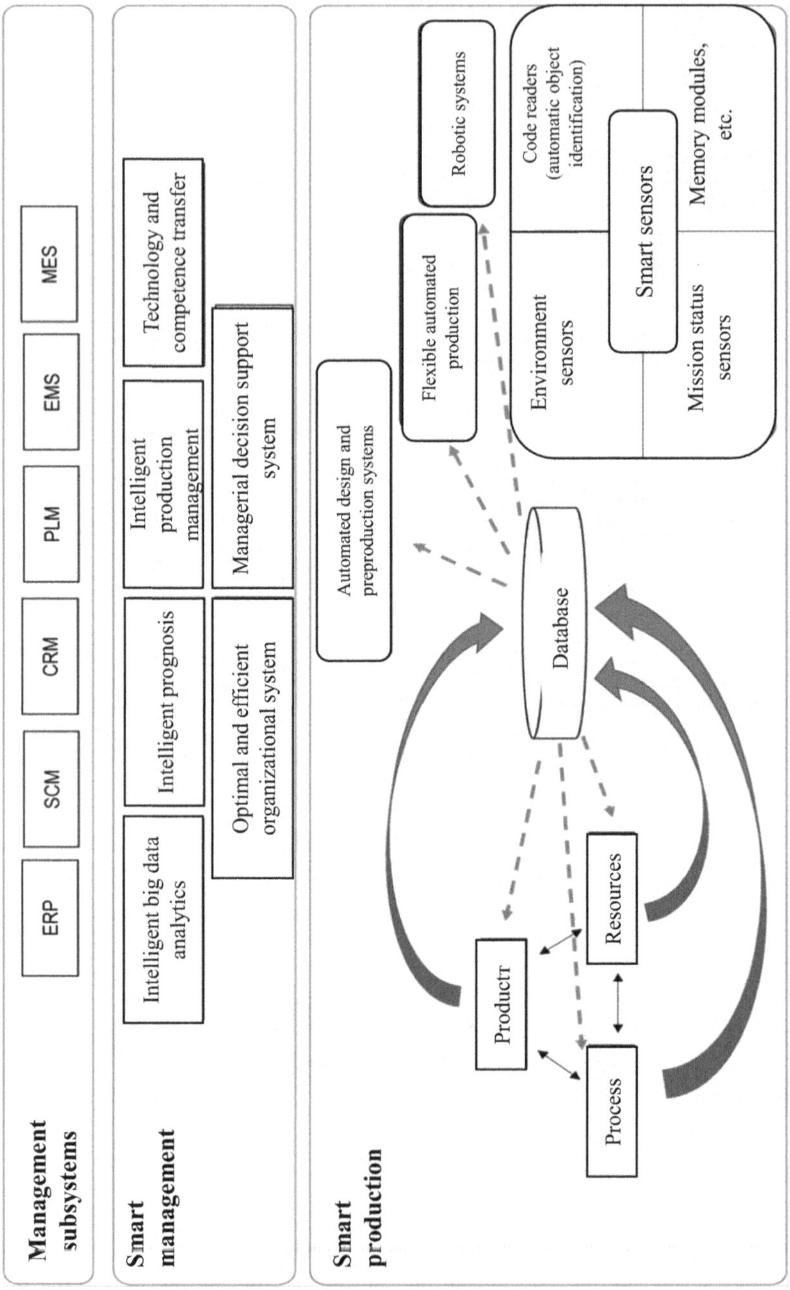

Fig. 8.16 A smart plant with an effective digital production system

Table 8.4 Differences between LM, ORM, and AM

Factor	Lean manufacturing (LM)	Quick response manufacturing (QRM)	Agile manufacturing (AM)
Strategic benchmark	Reduced costs	Execution speed	Effective work in unstable economic environments
Type of manufacturing	Large-scale and mass production	Medium- and small-batch production	Small-batch and one-off production
Product and service customization	Low–medium	Medium–high	High
Use of resources	100%	80%	>100% (use of numerous third-party resources)
Innovative	Low–medium	Medium–high	High

number of add-ons) and continuous modernization (ultimate changing/modifying existing subsystems' functionality and elements of a single system).

Up to a point, the structure of a technical system may expand due to incorporation of ultimately new elements. This requires large investments, use of one-of-a-kind competences, which can solve the problem of developing the system's functionality or transformation of existing systems, which are functionally similar, through their integration into a complex unified technical system. This should improve their functionality and quality of solutions to problems, which they are tailored to solve. All this can be implemented with the use of cutting-edge adaptive digital technologies that can help launch an Agile project.

In turn, a system's modernization leads to its evolution, raises the quality of existing, or adds new functions (for example, in the time of global digitization, every technical system should have parameters like data accessing speed, accuracy of data processing and analytics, informational and digital data transfer security, etc.) When modernizing a system, it is necessary to rely on policy objectives that are or should be reached with the help of this system, and on effectiveness analysis results demonstrated by the existing system; a negative dynamics of these signals a necessity of making new decisions concerning its modernization.

An important trait of technical system modernization process is that their dynamic progress implies a growing importance of recording complexity factors that accompany existing and future systems and complexes, which influence their effectiveness.

The effectiveness of any technical system is determined by approaches to its management concerning optimal use of material, financial and manpower resources, data flow regulation, improving the speed of the decision-making process, decision's quality, etc. Therefore, any technical system, especially a complex one, requires creation of an organizational and economic system with a functional including information backup, management, functionality control, regulation of business processes involved in ensuring effective functioning of the technical system.

The system's modernization will inevitably necessitate a transformation of the organizational and economic system, which manages it. A broader functionality (thanks to transition to an automated mode and elimination of redundant operations, adding new functions related to the rapid technical progress and changing market needs, etc.) simplifies the organizational and economic system's structure and, consequently, helps reduce costs. This raises a question: how significant are the changes resulting from the modernization of a complex system functioning within an economic system?

So far, there is no description in scientific literature of the relationship between the development of a complex technical system and organizational and economic system, which manages it, although improvement and modernization of technical systems changes organizational and economic systems' structure, as the management process is optimized, becomes more effective and less labor costly.

For example, transformation of a system that governs a complex technical system (when technical systems are integrated into a large one), by an organizational and economic system, may happen as shown in Fig. 8.17.

The pattern shows that chaotically structured technical systems, which lack an organizational structure, require a large number of budget sources, and often have redundant missions and developments, which take money.

Integration of isolated technical systems into a large and complex one, which should be governed by an organizational and economic system, produces pronounced effects resulting from cost reduction due to elimination of redundant elements and management processes, use of standardized basic structural elements, complex modernization planning with account for progressing digital technologies, reduced labor use in system governance, etc.

The center of a complex organizational and economic system (with a verified staff structure) solves the following problems related to its functioning as an integrated system:

- Choosing directions of advanced system development
- Technical and economic planning
- Real-time system management
- Planning, recording, and analysis of resources necessary for development, modernization, and proper functioning of the system
- Compiling scientific and research information about the system
- Management of scientific and research activity within the system
- Categorization of management processes taking place within a large integrated technical system with the help of advanced digital and artificial intelligence technologies

This is a description of the formation of a complex organizational and economic system, which requires using new approaches to operation of a large technical system.

The problem of finding new approaches and improving existing ones in the economic theory and practice is multidimensional, and it requires using a comprehensive and targeted approach. Particularly, it focuses on the development of analytical effectiveness research methods, as well as organizational and economic methods, which analyze any systems (technical, etc.) from the perspective of economy and transform them into complex organizational and economic systems,

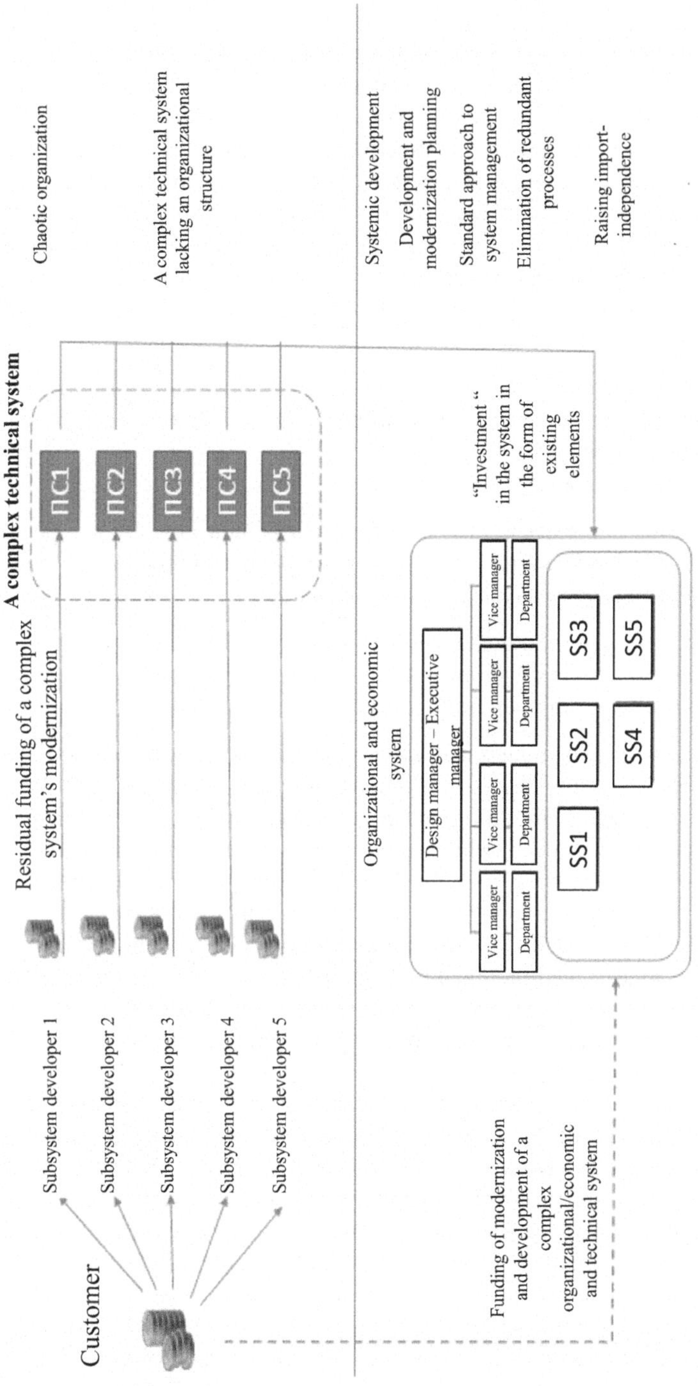

Fig. 8.17 Transformation of a system that controls a complex technical system as a unified organizational and economic complex (SS means "subsystem")

the management of which includes both economic and technical/technological steps, which raise its effectiveness.

The current state of research of modernization and organizational and economic systems, which focuses on their effectiveness, is such that so far there is no generally accepted terminology for this issue. Prior to directly addressing the problem of generating an integrated economic effectiveness indicator for a complex organizational and economic system, it is necessary to briefly explain the term.

A complex organizational and economic system is one, the analysis of which requires combined use of diverse models, theories and, in some cases, science disciplines (interdisciplinary research). In order to distinguish a complex organizational and economic system from other economic systems, several complexity aspects are considered:

- Structural complexity
- Complexity of the functioning (operation)
- Modernization complexity
- Development complexity

Therefore, the integrated economic effectiveness indicator of a complex organizational and economic system *B* is going to be a combination of generalized economic indicators of the system:

I_1—a generalized system indicator (million rubles)
I_2—a generalized system development indicator (million rubles)
I_3—a generalized system operation indicator (million rubles)

The goal of building integrated economic effectiveness indicators is ultimate when it comes to making strategic decisions and modernization for the foreseeable future. The prospect of building integrated quantitative economic effectiveness indicators should be solved while relying on new economic and mathematical integrated indicator building models. A standard integrated indicator computation mechanism is the summation method, which uses weighing coefficients. In different variants, this approach helps narrow multicriterion optimization down to standard scalar optimization of the functionality. However, the use of weighing coefficients or functions turns out to be a rough approach. Another serious drawback of using these is their subjective nature, because choosing a weighing coefficient is a challenging and hard-to-formalize task.

Integrated economic effectiveness indicators of particular systems help make a complex assessment of these new methods. It should be noted that integrated indicators imply generalization, categorization, and overall evaluation of different aspects of systems' modernization, development, and operation. Full and complete assessment of a system's economic effectiveness is a nontypical and challenging task. This poses a necessity to equal the new evaluation criteria with quite so stable and reviewable structures, which enable revealing significant characteristics in these systems, i.e., formalization of integrated indicators based on evaluation criteria.

Generalized indicators are a sum of costs referring to each of the cost groups that are characterized by such an indicator. Thus, the economic idea of an integrated economic effectiveness indicator as applied to a complex organizational and economic system lies in generalization of all modernization, development, and operation costs:

$$B = I_1 + I_2 + I_3.$$

To determine a complex economic system's economic effectiveness, it is necessary to calculate an annual integrated indicator with the help of statistical data, in the current situation (B_0 is the original state) and with ongoing modernization and development of a complex organizational and economic system (B_1). The economic effect—Δ (million rubles) is calculated in the following way:

$$\Delta = B_0 - B_1 - R,$$

where R is the costs resulting from the use of a complex organizational and economic system (personnel, etc.).

It is necessary to discuss the formation of the generalized indicator of a system's modernization (I_1) in detail. In this context, modernization implies significant improvement of separate components or whole subsystems within a complex organizational and economic system with account for modern technical and functionality standard. The main characteristic that influences the generalized modernization indicator is the share of modernized elements in subsystems and the number of subsystems, which are to be modernized within a given period of time. Individual factors, which influence the generalized indicator, are the extent of standardization, unification, components' series production, etc. Once activities that are subject to standardization/unification are verified, a known volume of work will can help calculate work costs. The hardest part of it is calculation of unification work costs. It can be determined from the perspective of experience, statistics, etc. Analyzing all aspects of unification can help calculate the cost of each of these aspects.

Assume there is a known list of modernization activities (for each subsystem), which are to be completed within a given period of time. These include ones that cannot be standardized. The cost of unification activities for each subsystem is A_i, where i is the subsystem's number. The remaining activities are subject to standardization. The biggest part of this work should be carried out one time, and the results should be applied to all subsystems, which are being modernized in this direction. Then the subsystem modernization cost for a complex organizational and economic system should be calculated as follows:

$$I_1 = \sum_{i=1}^{n} A_i + \sum_{j=1}^{m} \left(S_j + \sum_{k=1}^{l_j} P_{jk} \right),$$

where A_i denotes mandatory work applied to all subsystems (n is the number of subsystems); S_j is the cost of the j standardized work (m is the total number of standardized activities); P_{jk} is the cost of the standardized activity j for a subsystem (l_j is the total number of subsystems, to which the work j must be applied).

If a complex organizational and economic system is in its original state, the respective formula has the following shape (no standardization):

$$I_1 = \sum_{i=1}^{n} A_i + \sum_{j=1}^{m} \left(l_j \cdot S_j + \sum_{k=1}^{l_j} P_{jk} \right),$$

where S_j is the cost of the non-standardized work j (m is the total number of such works) for each of the l_j subsystems.

The economic effect of modernization within a complex organizational and economic system can be expressed as:

$$\Delta(I_1) = \sum_{j=1}^{m} (S_j \cdot (l_j - 1)),$$

where S_j is the cost of the j standardized work, l_j is the number of subsystems, in which it is necessary to do the work j.

Generalized development and operation indicators can be obtained in similar ways.

The economic effect of activities carried out within a complex organizational and economic system can be expressed as follows:

$$\Delta(I_2) = \sum_{i=1}^{q} A_i,$$

where q is the number of factors contributing to a complex organizational and economic system's competitive advantages; A_i is the acquired benefit of developing a complex organizational and economic system, which has resulted from the influence of competitive advantage factors.

The functioning of a complex organizational and economic system as a single whole implies less labor-costly subsystem servicing thanks to:

1. Sequential management of all operation processes with incorporation of smart management approaches. Single-type functions, which are performed by subsystem developers, are grouped and used by a management unit that is featured in a complex organizational and economic system, which can play the role of a computation and analytics center, and they ensure automated control of the system.
2. Introduction of advanced digital economic technologies into the system management process. For example, many types of activities can be fulfilled with the use of artificial intelligence, which can help automate a specific group of processes.

Eventually, it helps reduce personnel costs and achieve the target cost of the product being created.

The economic effect that occurs when operating a complex organizational and economic system can be expressed as follows:

$$\Delta(I_3) = \sum_{i=1}^{w} U_i,$$

where w is the number of factors, which help reduce operation costs resulting from the use of a complex organizational and economic system; U_i are the saved operation costs, which is thanks to the influence of each of the operation cost reduction factors.

The economic effect Δ can be calculated in the following way:

$$\Delta = \Delta(I_1) + \Delta(I_2) + \Delta(I_3) - R,$$

where R are the maintenance costs incurred by a complex organizational and economic system.

Thus, it is possible to calculate the economic effectiveness indicator of a complex organizational and economic system. Monitoring of changes in this indicator can help evaluate its dynamics; should there be a decline in the system's effectiveness, additional research can be held to determine the causes of the decline. On the one hand, it can be caused by the ineffectiveness of a technical system (due to functional depreciation, poor automation, failure to use advanced technologies and artificial intelligence), which is governed by the organizational and economic system; on the other hand, this can be due to the ineffectiveness of the organizational and economic system itself resulting from improper use of the managerial staff and material/technical resources engaged in the managerial process. Both cases require restructuring of the organizational and economic system, which will depend on the modernization of the technical system or on the necessity of structural optimization to stimulate the expected effect.

It is monitoring of the quantitative effectiveness indicator that helps determine measures aimed at costs needed for maintaining the organizational and economic system's functioning, raising its effectiveness while providing for high quality of solutions to problems posed by the system.

The economic effect of setting up and using a complex organizational and economic system as a single system consists in the following:

1. Ensuring progressive advance and planning of an organizational and economic system's work to maintain a product's and a whole organization's competitiveness in future, through integration of separate and functionally similar systems and applying a universal development pattern to these.
2. Elimination of redundant development and test costs through implementation in new subsystems of standardized components, software utilities, programmed subsystem building processes that are based on advanced artificial intelligence

techniques. This can reduce the labor cost of a complex organizational and economic system's components.

3. Reduced modernization costs and a smaller number of newly created subsystems in future by improving the complex organizational and economic system's technical characteristics and recalibration of existing initially high-tech systems to new objectives, which can predetermine a setup of an Agile manufacturing system.

However, the economic effect of such step-by-step modernization of the technical and the organizational and economic system, which manages it, has a limited potential, because at some point modernization becomes impossible or ineffective from the perspective of manufacturing advanced competitive goods that are supposed to win market dominance. This can be explained by the rapid technical and technological progress, emergence of new principles and methods of production, which necessitate using new technical and economic approaches to production of unique and highly competitive goods. This poses a necessity to create new technical and large organizational and economic systems, which rely on in-depth automation, industrial robotization, and increasingly versatile manufacturing facilities.

A pioneering system was built in 1981 by the Fanuc Corporation, Japan. The company built a flexible automated robot manufacturing system—Fuji FN-2, which predetermined the development of new production and production management patterns, such as QRM and Agile production, which are described in Table 8.4.

It should be noted that the evolution of the Agile method started with its use in software development. That was a transition from the standard step-by-step process, which is implemented throughout the lengthy product development and marketing period in the very end of the life cycle, to the new Agile method, which helps put a baseline product in the market and later add new characteristics in its future versions, reflective of customer feedback. The organization continues to receive feedback from customers and end users throughout the product life cycle, so any changes in customer and user reviews are included in the production plan.

Every new change dictated by the market demand is prioritized when designing a new product version; they can be applied when fulfilling the following cycle of designing the new version. The main characteristics of the Agile method include adaptability, effective use of customer feedback, and quick response to new market requirements. These fundamental principles define approaches to effective structuring of an enterprise to satisfy the rapidly changing market requirements and future needs.

Agile Manufacturing is a kind of management strategy, which should make a business more resistant to crises, demand fluctuations and other unpredictable phenomena.

Typical of organizations that have set up Agile businesses is an ability to promptly reconfigure labor and material resources, so they do not miss a chance to make a profit and avoid trouble. The concept's greatest advantage is an opportunity to quickly adapt to a changing situation and run a business in uncertain market environments. Agile manufacturing is well suited for branches with a high level of uncertainty (IT, electronics, etc.)

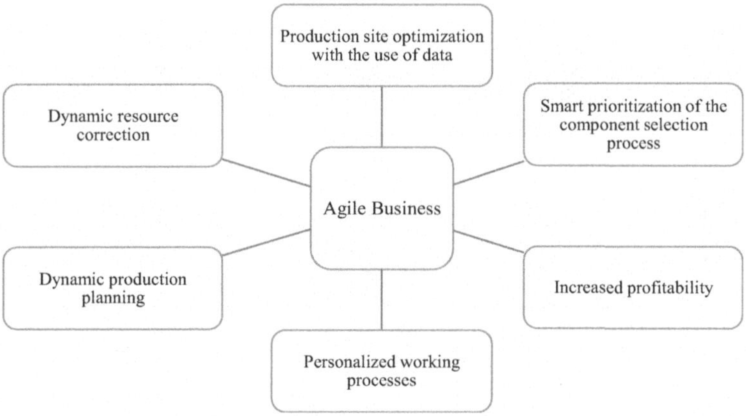

Fig. 8.18 Benefits of an Agile business

The use of the Agile method in development of advanced products relies on a concerted transition to digital technologies throughout the chain—from the design through shipment stage. Removal of design and production limits can give a business a competitive advantage in promoting its existing and advanced projects; besides, it can generate ultimately new profit channels with the help of Industry 4.0 technologies, which include big data, automation, the Internet of Things, etc.

Agile Manufacture implies the use of modern digital technologies and software solutions and solving important problems. This helps a company stay within specified cost limits while designing sophisticated future products (Fig. 8.18).

Production site optimization (equipped areas, where a company fulfills production) with the use of big data adds transparency to the process and speeds up the data flow throughout the production process.

Smart prioritization is carried out with the help of a digital mechanism of preparation for and recommendations concerning the choice of parts and components during assembly. The process relies on several criteria, including the type of material, delivery data, and availability of assembly equipment.

The goal of increasing profitability is directly related to an opportunity to maintain a high machine load and deliver products to the market or individual customers on time. Smart prioritization helps determine the most optimal production routes with account for the choice of material, equipment adjustment and processing. The production of each item is automatically organized into a sequence of operations, which can be implemented with a maximally effective machine loading.

Within a business, hundreds of operations are performed with the use of a variety of resources (materials, technologies, competences, personnel, etc.), which enables Agile businesses to create production (process) templates for each type of product. These templates can be assigned automatically with the use of existing methods, or they can be calibrated to each separate part with the use of smart methods. This approach helps carry out dynamic production planning.

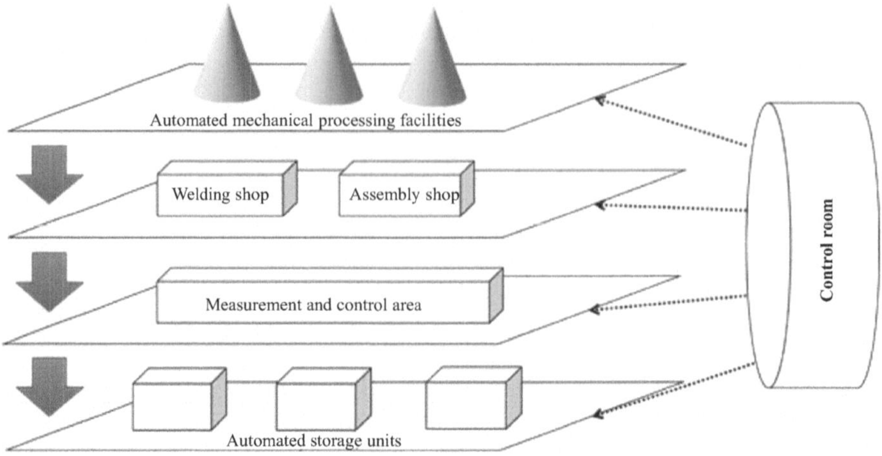

Fig. 8.19 A flexible automated industrial robot production plant

Dynamic planning consists in optimal equipment- and man-loading depending on market needs and the time of putting a new product on the market.

Dynamic resource correction implies automatic reallocation of jobs between units, should a separate system element break down or stay idle.

Thus, it can be concluded that Agile manufacture presents a combination of two ultimately new dynamically evolving (with account for new scientific discoveries, technical and technological progress) systems: a complex technical and large organizational and economic one, which manages the technical system. It is the effectiveness of the functioning that influences a business's ability to achieve the target prime cost of a new product.

The Agile systems discussed above have laid the ground for the manufacturing pattern used by Fanuc Corporation. A new versatile automated industrial robot production facility was built according to the Agile method, and is structured (it should be noted that the factory's structural pattern has not changed; it is only the scale and process management systems that changed with the advent of new technologies) as described in Fig. 8.19.

The industrial robot production system uses industrial software, which controls a variety of systems, and the machinery interacts with the help of the Internet of Things. The company develops high-quality products that boast "intelligence," "super-accuracy," and "high functionality," and it keeps to one key criterion of successful robot production—high reliability.

Right now, Fanuc employs more than 1000 robots. The company continues to extend the use of smart robots for more effective processing and assembly. This helps achieve higher technical characteristics and optimize the cost component through in-house automation.

Fanuc's tremendous experience in in-house implementation of state-of-the-art advanced products demonstrates that setting up a smart, automated, and robot-

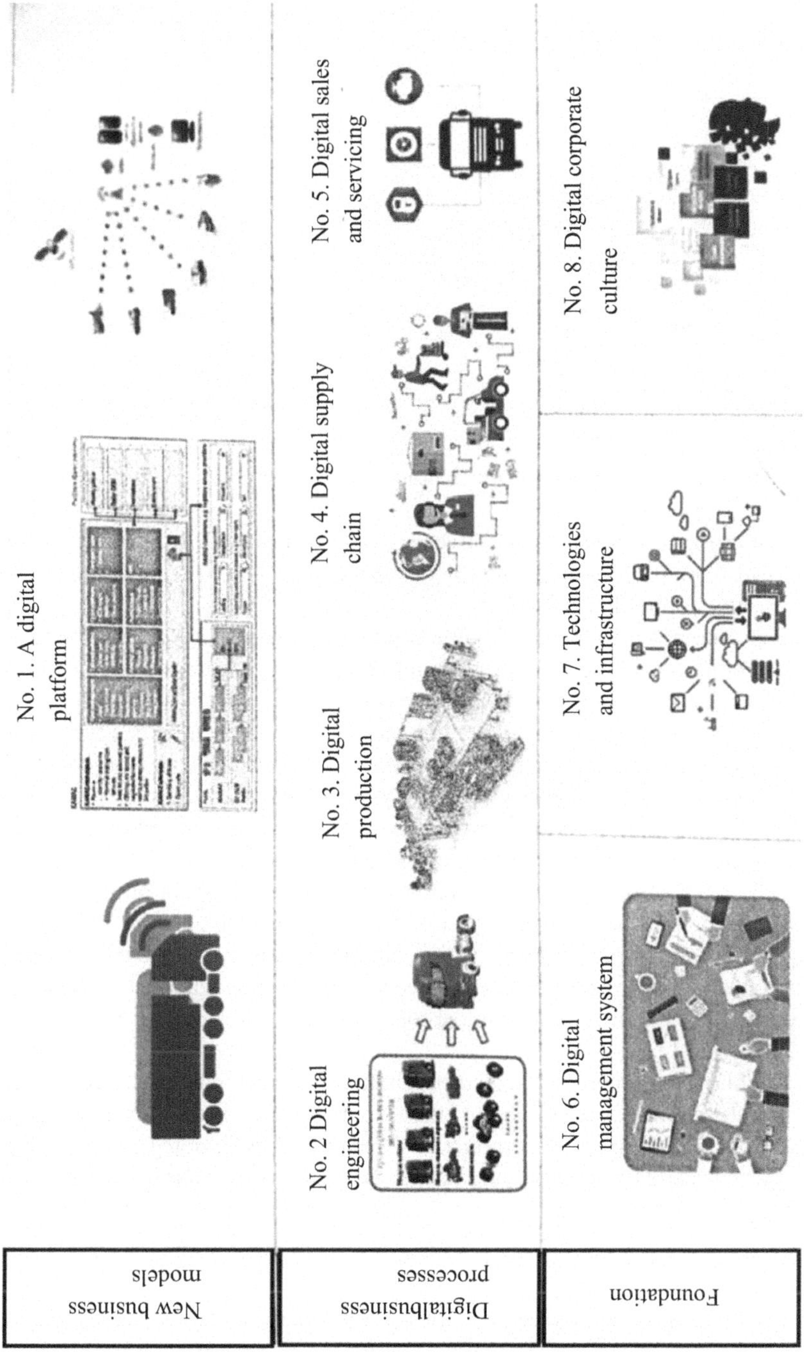

Fig. 8.20 The main Agile manufacturing components

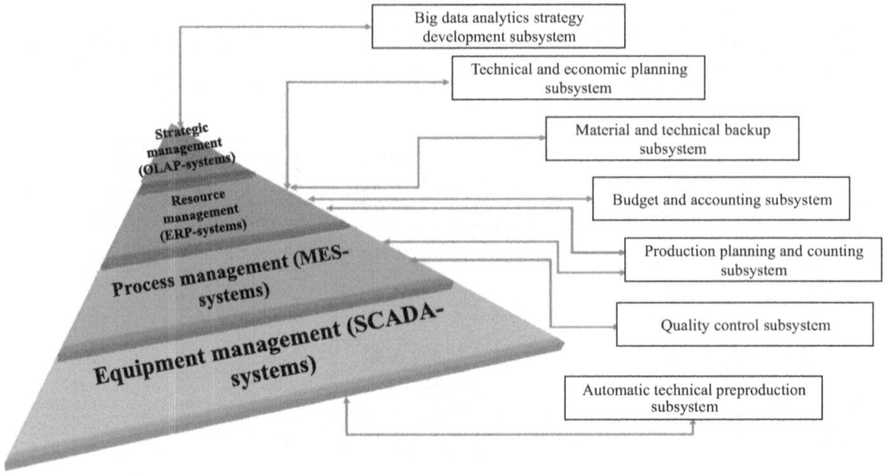

Fig. 8.21 An integrated automated production management system

powered enterprise ensures high effectiveness and low production costs. Theoretical and practical work in designing smart plants has contributed to the implementation of smart production tools in Russia.

Building automated facilities, flexible production lines and Agile solutions are extensively practiced in the Russian automobile industry. Figure 8.20 demonstrates the use of Agile technologies at one of such enterprises.

The circle of companies implementing the Agile method is expanding. It has spread to machine building in Russia as well, as subsystems and processes are divided into modular components, which can be modified and recalibrated when necessary. There is a network of trustworthy suppliers, which optimizes the cost of components and helps modify delivery and client schedules.

This is how a manufacture process is transformed on the basis of progressing and developing techniques and technologies. As a result, smart plants are launched on new Agile Manufacturing systems, which are controlled by respective integrated automated manufacturing systems, which work as large organizational and economical systems. They comprise a number of automated subsystems, as shown in Fig. 8.21.

The use of an integrated automated production management system results in a transition of an entire business onto an end-to-end digital platform, which integrates automated design, equipment-, production-, resource-, and strategic management systems, which contain separate subsystems that focus on a number of important goals:

- Design and manufacture products to a fixed cost and planned competitiveness level
- Reaching a highly competitive quality level through continuous search for and incorporation of cutting-edge scientific and technical projects

- Achieving price competitiveness through reduction of costs and losses at each product life cycle stage
- Sustainable growth through production line updates with the use of digital technologies and the potential of Agile Manufacturing is achieved by combining rapidly evolving techniques and technologies used in these systems

Integrated automated production management systems are built to ensure optimal control of the processes of design and production of advanced goods, which is demanded at all life cycle stages, with the use of advanced smart cyber-physical and cyber-economical systems. These systems make up an integrated digital platform that backs up the making of effective managerial decisions during life cycle management, which provide for effective informational backup of a life cycle management system within a business; also, they focus on creation of products, which can capture and maintain an good share of consumer market for a long period of time, thus prolonging the product's life cycle.

Chapter 9
Product Life Cycle Management

9.1 Integrated Digital Platform Supporting Effective Managerial Decision-Making in Product Life Cycle Management

Influenced by the digital economy and extensive use of advanced digital technologies that are part of all corporate business processes, production management processes are changing rapidly. Digital technologies stimulate creation of new business models intended for designing and producing future goods, which can help a company launch the rapid product development process. New market players coming up with ultimately new products make it necessary to use a production management strategy, which should help produce highly competitive goods and services with the highest technical and quality characteristics at the lowest cost.

Transformation of a company into a digital one is the only solution. The basics of building one are described in the previous chapters. Data and analytics become these companies' main assets.

According to Gartner, in 2018 there were more than 7 billion objects connected to each other with the Internet of Things.

The distribution of devices connected to a unified interactive network has changed corporate work methods in all spheres. Smartphones, laptops, tablets, and PCs ease the data analytics process and help companies automate and improve all operations within any unified digital organization relying on the use of respective technological toolkits that combine various cutting-edge technologies. There are three main complex tools, which are widely used by industrial companies. These include data management tools, which act as an important corporate asset; process support tools; tools that coordinate interaction with the outer environment. The whole toolkit is shown in Fig. 9.1.

Figure 9.1 shows that the toolkit includes technologies like artificial intelligence, fog computing, end-to-end, quantum, supercomputer, and identification technologies, Blockchain, and neural networks. A combination of these ensures effective

A. Tyulin, A. Chursin, *The New Economy of the Product Life Cycle*,
https://doi.org/10.1007/978-3-030-37814-1_9

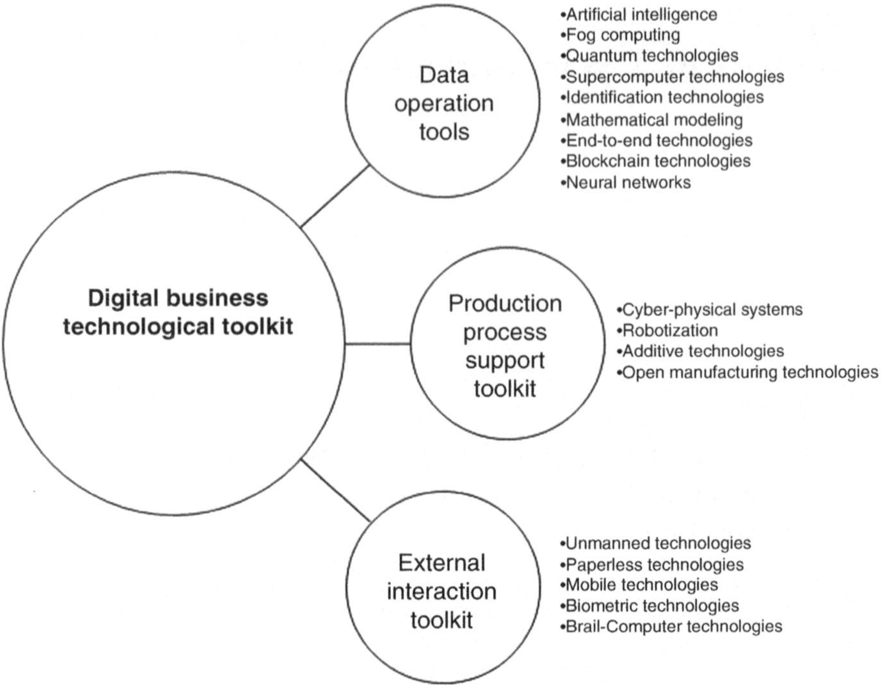

Fig. 9.1 A digital company's technological toolkit

management and use of big data in the corporate and outer environments, as the data helps construct an effective managerial decision-making process. This toolkit helps use past experience in developing future solutions while relying on advanced self-learning algorithms. Besides, decision-making models are capable of adapting to new conditions once new data is obtained. An important advantage of using these technologies in data management is an opportunity to achieve high processing accuracy and minimize the occurrence of human-factor-related errors.

The process support toolkit includes cyber-physical systems, 3D (printing) technologies, robotic, additive, and open manufacturing technologies. Practical use of this toolkit helps businesses arrange the production process in the most effective fashion and create a *high quality–cost* balance to ensure competitive rapid product development. In this case, all technologies are used with the purpose of minimizing the product life cycle length and costs at all stages while retaining high technical parameters and creating a competitive functionality.

The toolkit that regulates interaction with the outer environment includes unmanned, paperless, mobile, biometrical, and Brain–Computer technologies. On the one hand, this part of the technological toolkit enables real-time monitoring of changes in the outer environment and customer preferences, to form innovative future product design concepts; on the other hand, it prolongs a product's market life cycle through organized servicing, which boosts a manufacturer's image on the consumer market.

For more effective use of the technological toolkit when designing advanced products, a company can create integrated digital effective managerial decision-making platforms while managing such products' life cycle.

An integrated digital platform should pursue the following goals:

- Organizing rapid product development processes that ensure products' competitiveness by the time they are sold.
- Maintaining the decision-making process while doing the life cycle management of advanced products with the use of intellectual methods.
- Maintaining the decision-making process with the use of extended data analytics.
- Automation of business processes and decision-making procedures that are part of business processes.
- Smart integration of technologies currently used by an organization and recombination of innovations.

Recombination of innovations implies smart integration of various technologies into a single one, which helps create new digital business assets. These assets are key tools that maintain real digital transformation of business models that are used by a rapid product development company. Recombinant innovation enables implementation of the company's technological and competence potential, which comprises a set of key technologies and competences. These can be reconfigured with the help of an integrated digital platform in new advanced fashions to generate new concepts.

A bright example of using this approach to running a business is a model and background built by the 3M Company. The organization uses 47 "technological platforms," based on which it creates innovative products relating to different spheres. Since its inception, the company has produced more than 55,000 items, i.e., more than 1000 items per technology.

An important strategic direction, in which the company works, is integration of existing technological platforms with the goal of designing innovative products, whose technical characteristics and customer appeal outrun those of similar products made in other parts of the world. A good example of such a product designed with the help of a group of technological platforms is a high-voltage cable described in Fig. 9.2.

As follows from the picture, the 3M Company employed five corporate technological platforms to design a new-generation cable. This is quite indicative of effective in-house communication between specialists and technological integration. Thus, the use of existing technological platforms helps generate new product design concepts and produce goods, which can capture a large share of the market or create a new consumer market segment.

Meanwhile, the formation of a new product design concept based on recombined innovations with the use of an integrated digital platform launches a design and engineering project realization cycle for a product, which can trigger a rapid development process managed by the same integrated digital platform.

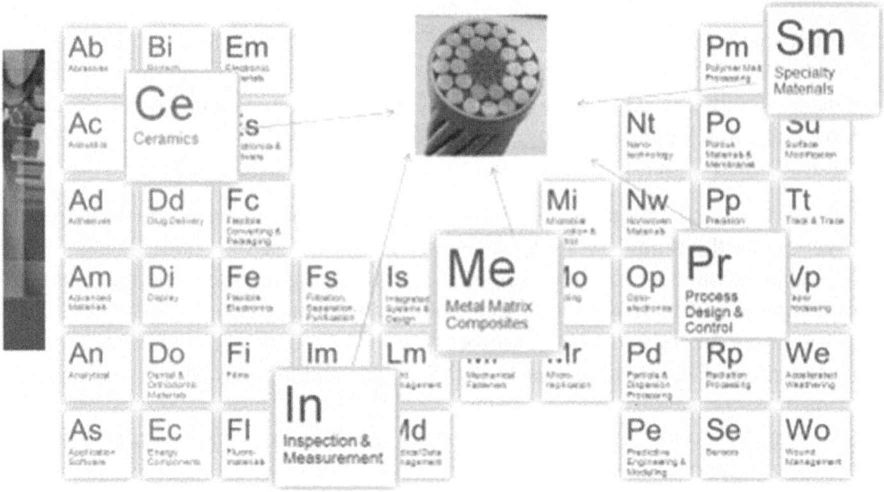

Fig. 9.2 Recombination of innovative technologies

As follows from the above, an integrated digital platform is based on PLM-systems, which help solve a bunch of product life cycle management problems using a modeling technique. Important tasks that are solved by such systems, which can become part of an effective integrated decision-making platform's functionality while doing PLM procedures, include:

- Maintaining the product life cycle management process until the utilization stage is over.
- General technological development achieved with the help of a flexible design system.
- Document and information model management.
- Providing funds for delivery and maintenance of products' operating cycle.
- Analyzing and testing working objects with account for their characteristics.
- Reliability checks that are part of technical maintenance and other functions.

However, that is not enough for an effective managerial decision-making support that should be provided while managing the life cycle of a rapidly developing product.

Therefore, there are standard requirements to the process of creating an integrated digital platform, which can solve the problem.

First, this digital platform should rely on automatic development of recommendations concerning management of a rapidly developing product's life cycle while referring to the analysis of the current situation and previous tendencies executed after big data processing.

In this context, automated support of the decision-making process means:

1. Providing reference data without automatic formulation of database inquiries.

2. Providing reference data with automatic formulation of database inquiries and referring to the problem environment.
3. Graphic visualization of received reference data and information about decision-making methods.
4. Providing decision-making recommendations.
5. Narrowing down the search of a solution by users.
6. Choosing and advising the most acceptable decisions with reference to rankings.
7. Modeling possible consequences of decision-making when carrying out life cycle management.

The second requirement is building an analytics block (a decision-making process support subsystem) as part of an integrated digital platform, which should help solve the life cycle management problem by providing recommendations in such a way as to ensure the product's categorization as a rapidly developed product by the time of its sale. This analytical block can be structured as shown in Fig. 9.3.

The decision-making support subsystem relies on analytical instruments, which help it formulate recommendations concerning the results of decisions made when solving life cycle management problems, such as:

- Evaluation of future market demand for a product depending on scientific, technical, technological, and customer expectation trends.
- Evaluation of a company's innovative, competence, and resource potential.
- Evaluation of the competitiveness of a product's technical and economic image followed by its monitoring at all life cycle stages with the use of low signal risk analytics.
- Evaluation of changes in the market situation to determine the market saturation point and shape a new advanced product.

Thus, the most important problem, which should be solved prior to deciding to design a new product is stimulating future market demands for specific product types, analyzing their technical and economic images' conformity with customers' requirements and desires and identifying their most important technical characteristics, which meet customers' expectations and make them\products competitive.

To solve this problem, the analytics subsystem should reflect smart customer expectation evaluation methods, the importance of specific properties, which are paramount for customers, and competitors' offers and projects, which prompt the subsystem to release adequate recommendations concerning the formation of a future product's technical and economic shape that ensures successful rapid development. Therefore, the analytical decision-making subsystem should include a customer expectation, characteristics, and properties analytics block, as these expectations, characteristics, and properties are exceptionally valuable for customers and characterize future needs of the market.

The second important problem, which is solved when deciding upon a future product's technical and economic image, is providing a sufficient amount of resources for production while relying on this image. The technical and economic image and its replication should rely on the company's respective innovative

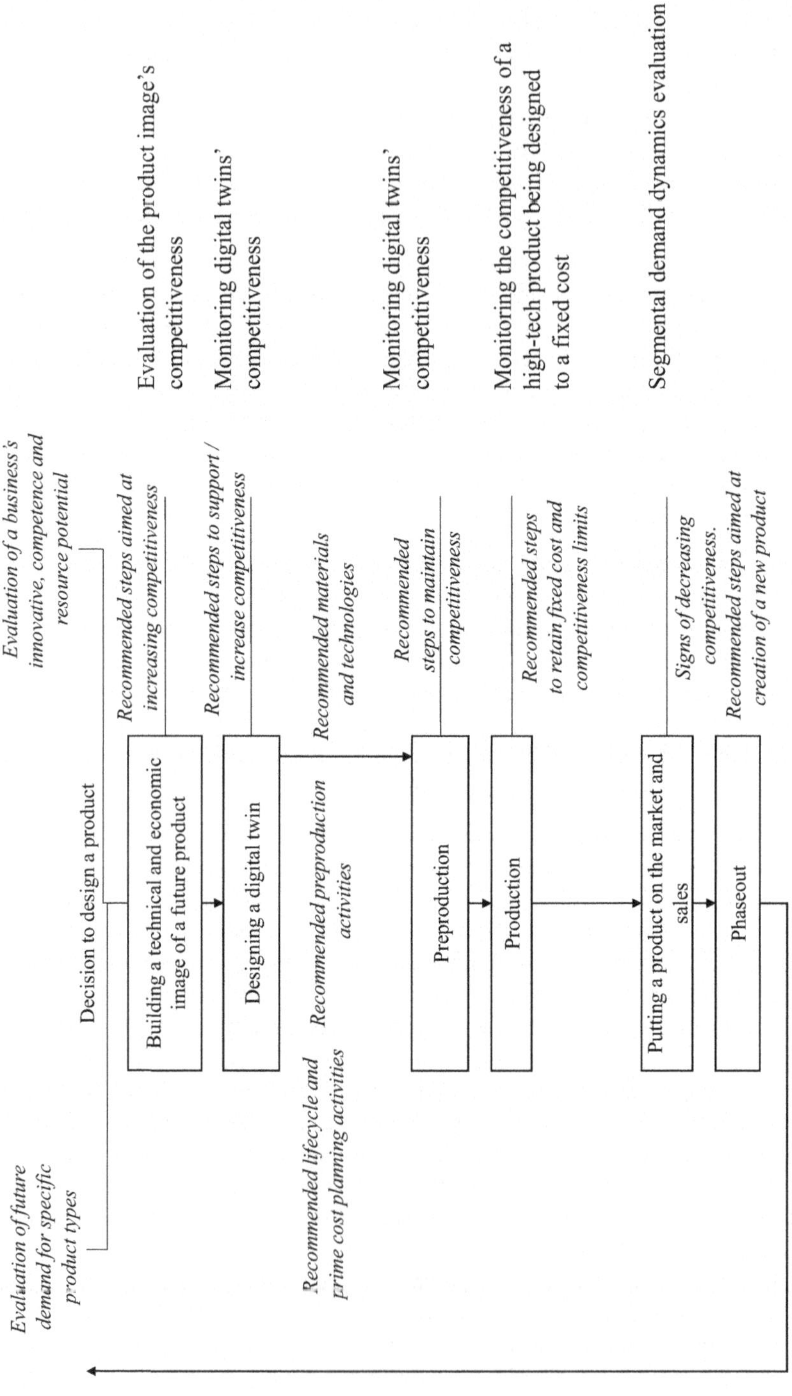

Fig. 9.3 Analytical block (analytical decision-making support subsystem) of an integrated digital platform

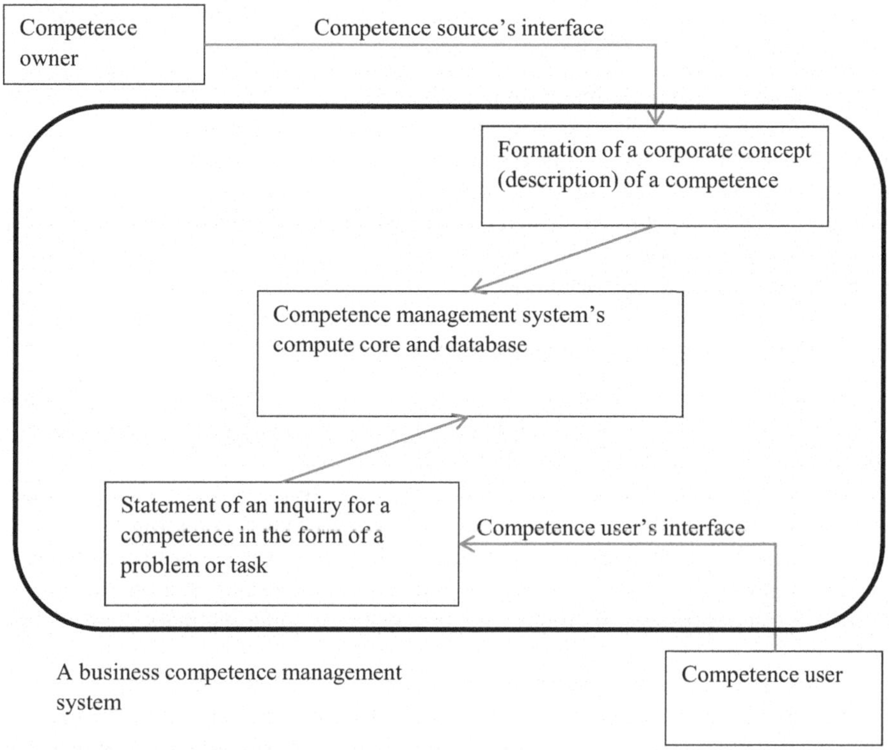

Fig. 9.4 A business competence management system

potential and resources that are needed for successful running of a project. This is the only scenario, which helps design a competitive product and put it on the market in due time. It solves two subtasks: management of material resources, which are formed as required by existing *resource and innovation potential formation models* that are used to design new advanced products; and management of nonmaterial assets (competences), which is carried out with the use of respective mechanisms.

Competence management mechanisms imply building a formalized idea of a competence, which can describe the characteristics and functional potential of an organization's competences and their uses, as well as describe the characteristics and functions of competences, which a company needs to use while designing advanced products. In other words, this mechanism should be effective, and the competence user (the advanced product manufacturer) should be able to describe needed properties and potential of a competence, which should help solve a current problem or task. The information system will browse for a competence upon a customer's inquiry.

The competence to customer route is shown in Fig. 9.4.

The general purpose of this system is formalization and classification of all competences available within an organization and enabling their distribution.

The segmentation of generalized competence management helps reveal the following basic processes within the mechanism:

- Revealing competences inside an organization.
- Formalization of competences according to specific criteria and forms.
- Classification and inclusion of competences in a knowledgebase.
- Checking the knowledgebase for competences according to specific criteria.
- Ranking of competences, which are stored in the knowledgebase and meet specific fixed criteria.
- Providing access to competences.
- Setting up interaction between sources of competences and competence users.
- Developing competence implementation methods.

The described approach accepts business competence management as part of the block that evaluates the innovative potential of an analytical decision-making subsystem that is part of the integrated digital platform.

After implementing these two blocks' smart algorithms, the subsystem generates recommendations, which help make a relevant choice of industrial projects based on market need and manufacturer potential analytics; also, it generates recommendations to improve the product's technical and economic image based on customers' expectations concerning the product that is expected to enter the market after a few years.

At this stage, analytical algorithms of the block that defines the product image's competitiveness and predicts it begin to work by the time of its entry into the market. They take into account the production cycle's length and external factors, which can deprive the product of its unique features and ability to stimulate rapid development due to rapidly developing competitors' innovative concepts introduced by major market players. This explains the necessity of introducing the competitiveness monitoring block into the analytical subsystem at all life cycle stages with account for internal and external influences to ensure optimal execution of the production process.

Therefore, the next problem, which is solved when shaping a future product, is evaluation of its competitiveness and monitoring it throughout the rest of the life cycle. Along with the algorithm, it is necessary to include in the platform's analytical subsystem the *product image competitiveness evaluation* and the *competitiveness dynamics evaluation blocks* at all life cycle stages.

It is important to determine economic (price) parameters, which should keep a product competitive in future.

The use of smart price prediction methods can make a product price competitive, as well as form possible prime cost options, which depend on how internal and external risk factors and uncertainties influence the production process.

Based on this data, the analytical decision-making subsystem generates cost (prime cost) threshold recommendations for a future product being manufactured according to the analyzed technical and economic image. These recommendations should be introduced and later used when designing a product to a fixed cost.

This block analyzes the opportunity to design a product with the use of planned design and engineering solutions while ensuring its price competitiveness. In case whereby a product fails to demonstrate price competitiveness, the subsystem recommends a more effective design and engineering solution to bring the price to a competitive level. These design and engineering solutions should be able to reduce an item's prime cost. Besides, the subsystem can recommend to take steps to maintain or increase competitiveness, optimize costs when fulfilling the production cycle, choose (recombine) materials, and technologies, which should help achieve rapid product development. Also, it can ensure the most effective preproduction with maximum automation and reduction of the current life cycle stage.

While further monitoring a product's competitiveness indicators at all life cycle phases, the analytical system will immediately respond to any threat to the prospect of global leadership or to fulfillment of the rapid development strategy.

Thus, the need to introduce this analytical managerial decision-making subsystem is explained by the importance of tasks that are solved in an automated mode with the use of intellectual big data processing and analytics algorithms. Meanwhile, this subsystem provides new added value tools when designing and selling advanced products and using cutting-edge digital solutions at all life cycle stages (Fig. 9.5).

Formation and activation of the tools shown in Fig. 9.5 with the help of the analytical decision-making subsystem's functions will produce the following economic effects:

- Reduced maintenance costs
- Shorter market launch periods
- More accurate customer behavior prognosis
- Reduced quality control costs
- Reduced storage costs
- Higher productivity
- Minimal downtime
- Fewer managerial errors and losses through minimal use of the human factor while justifying decisions

As follows from the above, the integrated effective managerial decision-making platform that is used in life cycle management comprises a group of large systems, which can maximally optimize the process of its creation with minimal costs, highest quality, and best available technical and functional product characteristics. These systems combine and interact with the integrated digital platform as shown in Fig. 9.6.

The integrated digital platform shown in Fig. 9.6, which makes effective managerial decisions as part of PLM, illustrates the interaction between its three main systems: the ERP-, PLM-, and an analytical managerial decision-making subsystem. The main problem, which these systems solve while being part of an integrated digital platform, is effective management of the product creation process at all life cycle stages. It should help the product become an advanced one.

The input data received by the integrated digital platform concerns a rapid product development project and a task of analyzing its potential successful

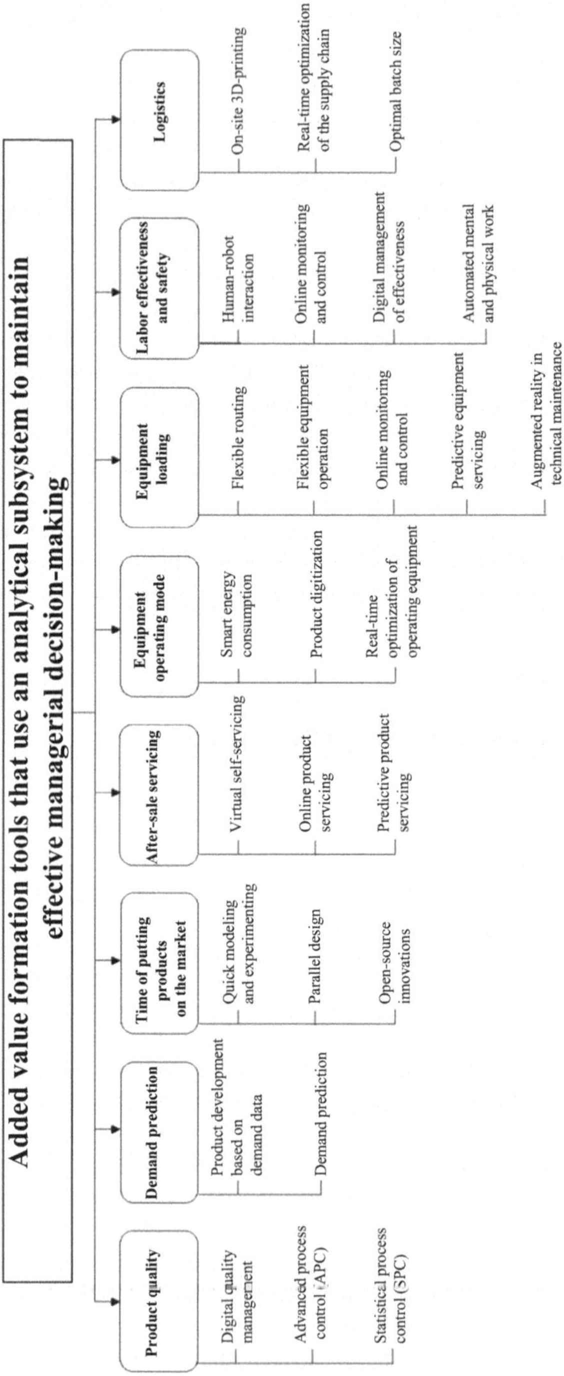

Fig. 9.5 Added value formation tools

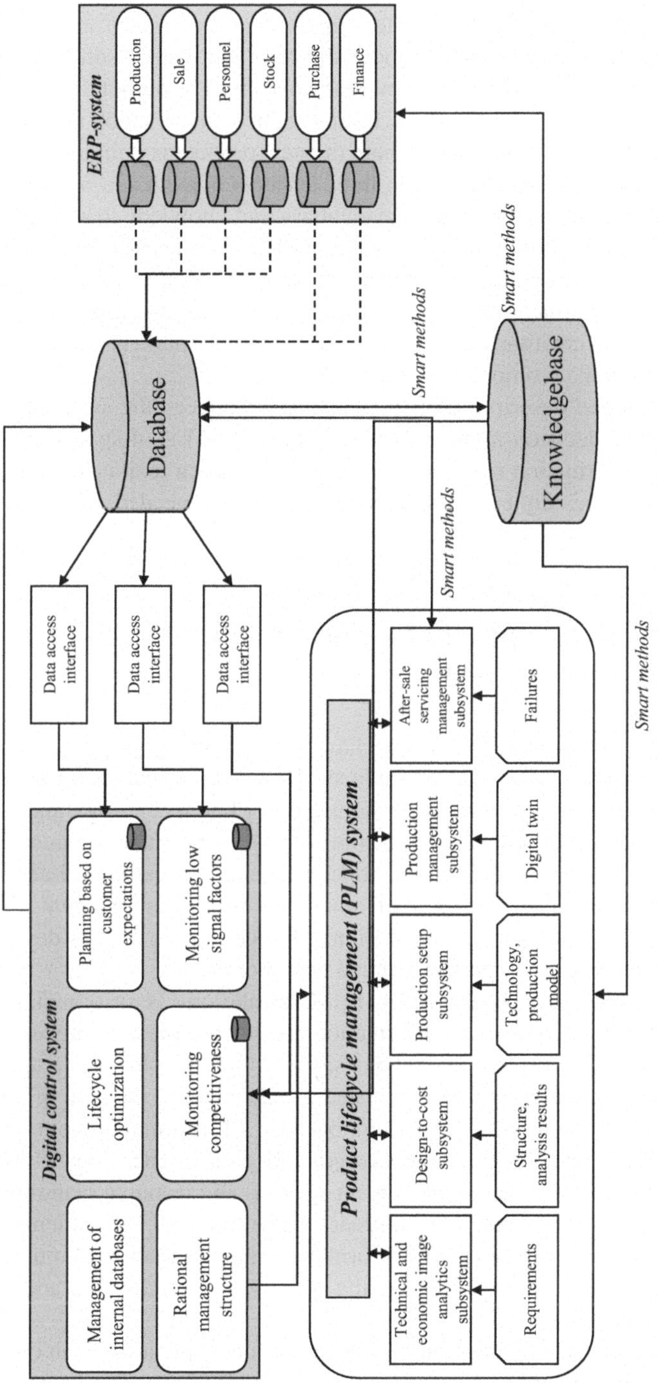

Fig. 9.6 An integrated digital platform, which makes effective managerial decisions at the product life cycle management stage

distribution. The next task is operating competitiveness parameters while relying on incoming information about the previous individual experience and data derived from the global information space about the best innovative solutions, effective management, business-running tools, mechanisms, etc. The output data accumulated throughout the life cycle highlights the most effective steps, which need to be taken to achieve leadership in all markets, where the new product is going to be presented.

The interaction between these main three systems is mediated by the integrated digital platform's two key elements: the database and knowledgebase. The database presents information required for design and production process management, while the knowledgebase provides an intellectual support of this process and builds algorithms of smart interaction between all systems of the integrated platform; their main goal is creating advanced products in all markets and bringing the manufacturer to global competitive leadership.

Making a detailed description of the interaction between the systems within the digital managerial decision-making platform during PLM is of special interest as it helps track the information support of integration between major systems' separate elements and how this digital platform helps achieve the needed effect.

9.2 Information Support of a Corporate Life Cycle Management System

To maintain competitiveness, businesses have to constantly and quickly launch new products and satisfy market needs. Reduced launch time and therefore increased effectiveness can be achieved through the use of a PLM technology and a communication system as an implementation tool. Usually, the main product life cycle processes include design, production, operation, and discarding. Each of these processes comprises a combination of other processes. For example, the design process includes scientific, research, design, engineering, as well as designing and engineering preproduction, and launching a new product.

The product life cycle management (PLM) technology is an organized technological system, which controls all information about a product and processes related to it throughout its life cycle—from the design through the phaseout stage.

Because the technology is directly related to information, the core of a life cycle management system is a unified platform, because all information contained in the PLM system makes up the new product's *digital life record*. Therefore, PLM relies on the integration of product data management systems and resource planning within a single communication system. This approach helps bring all departments and divisions to cooperation within a unified information space. Creation of a unified environment for effective interaction makes the activity of each division and the whole business more transparent.

This is followed by a description of a PLM platform operating in an organization and its information support.

The previous chapters describe a comprehensive methodological network, which helps maintain effective production of highly competitive products. The methodology relies on economic and mathematical models, which use various types of original data describing products' technical and economic characteristics, as well as market conjunctures, manufacturing environments, etc. Therefore, specific data, which must be used to make the methodology effective, is diverse, and their effective implementation is related to building universal cyber-economic systems.

A unified corporate digital life cycle management platform, being a cyber-economic system, helps fulfill an important task of standardizing original data and tools when implementing methodical and applied instruments. A digital platform's information structure is built with reference to standards applied to calculation instruments while solving economic and mathematical tasks in a science-intensive industry. Here is how a schematic structure of a digital platform's information core can be presented:

- A digital platform's low-level framework
- Original data structure developer
- A library of different methods
- Computational conveyor
- Universal database

These structural elements need a closer look. The platform's low-layer framework is a set of API functions intended for unification and operation of the main types of data. The module provides a service for the digital platform's main blocks. It should be noted that this block should be implemented effectively, because its effective implementation predetermines the whole platform's effective implementation.

The original data structure processor is a library of specific original data processing methods. Because economic and mathematical methods that are used in high-tech industries operate original data, which is structured in different ways, specific procedures should be applied to process and unify this data.

Processing unstructured original data involves the use of abstract data structures. Bringing different original data down to a common shape helps effectively unify and standardize input and output data for various methods. On the one hand, there are methods, which are used to bring big data volumes down to an abstract (common) shape for unification; on the other hand, a digital platform should feature advanced tools for processing complexly structured data. Therefore, the original data structure processor uses object- and application-oriented programming technologies.

Object-oriented data structures represent data of any complexity. However, complex object-oriented data structures are based on universal abstract data structures. This approach helps achieve effective fulfillment and high flexibility when processing original data structures.

The library of common methods has its own set of service functions, which help implement needed mathematical methods, which are parts of economic and mathematical methodologies used in high-tech industries.

Along with standard mathematical and statistical functions, this block contains quite a number of specific functions, which are used in methodologies. The library of common methods has one most important property: it is constantly replenished by new functions. The library's architecture is designed in such a way as to enable inclusion of new methods, which emerge when implementing new methodologies. Thus, the digital PLM platform, apart from being used for various existing methodologies, naturally evolves when new methodologies are applied.

Data flow computing relies on a computational conveyor, which controls the computation process according to a program set by the methodology being used. This computation pattern helps effectively carry out computing experiments while using economic and mathematical methods.

Because many economic and mathematical methodologies rely on simulation modeling, the digital PLM features a computing conveyor, which is intended for simulation modeling as part of economic and mathematical calculation methods used at different life cycle stages.

Closely related to simulation modeling is the presence in the computing conveyor of calculation tools that rely on agent-based and game theory modeling.

A specific role in the structure of a digital platform belongs to a universal database. As noted above, effective use of a PLM platform in a company depends on an opportunity to integrate different economic and mathematical methodologies. To provide a unified communication and computation environment for mathematical methodologies, a universal database should be created. The database should be abstract enough to reflect data with different structures. This can be achieved by using the original structure of an abstract database. The needed level of abstraction is provided through the use of primitive data types with the goal of storing information in the database. However, the abstract database should feature specific mechanisms enabling its operation with the use of object-oriented mechanisms.

The digital PLM platform has an effective scalable structure, which enables stream processing of information analysis methods that are used in production. These economic and mathematical methodologies help carry out simulation-modeling-based computing experiments from unified positions. In economic and mathematical methodologies, simulation modeling helps analyze various scenarios with reference to the changing economic situation.

Meanwhile, the digital platform relies on the model of a database that would be intended for storing original data and computation results. According to a chosen type of architecture of the given universal computational platform, the database core plays an important role in stream computation and data obtained after data computation.

This digital platform system that relies on the database model with an integrated interface between the computation conveyor and database core, enables planning and execution of multi-layer computing patterns that rely on economic and mathematical methods.

Next thing we do is formalize the scheme of the given unified information approach to simulate evaluation of product life cycle parameters while relying on information provided by the digital control system. The main parameters, which characterize different product life cycle stages, are usually presented as integrated indicators. Each integrated indicator is built on the basis of a variety of individual indicators. There number of individual indicator integration options may differ depending on the breakdown of parameter representation. This approach helps calculate and evaluate the main life cycle parameters from a unified position.

The information basis for implementation of mathematical evaluation of the main product life cycle parameters is a database, which is integrated into a unified digital PLM platform. With its help, it is possible to structure computational procedures from a unified position. On the other hand, using a single database that is integrated into a digital platform makes it possible to use existing data as original data in completing mathematical evaluation procedures for the main life cycle parameters.

The mathematical statement of evaluation of the main product life cycle parameters is described below. Assume that the main parameters are:

P_1 is the first evaluation parameter
P_2 is the second evaluation parameter
...
P_N is the N evaluation parameter

Every main parameter is evaluated according to its ability to reach a fixed value. Fixed values are original data for the economic and mathematical model, and they should be specified in the main documents describing these parameters.

This mathematical model uses a nondimensional approach. Assume that each main parameter that characterizes a particular life cycle stage is an integrated batch of individual parameters. This hierarchical structure will be represented as a directed connected graph. Here is a formal definition of the graph that is used in the mathematical model.

Assume that there is a fixed main product life cycle fulfillment parameter (P). According to this approach, the parameter evaluates the degree of achieving goals of a new product design project. The parameter will depend on a multitude of individual ones. These are:

$\{S_1, S_2, \ldots, S_M\}$ is the multitude of individual parameters.

Thus, the main parameter comprises a M number of individual parameters. Each integral parameter S_j has an information element, and it presents the following values:

- [Name^j] is the name of the parameter.
- [Pointer^j] is reference to an integrated parameter, which includes this particular parameter.
- [R^j] is the degree of reaching the planned parameter at a certain time point.
- [α^j] are internal factors occurring at a particular life cycle stage.

As usual, it can be assumed that if a given parameter is the main one characterizing a life cycle phase, referring to a parameter, which includes this parameter as an element, has a NULL value. Assume that the graph is an information tree, in which only one vertex has a null reference to the life cycle stage, which is characterized by the parameter being used.

The information structure built in this fashion is a directed graph, which is a tree. Use of these structures is traditional while presenting the hierarchy of indicators. This model is also a connected directed graph.

A connected directed graph is one, in which each vertex and edge has a numerical function. These numerical functions can be statistical and dynamic. In a statistical directed graph, functions ascribed to vertices and edges are not time-dependent. The information model described above presupposes the following functions (functional values):

$$F : \text{Pointer}^j \to \mathbb{R},$$

where $\mathbb{R}$ is the real axis. The meaning of the F value is that it demonstrates the indicator's weight (significance) at a particular life cycle stage. These ratios can be designated as follows:

$$\beta_j = F\left(\text{Pointer}^j\right).$$

The above model has the value R^j, which denotes a project's readiness at a given point of time. This parameter can take on values within the $R^j \in (0, 1)$ range.

Therefore, the fulfillment of planned parameters ranges from 0 to 1. Assume that this value is above zero and therefore it is possible to evaluate the fulfillment of a life cycle stage. The 1 value means that the project is fully finished.

On the one hand, this model reflects the internal fulfillment of planned parameters (R^j); on the other hand, it is necessary to focus on functional fulfillment. The difference between these concepts is that the internal degree of fulfillment is related to temporal and material resource spending when completing a life cycle stage, while the functional fulfillment is determined based on the percentage of life cycle completed by the current moment of time. The division is explained by the inevitable lag in the functional completion, which is observed at the beginning of life cycle stage fulfillment, compared to the internal fulfillment of planned parameters.

To model the dependence of the functional and internal fulfillment, a logarithmic law of dependence should be used:

$$f^j\left(1 + \alpha^j \ln R^j\right).$$

This functional dependence is expressed in coefficients α^j, which describe the nodes of the connected directed life cycle hierarchy graph. An approximate diagram of this function is shown in Fig. 9.7.

The X line reflects the internal fulfillment of planned indicators, while the Y line—the functional fulfillment of planned indicators. As follows from the diagram, at the initial stage functional fulfillment can be negative. To avoid that, the following formula should be used to calculate functional fulfillment of planned indicators:

$$f^j\left(1 + \alpha^j \ln R^j\right)_+,$$

where

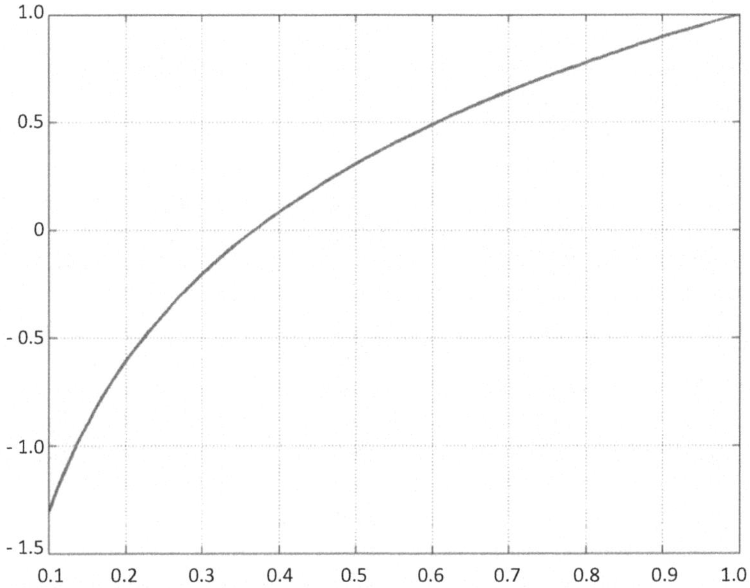

Fig. 9.7 Oriented indicator hierarchy graph for a specific life cycle stage

$$(x)_+ \begin{cases} 0, x \leq 0; \\ x, x > 0. \end{cases}$$

This cutoff function is used because formally obtained negative values are replaced by zero values. This approach helps use these mathematical structures when building a mathematical life cycle state fulfillment models when launching a new production.

The function, which is used to calculate functional fulfillment of planned values, is logarithmic. It has just one internal parameter—α, which is a linear part of the function. However, as long as we use a cutoff function, which zeroes negative values, the functional fulfillment function becomes highly nonlinear according to the α parameter. To research this nonlinearity, the dependence between this parameter and the threshold value should be found. Here is an equation:

$$f(\alpha, R) = 0.$$

Evidently, with $R = 1$ the equation is solved in a trivial way; therefore, it should be assumed the $R \neq 1$ statement is fulfilled. In this case, we have the following equation:

$$(1 + \alpha \log R) = 0,$$

from which follows

$$\alpha = -\frac{1}{\log R}.$$

The dependence of α on R is shown in Fig. 9.8.

The diagram shows that the nonlinearity becomes observable with low values only. The fulfillment of planned indicators should be high (0.5 or higher).

The obtained dependence equation demonstrates the internal indicator's α influence on the functional fulfillment of planned values.

The mathematical algorithm of using the mathematical life cycle evaluation model, which defines the main parameters, is described in Table 9.1.

After using this algorithm, only one vertex remains, which describes a life cycle stage. The vertex is going to contain the value R^0, which expresses the fulfillment of planned indicators at a particular life cycle stage. Based on the resulting value, it is possible to evaluate the main indicators of the implementation of innovative projects.

This algorithm is based on the procedure of original graph's factorization. This procedure can be illustrated. Assume that there is a graph tree with indicators, which characterize the product life cycle stage shown in Fig. 9.9.

Step 1 of the factorization procedure results in a graph, from which the bottom level D has been removed, and the next level C parameters are calculated according to formulas shown in Table 9.1. The resulting situation admits the use of a similar procedure for level C, which is schematically shown in Fig. 9.10.

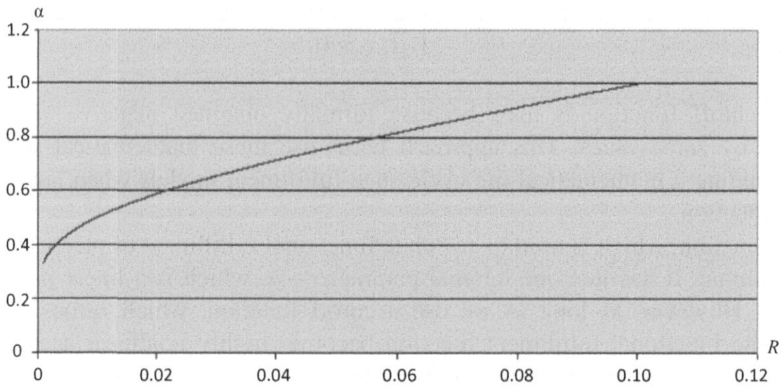

Fig. 9.8 Dependence of α on R

Fig. 9.9 The hierarchy of indicators characterizing a product life cycle stage

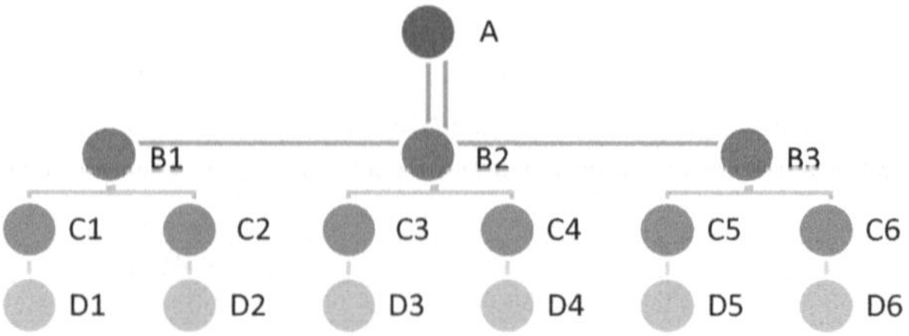

Table 9.1 The mathematical algorithm of using the mathematical life cycle evaluation model, which defines the main parameters

Step 1	Building a hierarchical life cycle stage indicator model.
Step 2	Choice of coefficients α^j, β^j and other mathematical model parameters.
Step 3	Carrying out the tree factorization procedure. It can be expressed in the formula: $R^i = \frac{\sum_{k=1}^{M} \beta^k \left(1 = \alpha^k \ln R^k\right)}{\sum_{k=1}^{M} \beta^k}.$ This formula calculates the internal fulfillment of planned indicators for all vertices that are referred to only by unreferred vertices.
Step 4	Removal of all vertices, to which there are no references and respective edges.
Step 5	If the remaining graph has only one head vertex, the algorithm ends. If not, it is transited to Step 3.

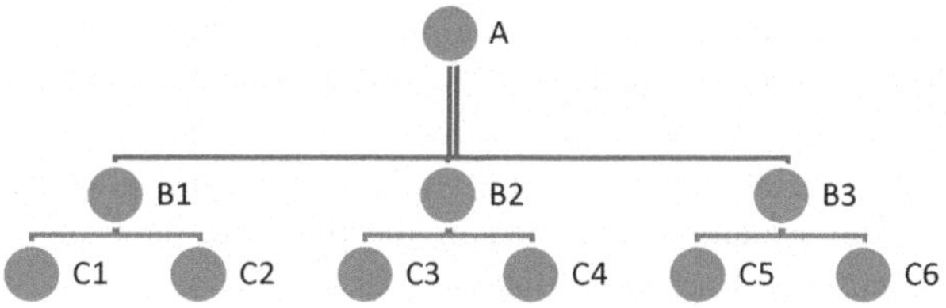

Fig. 9.10 A hierarchy to which a single factorization procedure has been applied

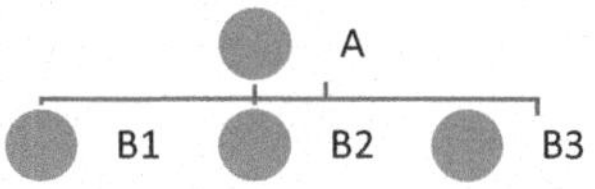

Fig. 9.11 A hierarchy, to which the factorization has been applied twice

Step 2 of the factorization procedure results in a graph, in which there is no level C either, and the C-level parameters are used for calculating B-level parameters. Again, it is possible to apply the same procedure to level B, as it is shown in Fig. 9.11.

The final step results in a graph consisting of levels A and B. In this case, the factorization procedure can be applied too, which will result in the removal of the bottom level. The resulting graph will degenerate, as it will consist of the head vertex only (Fig. 9.12). Actually, the factorization operations resulted in parameter data calculation taking place at every next level.

A theoretical ground for the use of simulation modeling when evaluating basic parameters of innovative project implementation is represented by Markov processes. The Theory of Markov processes is a powerful tool for analyzing process dynamics influenced by occasional factors. The main statement of the theory is that the likelihood of a system's transition from one position to another depends on the current situation only, without any reference to history. Herein, the Markov processes are going to be applied to evaluate changes in the parameters at different product life cycle stages. As a rule, it is economic uncertainties that act as occasional factors in this context. The flow of risks for these factors can be accurately described from the perspective of the Theory of Markov processes.

It is structured in the following way:

1. There is a finite set $S = \{s_1, s_2, \ldots, s_m\}$, which are a system's states.
2. The finite set $X = \{x_1, x_2, \ldots, x_n\}$ is fixed, which it is an input set.
3. The finite set $Y = \{y_1, y_2, \ldots, y_k\}$ is fixed, which is an output set.
4. A specified function: $T : S \times X \rightarrow S$. It is referred to as a *transition function.*
5. A specified function: $\Omega : S \times X \rightarrow Y$. This function is called an *output function.*

Assume that there are fixed functions $s \in S$, $x \in X$, then the state $s' = T(s, x)$ interpreted as the next state of the system that results from the influence x. The mathematically described structure is called the *finite automation* or *machine.* The multitudes X and Y are called *alphabets*, an *input* and *output* ones, respectively.

Within this task, this scheme has the following interpretation: different states are the main product life cycle stage values. The input alphabet is the influence of external factors. The output alphabet describes the cost resulting from these factors' influence. Later, the cost of the influence can also be taken into consideration while evaluating the main parameters of life cycle stages. The transition function is a reflection of external factors' influence on the parameters of life cycle stages.

The structure described in the previous paragraph is a deterministic machine because the transition function within it is one-values and nonrandom. However, as noted above, when describing economic parameters, which influence the mail life cycle parameters, the influence and its result should be viewed as random (uncertain). Particularly, the transition and output functions should be random.

The structure of a stochastic machine described above is an occasional model intended for evaluation of random parameters. In simulated progress parameter evaluation models, Markov's chains can be used, and they also can be viewed as generalized finite automations. Because the model focuses on a limited number of states, the transition function of a Markov chain can be expressed as a square matrix:

$$P = \begin{pmatrix} p_{11} & \cdots & p_{1n} \\ \vdots & \ddots & \vdots \\ p_{n1} & \cdots & p_{nn} \end{pmatrix}.$$

The idea of the matrix is the element p_{ij} describing the probability of the system's transition from the state j to the state i. Because elements of the transition matrix are probability values, it should meet the following:

Fig. 9.12 A degenerate graph consisting of the head vertex only (after triple factorization)

$$p_{ij} \geq 0, p_{ij} \leq 1,$$

and the following condition:

$$p_{i1} + p_{i2} + \ldots + p_{in} = 1,$$
$$i = 1, \ 2, \ \ldots, \ n.$$

Matrixes, which meet these conditions, are called *random* or *stochastic.*

In simulation life cycle evaluation models, Markov processes are used as described below. In evaluating a set of main life cycle parameters, the following vector can be used:

$$S = \begin{pmatrix} s_1 \\ s_2 \\ \ldots \\ s_n \end{pmatrix}.$$

Assume that the model features a set of random factors, which influence the main parameters. These factors are expressed by the vector

$$F = \begin{pmatrix} f_1 \\ f_2 \\ \ldots \\ f_m \end{pmatrix}.$$

To evaluate each parameter, it is advisable to use a Markov chain, which refers to random external factors' influence. Therefore, these factors' influence depends on the current life cycle parameters.

To calculate probability characteristics of final life cycle values, simulation modeling should be used, which can be described by the pattern shown below. Assume that there is a system with parameter *S,* and it is influenced by random parameter *F.* With account for a Markov chain, these factors' single-time influence can be described by the following pattern:

$$\widetilde{S} = T(S, F).$$

The concept of simulation modeling lies in analyzing of a large consecutive group of random influence scenarios to understand their influence on the system. The result is an iteration formula, which should be used during simulation modeling. The pattern can be described by the following formula:

$$S^{t+1} = T(S^t, F),$$

where S^t denotes the value of the parameter vector and a time point t. This formula can be expressed in a coordinate form:

$$\begin{pmatrix} s_1^{t+1} \\ s_2^{t+1} \\ \dots \\ s_n^{t+1} \end{pmatrix} = \begin{pmatrix} F_1\left(s_1^t, f_1\right) \\ F_2\left(s_2^t, f_2\right) \\ \dots \\ F_n\left(s_n^t, f_n\right) \end{pmatrix}.$$

The existence of life cycle evaluation limits is very important:

$$\lim_{t \to \infty} S^t = S^*.$$

The existence of such a limit can be confirmed with the help of so-called *ergodic theorems* that work for Markov processes. With a mathematical model, which uses ergodic Markov processes it is possible to determine clear S^* values. However, to execute simulation modeling, mean type methods will be used herein to average random factors' influence. For that purpose, the digital platform uses a specific simulation module to carry out simulation modeling and to process average information.

Full-scale use and implementation of this mathematical model is possible with the help of digital platform's functions. The estimates' accuracy will depend on that of original data, which necessitates the use of a database that is integrated into the platform. The use of this database with the advanced project implementation evaluation mode activated can help fulfill the dynamic method of evaluation of the main parameters.

Using the economic and mathematical modeling device for evaluation of the main product life cycle parameters implies the use of the simulated modeling concept with the goal of obtaining indicators; this concept characterizes the life cycle stage at the current moment and ensures effective prognosis of these indicators' dynamics in future. When determining these indicators, different information, which describes particular life cycle indicators, should be used.

When processing big data with the goal of building economic and mathematical product life cycle evaluation models, it is necessary to use a special kind of information and analytics system. It is a digital platform featuring a set of necessary tools. A database management system should be the core of this system. Therefore, when designing the database management system, all requirements applied to the platform should be taken into consideration.

A full and complete implementation of this digital platform should be based on an abstract database designed and implemented with the use of a modern database concept. The use of a large database as a key element of the digital platform helps implement a huge number of economic and mathematical methods, which ensure effective production of new types of goods. The results of mathematical modeling will be stored and processed in a unified database while relying in a unified interfacing protocol.

9.3 The Theoretical Basis of Creating Future Goods, Which Put a Business onto the Path of Rapid Development

Modern science-intensive businesses have to plan their development pathways with account for the importance of rapid product development because of the toughening competition coupled with increasingly strict requirements to science-intensive products. Consequently, to ensure sustainable competitiveness of a product and/or business, advanced products should be designed and manufactured, which should put organizations in a rapid development mode.

Rapid development of a science-intensive business stimulated by the manufacturing of advanced products is driven by the information support of the product life cycle management system described in Sect. 9.2. Also, to maintain rapid development, businesses should continue to modernize their own management and decision-making systems, because the economy has entered the stage of rapid changes, which include in-depth digitization and extensive use of unique technological competences. The changes are extremely quick, so they require that the traditional management system used in science-intensive companies be adaptive to these changes and modern economic trends.

The biggest problem that accompanies designing and manufacturing of science-intensive goods is a lengthy production cycle, since it takes quite a while to make this kind of product; therefore, there is a serious risk that by the time it is released into the market, its technical and economic characteristics will have fallen behind. To solve this problem, it is necessary to rely on an approach that is based on designing and manufacturing of advanced products.

When designing advanced products, both existing and future standards should be considered. While it is hardly possible to look into future, it is possible (and necessary!) to try to predict future market needs when designing advanced products and relying on rapid development methods.

The ultimate task is building a product life cycle management system in such a way as to ensure designing and manufacturing of products while relying on the rapid product development concept. To solve this task, it is necessary to rely on postulates that provide a theoretical basis for designing advanced products, which help a business to progress rapidly.

Postulate 1 To ensure flexible development of science-intensive products, it is necessary to rely on digital twins of both goods and production cycles.

Postulate 2 When building advanced products' digital twins, it is important use methods that rely on flexible and multilayer development processes.

Postulate 3 At each product life cycle stage, it is necessary to rely on prognosis of technological and economic market needs.

Postulate 4 While carrying out a prognosis of future needs in high-tech markets, it is necessary to use a flexible planning horizon.

Postulate 5 To ensure permanent competitiveness of a product, it is necessary to constantly modify and modernize the future product's image.

These postulates need a closer look as applied to the goal of creating advanced products and ensuring an organization's rapid development.

Postulate 1 states that designing and manufacturing science-intensive goods of future requires the use of digital twins and thus provides for a flexible design process. Flexible product design and project launching implies that the current project execution schedule be constantly modified as information about the project's current status continues to arrive. For science-intensive and high-tech products, changes in the project, as a rule, mean serious financial and time costs; not infrequently, these changes are not possible at all. Therefore, the digital twin model's products and processes are vital tools, which ensure flexible development of science-intensive projects. A digital twin is a digital copy of a physical object. A digital model is very easy to modify. Also, when designing a future product, it is necessary to use a multivariant design approach, which allows to consider several technical solutions. With the help of digital models, it is possible to not only mobilize tools and launch a parallel process, but also foretell products' technical characteristics.

According to Postulate 1, a digital twin is a kind of thing that really helps create advanced products, which can meet future consumer needs and market conjuncture. Automated project management systems play a crucial role in the process. The information support of a product life cycle management system helps obtain real-time data and make managerial decisions to ensure flexible correction of the project execution process. Although automated decision-making systems, which use real-time data, are not always applicable in real production processes, they can be effectively used in a digital twin of a product or process. Once the results of using automated solutions in the digital environment are analyzed, these solutions can be used in real processes.

Figure 9.13 shows a generalized scheme of the process of using digital twins while designing science-intensive goods. The arrows denote data flows. According to Postulate 1, the design process starts with creation of a product's or process's digital twin. These twins generate a secondary data flow, which, according to the information support concept, is analyzed in the Data Analytics block. The block can also be implemented in an automated mode while relying on smart data analytics.

The results of the analysis digital twins' functioning are automatically directed to the automated digital decision-making system, which uses artificial intelligence and machine learning techniques and automatically makes decisions, which are instantly implemented in future products' and process's digital twins. This creates a kind of information support cycle.

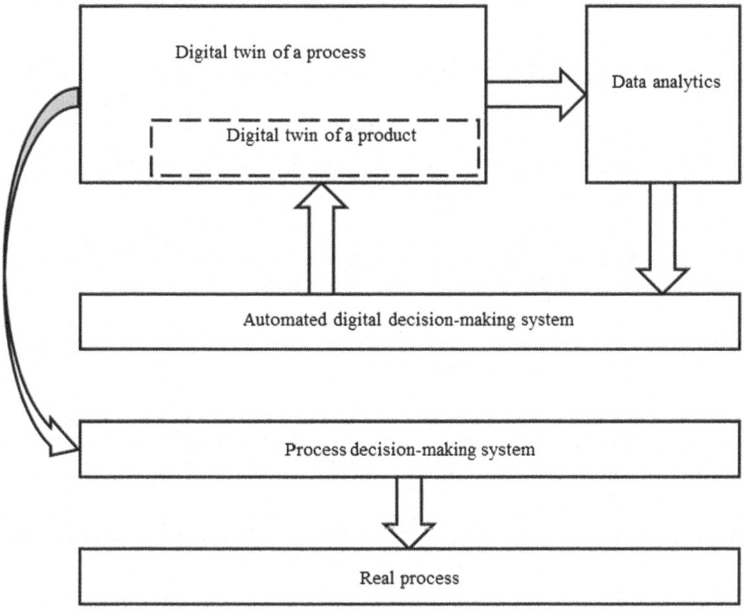

Fig. 9.13 Use of digital twins

With the help of information describing the digital production and automatically generated decisions, managerial decisions are formed and then used in real processes.

This information support pattern lays a theoretical ground for designing future products to help a company launch a rapid product development process with the use of digital twins.

Postulate 2. Creation and use of digital twins is a challenging task, because thorough modeling of complex objects requires analysis of huge volumes of information concerning objects' states and characteristics. It is not only about duplicating complex systems, but also about using multivariant development methods. Multivariant development of a digital twin implies simultaneous development of several technical solutions. This should be done to analyze a number of alternative solutions and choose the best one that should satisfy future market and customer needs. This flexible method is feasible thanks to the use of digital twins as stated in Postulate 1. Multivariant digital twin design can seriously boost the economic effectiveness of designing advanced products.

Figure 9.14 presents a scheme of using a multivariant approach to creating digital twins according to Postulate 2. It focuses on several digital twin variants. Simulation modeling helps analyze digital twins' properties. Data obtained from the analysis of each variant should be compared to the product's predicted future. On the one hand, the prognostic data is used to analyze and select different digital product variants; on the other hand, it is possible to predict the future contour of the product based on digital and simulation modeling data.

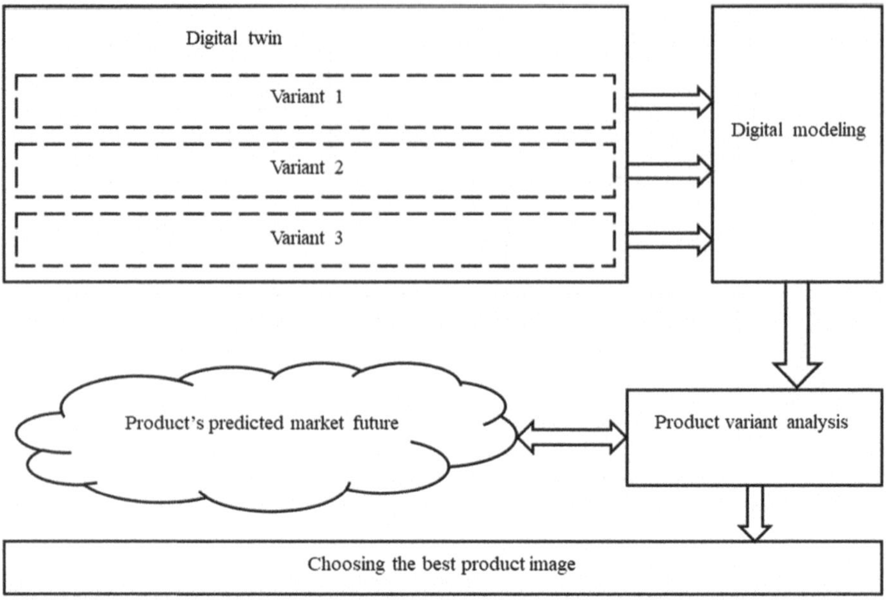

Fig. 9.14 Using a multivariant digital twin

Postulate 3. Crucial to successful creation of a future product is the use of a predicted technological and economic contour of the product being analyzed. According to Postulate 3, it is necessary to make prognosis with the use of existing data at each product life cycle stage.

According to Postulate 1, a product is designed with the help of a digital twin, which is developed with account for multivariant modeling. Therefore, digital twin analytics is a vitally important source of information, which should be used to predict the product's behavior in consumer markets. On the other hand, designing an advanced product requires using external data as well, which can shed light on the product's conformity with the economic reality and customer expectations.

Prediction of future technological and economic market needs is a challenge. However, the concept implies the need of full-scale modeling of not only future products, but also building a similar digital twin for target markets.

A digital twin of a market is a model simulating customers' and companies' behavior, which is centered around the digital twin of a future product. As stated by Postulate 2, creation of a digital twin requires using a multivariant approach.

Figure 9.15 describes the digital twin of a market intended for modeling future products. Based on a digital model of a market, different product twin variants can be analyzed to predict a product's behavior in future markets and generate advanced solutions in the sphere of advanced product design.

It is possible to use different approaches to simulation modeling while creating an economic and mathematical digital twin model. Another way to implement a digital

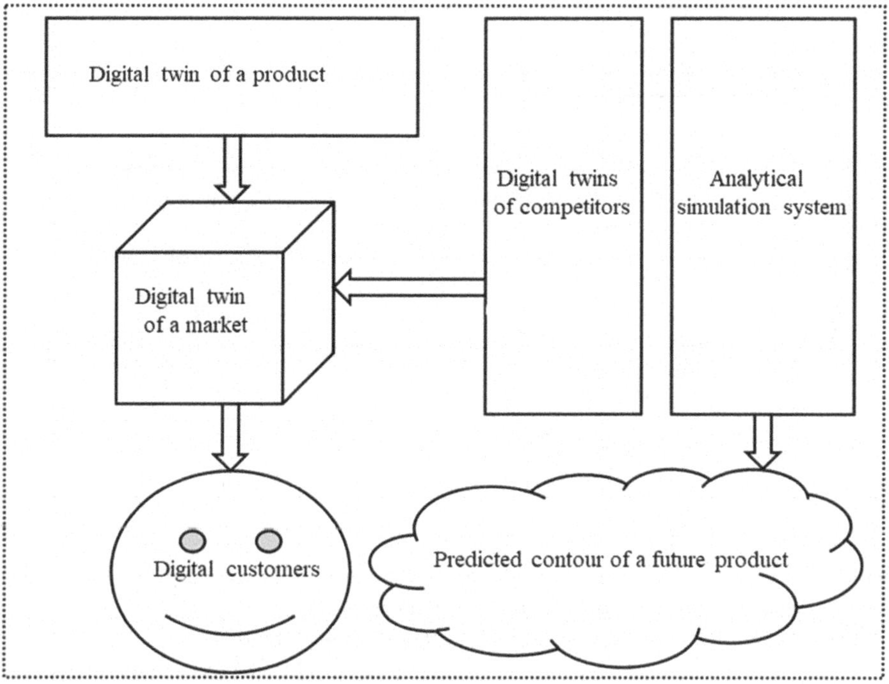

Fig. 9.15 Digital prognosis of future products' behavior

twin of a market is using the theory-and-game approach to testing possible market twin behavior scenarios.

Postulate 4. According to Postulate 3, a crucial part of creating a future product is predicting product digital twins' and other agents' behavior. As a rule, prognosis implies setting up the so called *planning horizon*, which reflects a possible prognostic depth. On the one hand, this horizon is set according to prognostic requirements, on the other hand, it depends on prognostic potential.

The use of digital twins according to Postulate 1, predicting the contour of a future product can pretty much extend the prognostic potential. This requires using a flexible planning horizon, which will depend on products' and companies' digital twins' potentials.

To implement a flexible planning horizon, it is advisable to use cascade forecasting with the use of digital twins. The cascaded prognostics method enables to extend planning horizons based on existing predictions. The concept of cascade forecasting is shown in Fig. 9.16.

Figure 9.16 demonstrates that the original digital twin marks the initial stage of predicting the contour of a future product. Once first forecasts are obtained, a new version of the product twin is built with account for predicted factors, which influence the product's image; then the new version of the twin can be used repeatedly for

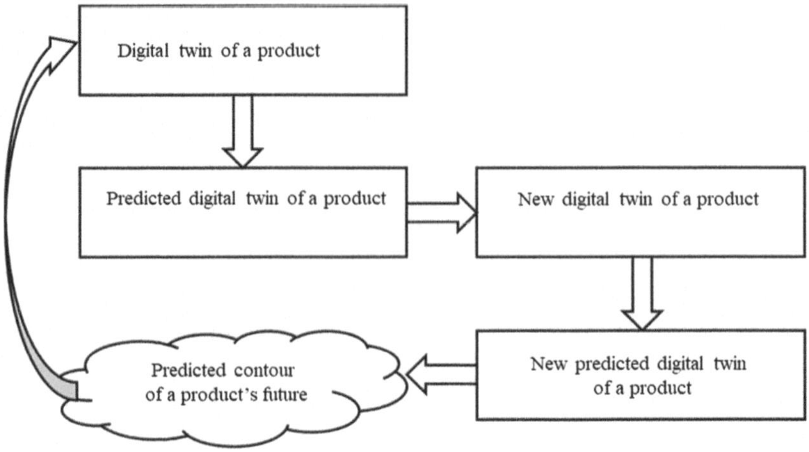

Fig. 9.16 Flexible planning horizon

building a forecast. This results in the appearance of an advanced prognosis of the future product's contour with account for market factors.

If necessary, this cascaded prognosis pattern can be extended.

Postulate 5. In order for a business to launch a rapid product development process, it is necessary to not only design/manufacture advanced products, but also manage the company's competitiveness on a regular basis. Therefore, science-intensive organizations must constantly modernize their products in keeping with customer needs and the market conjuncture.

As stated by Postulates 1, 3, and 4, a rapidly developing business should use digital twins of its products and take regular steps to predict their contours. Apart from the fact that the prognostic process improves a product's digital model, to ensure a greater conformity of products' consumer properties to future market needs, it is necessary constantly modify product samples at the design stage.

In some cases, it is possible to use predicted market parameters when designing next-generation products, which should demonstrate a greater conformity with predicted market needs. Recommendations and requirements to new product versions can be generated automatically and with the use of digital twins of products and future markets. Automated expert systems, which rely on knowledgebase provided by product and market digital twins, can be used for this purpose.

Based on these postulates, according to which creation of advanced products ensures rapid product development for a science-intensive company, a mathematical model of this approach can be built.

Computational mathematics describes a dynamic computation method known as the *predictor-corrector method*, which is used in solving differential equations, as well as for interpolation/extrapolation of functions. This method's concepts can also be used for describing the product creation process.

The main concept of the predictor-corrector methodology lies in making a brave move (the predictor) at the prognostic stage followed by correction according to reality (the corrector).

Here is a discrete time scale:

$$t_0 < t_1 < \ \dots \ < t_T,$$

where T is the total number of time spans. A product's state observed at each span is described by the vector:

$$Q(t) \ = \ (q_1(t), q_2(t), \ \dots, \ q_N(t)),$$

where N is the number of vector components $Q(t)$.

For simplicity, it can be assumed that

$$q_n(t) \geq 0, \ \ n = 1, \ 2, \ \ \dots, \ N.$$

Along with the real object, which is a future product being designed, its digital twin will be analyzed (as stated by Postulate 1):

$$S(t) = (s_1(t), \ s_2(t), \ \dots, \ s_N(t)),$$

where the digital twin's dimensions coincide with those of the real product.

Unlike the vector $Q(t)$, which is known at the current moment only, vectors $S(t)$ have been known before, because the complete digital twin embraces the pre-design period.

A prognostic step is made at every moment of time t_k:

$$[Q](t_k) = P(Q(t_{k-1}), \ S(t_k), \ S(t_{k+1}), \ \dots, \ S(t_T)). \tag{9.1}$$

Here, $[Q](t_k)$ denotes the predicted value of the vector $Q(t_k)$ at the moment t_k, and P denotes a forecast operator (the predictor), which makes a prognosis based on the value of the vector Q observed at the previous step and digital twin values observed at all following steps.

However, the real value of the vector $Q(t_k)$ is calculated according to the following formula:

$$Q(t_k) \ = \ K(Q(t_{k-1}), \ [Q](t_k)). \tag{9.2}$$

Here K denotes an operator (the corrector), which activates the following step in building the future product, also with account for a prediction calculated according to the formula (9.1).

Equations (9.1) and (9.2) are recurrent ratios, which help model the future product's market trajectory:

$$\Gamma(Q) = \{Q(t_k) : \ k = 0, \ 1, \ \ldots, \ T\}.$$

Apart from the future product's trajectory, there is its digital twin's market trajectory model:

$$\Gamma(S) = \{S(t_k) : \ k = 0, \ 1, \ \ldots, \ T\},$$

which can help evaluate the accuracy of the future product's trajectory:

$$\Delta = \sum |Q(t_k) \ - \ S(t_k)|,$$

with summation based on all $k = 0, 1, \ldots, T$.

This value can help make a quantitative estimate of a future product's conformity to its digital twin.

The theoretical basis of creating advanced products, which lays the ground for rapid product development, can also be used for creation of product design and manufacturing tools, which will form the main driver of rapid development and progress of a science-intensive business.

Chapter 10
Rapid Development of an Organization on the Basis of Product Life Cycle Management

10.1 Basic Management Tools for the Development and Production of Future Products in Order to Ensure an Organization's Rapid Development

The development and production of the products of the future ensures rapid development of an organization if an effective process for managing the production of new highly competitive products is built. The effective functioning of this process is ensured by a whole range of tools aimed at managing the competitiveness of the products depending on the use of certain technologies and competencies in the process of their development and production, selecting the most effective innovative technical solutions and assessing an organization's readiness for their implementation to create products of the future, which will lead to the rapid development path.

The production of new highly competitive products begins with the analysis of the main consumer expectations for the prospective technical and economic product image, resulting in the determination of the technical, economic, environmental, ergonomic, and other characteristics.

The initial data for assessment of the formation of the technical and economic product image for new products are the results of marketing research that determine consumer expectations, the importance of certain characteristics of products, as well as proposals and projects of competing organizations. Wherein:

1. The development of new products is carried out taking into account the characteristics of new markets (e.g., foreign markets). The process of studying foreign markets is not much different from the study of national markets. However, analysis of market chances and risks in international marketing activities require a more thorough collection of information about the potential of international markets and the possibilities of the external environment, including restrictions on imports, international law, etc.

A. Tyulin, A. Chursin, *The New Economy of the Product Life Cycle*,
https://doi.org/10.1007/978-3-030-37814-1_10

2. To organize an effective process for managing the development of new products, it is necessary to have sufficiently accurate marketing information about the markets in which updated products will be presented. Practice shows that desk research of foreign markets on the basis of a thorough analysis of the volume of secondary information in most cases is sufficient to achieve the goals set for the study of the foreign market of high-tech products. Conducting field research of foreign markets is one of the most expensive of the existing methods of marketing research, but it is able to give a maximum of fresh and reliable information in the case when it is necessary to include it in the research procedure to solve the tasks.
3. The life cycle of most high-tech products, the severity of competition, and the high degree of closure of the most solvent global markets for the import of foreign high-tech products welcome a study not only of potential markets, but also of know-how markets, leading achievements of scientific and technological progress, fundamental scientific groundwork in the relevant direction, i.e., marketing research of the innovation market.c

Technical indicators of a product are the source of its competitiveness. They need to be specified depending on the nature of the product purpose and the conditions of consumption and exploitation.

Economic models for assessing competitiveness should be built on the basis of a minimally necessary, but sufficiently representative group of technical and economic characteristics of products that ensure their sale on the market. Under these conditions, the achievement of competitive technical and economic indicators on the market is possible through the use of innovative technologies in the development, production, and marketing of products.

Nominally, all new types of promising production of an organization can be divided into several main types:

- New among the traditional markets
- New segments of the main market
- Creating entirely new markets

Work in all three focus areas is necessary for the stable economic development of an organization. At the same time, the development of new types of products for traditional market segments is necessary to maintain the current level of an organization's competitiveness. Creating products for a market segment in which an organization is not yet represented allows to acquire new competitive advantages and increase competitiveness. Finally, development aimed at creating new markets ensures rapid development of an organization (Fig. 10.1).

Each of the three areas involves the creation of new products. However, the process of creating a product can vary significantly. To enter existing markets for an organization it may be sufficient to create a new product line in which products will be developed on the same physical principles as in traditional segments, mainly due to already mastered technologies based on existing key competencies. However, in order to achieve rapid development an organization needs to create a completely new product with unique characteristics achieved through the use of new innovative

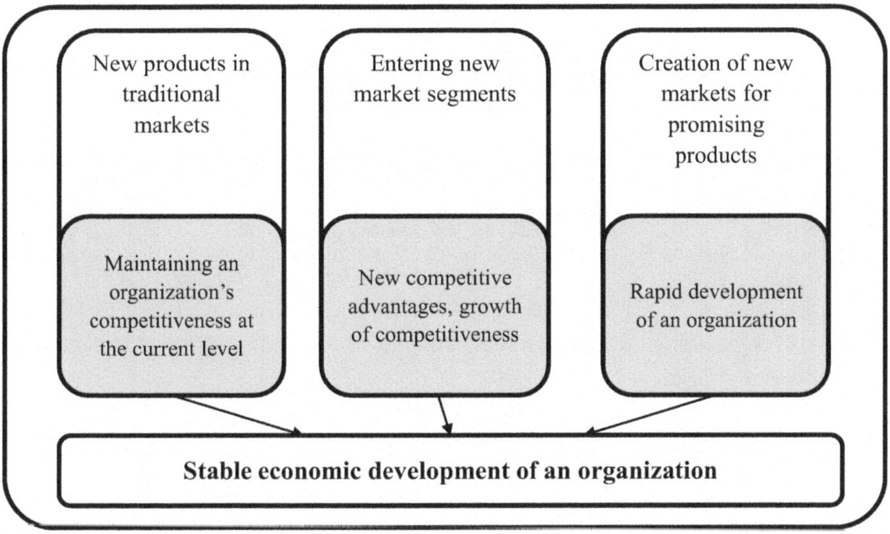

Fig. 10.1 Three areas of new product development

technologies, etc. Also, access to new markets can be achieved through a new market positioning of existing equipment by refining it for the tasks of a particular segment. In turn, maintaining an organization's competitiveness in its traditional markets can be realized both by creating a new product for this segment, and by modifying the existing product or radical processing of the product, its modernization or adaptation to the conditions of the current development of technologies (Fig. 10.2).

For stable economic development of an organization, passing into rapid development, it is necessary to make effective management decisions to determine the technical and economic image of promising products of each segment. Under technical and economic product image we mean a set of technical and economic indicators determining the competitiveness of products on the market, characterizing the properties of the goods and used to assess the competitiveness of the goods. Technical indicators of products are: key technical characteristics, indicators of product use and manufacturability, ergonomic, esthetic, environmental indicators; economic indicators: purchase price, commercial use price, delivery terms, delivery times, types and methods of settlements with the client. If economic characteristics are more or less traditional for different types of products (with the exception of a small number of parameters), the set of technical characteristics is unique for each type of products (or for group of products of the same type). At the stage of formation of the technical and economic product image, a manufacturer's task is to convert the abstract requirements of a consumer, representing a list of their wishes, into the integral value of the product, consisting of its specific technical and economic characteristics. That is, the requirements of a consumer should be directly correlated with general characteristics of the product, in other words, this requires an economic instrument for the quantitative assessment of compliance of created

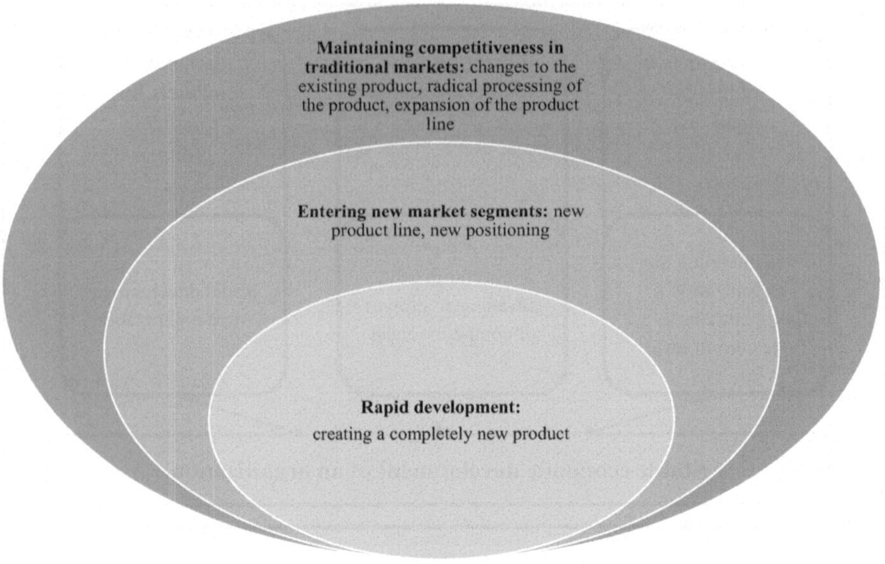

Fig. 10.2 Ways to create new products to ensure rapid development

products with a certain technical and economic image, consumer expectations, given the forecasting of the development of the needs of organizations and society (i.e., consumers) in the products of the considered type.

The present stage of the formation of the technical and economic product image can be implemented in the following sequence:

1. The requirements of consumers based on marketing research are determined. It should be borne in mind that a client is not always able to clearly formulate their requirements.
2. Consumer requirements are ranked by importance. At this stage, it is possible to use expert assessments of importance, for example, on a 10-point scale or complex expert assessments, for example, using a matrix of paired comparisons of consumer requirements.
3. The list of technical characteristics of products that affect the fulfillment of a consumer requirements is formed (at least two characteristics that affect each requirement).
4. The matrix of interaction of technical specifications and customer requirements is filled.
5. Customer satisfaction with your products and products of your competitors for each requirement is studied.
6. Analysis of competitor's products and benchmarking studies are carried out.
7. Target values of technical characteristics are determined and the relative technical difficulty of achieving each technical characteristic is estimated.

8. Interrelations between technical characteristics are established. This information is used in the future to study and overcome possible contradictions when modifying the technical characteristics.
9. Technical characteristics are determined, because their improvement has a positive impact on meeting the needs of the consumer (and vice versa).
10. The order of implementation of necessary modifications of technical characteristics is defined.

The algorithm of forming a list of technical and economic characteristics of products that meet consumer expectations is shown in Fig. 10.3.

On the basis of the formed technical and economic image, the fundamental possibility of creating new products should be determined considering the identified consumer expectations based on analysis of the current development of equipment and technologies. To develop a product with competitiveness indicators corresponding to the technical and economic image of promising products established on the basis of consumer expectations, it is necessary to control these indicators at all stages of design development (development of a pilot project, preliminary design, prototype, etc.).

Using special optimization economic and mathematical models and methods built on their basis for determining the amount of resource support sufficient to design products for a given cost and competitiveness (i.e., for a given technical and economic image), future costs for the production of the obtained design option can be defined for each design stage with a high degree of accuracy. If the achieved values of the competitiveness indicators deviate from the set values, it is necessary to use tools at the next design stage to improve product performance and thereby increase competitiveness indicators to an acceptable level.

Economic models for assessing competitiveness should be built on the basis of a minimum necessary, but sufficiently representative group of technical and economic characteristics of products that ensure their sale on the market. Under these conditions, the achievement of competitive technical and economic indicators on the market is possible through the use of innovative technologies in the development, production, and marketing of products.

The result of the assessment of an organization's capabilities is the calculation of the cost of production, given the need for the development of new technologies, development of competencies, etc., as well as an assessment of the time for product development (from idea to market entry). When the design of products is created in the light of a given cost, it is necessary to assess the ability of an organization to produce the products in a reasonable time. The time factor when developing products on the market plays one of the decisive roles, because the lost profit due to delays in the development of products significantly exceeds the possible additional costs necessary to accelerate the development process.

The main components of the economic and mathematical model for assessing the effectiveness of decisions to determine the technical and economic image of promising products are as follows:

- Technical and economic parameters of promising products $X = (x_1, x_2, \ldots, x_n)$
- Unit cost of promising products S

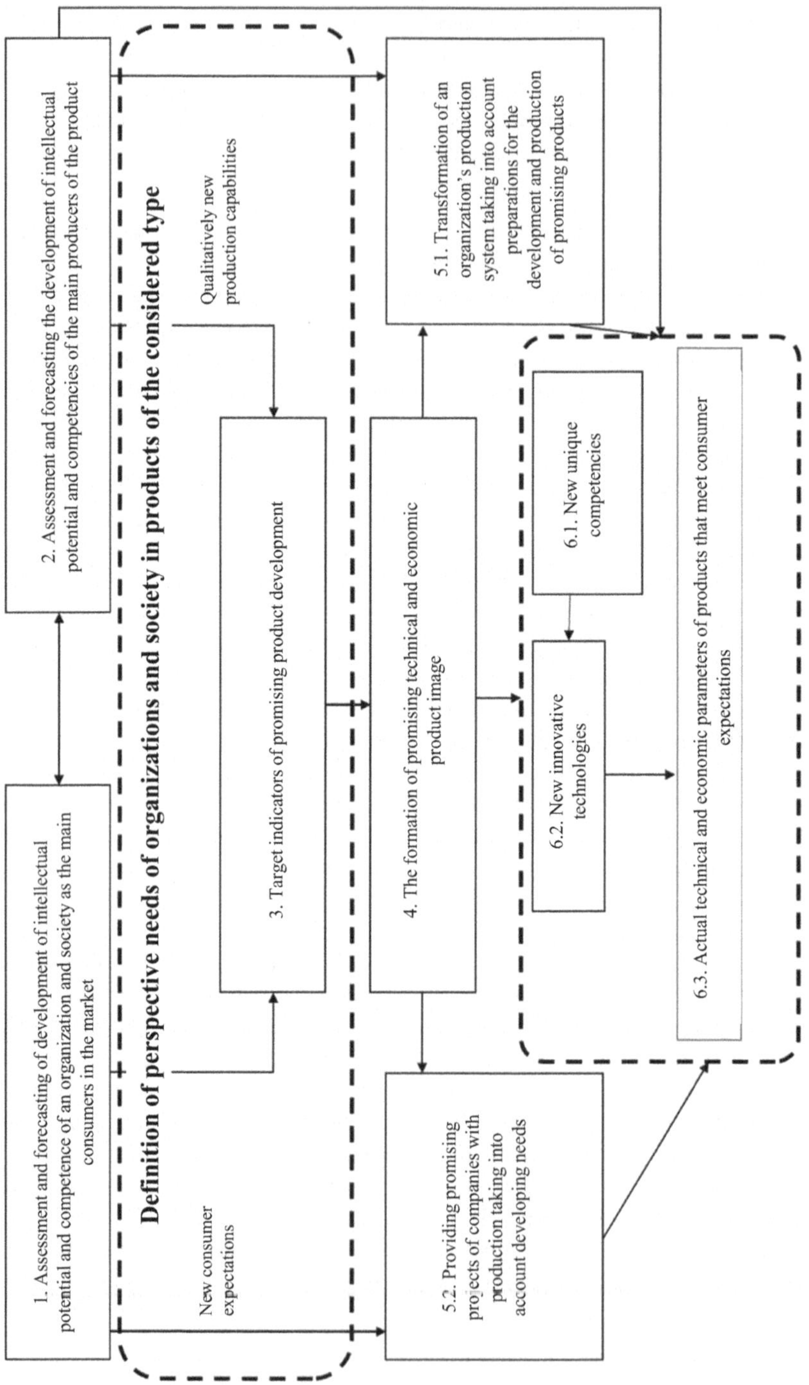

Fig. 10.3 Economic tool for determining the technical and economic parameters of new products

- Time shaping idea to market entry T
- Organization's capabilities (possession of innovative technologies, unique competencies, manufacturing capabilities, etc.) R
- Effectiveness of management decisions to determine the technical and economic appearance of promising products E
- Competitiveness of promising products Q

The expression connecting the listed values is:

$$Q = E(X, I, S, T);$$

Thus, the competitiveness of promising products is determined by the effectiveness of decision-making on the formation of their technical and economic image (i.e., a product whose technical and economic parameters are not worse than the given vector X, the cost price is not higher than S, and the development time is not more than T).

The main economic and mathematical models and procedures underlying the assessment of the effectiveness of management decisions to determine the appearance of promising products are as follows (Fig. 10.4):

- Processing and analysis of the results of marketing research to determine the lower limit of the space parameters of competitiveness of promising products depending on the target market (existing, new, rapid development)
- Assessment of innovative potential, competencies, production capabilities of an organization
- Assessment of the impact of innovative technologies (existing and promising) on the competitiveness of products. Assessment of an organization's competencies and the need to develop new competencies for the development of new technologies
- Determination of acceptable unit cost and time of its development (from idea to market entry)
- Assessment of the competitiveness of new products in the target market as a measure of the effectiveness of management decisions to determine the technical and economic appearance of products
- Simulation modeling in order to determine the optimal level of competitiveness, taking into account dynamically changing factors and risks

In the case of the development of multicomponent products, their technical and economic image consists of the technical and economic aspects of the components. On the basis of the analysis of innovative technologies, due to which the characteristics put in technical and economic image of production are reached, it is possible to define components whose development/completion is possible only as a result of applying completely new technologies requiring innovative development.

If it is necessary to master a completely new technology, its successful development is possible if an organization has a sufficient level of innovative potential. This issue was discussed in detail in Sect. 4.1, where, for a formal description of an

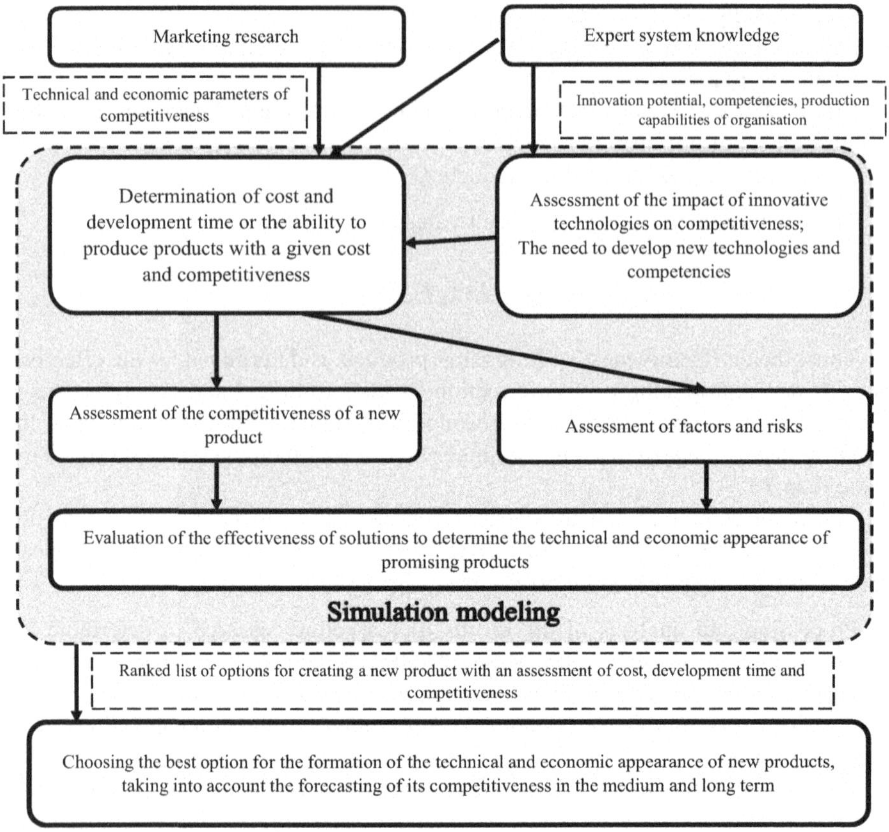

Fig. 10.4 Decision-making scheme to determine the technical and economic appearance of new products

organization's potential for highly competitive products, an economic and mathematical model was proposed for assessing the sufficiency of an organization's innovative potential in the face of the fast appearance of competing innovative solutions and the expansion of the global information space. The management of each component of the innovative potential of an organization is associated with the development of relevant competencies due to resource costs aimed at this process. An organization which seeks to dominate the market should launch a self-replicating process of improving competencies and resource efficiency. In the absence of competencies necessary for the production of advanced development, an organization should search for them in academic science, other industries, and special competence exchanges. Due to such competencies, innovative technologies can be created and the necessary technical and economic characteristics of the products will be achieved. In the context of digital transformation and the fast appearance of new consumer expectations in the market, organizations are required not only to possess certain knowledge, but also to have dynamic organizational skills to quickly adapt to emerging new needs and update knowledge through a system of lifelong learning, development of cooperation, etc. This will allow to constantly develop organizational knowledge base and to support it at the level not lower than the main competitors.

As we can see, the process of formation of new competencies and development of innovative technologies on their basis require significant resource costs. In the context of high financial costs for the development of new innovative technologies, special attention should be paid to the effective use of all types of resources. Based on the methods of assessment of key competencies and selection of innovative technologies, it is possible to build a program of activities of an organization and determine the most relevant areas of financing the development of new competencies and technologies needed to create new products. For this, the mechanisms for managing the development of competencies should include the need for their replication in order to create innovative technologies in various areas of technology development. That is, the possession of technology and its successful use in one of the possible directions does not yet indicate the possibility of its rapid implementation in another direction.

The process of transferring competencies and the formation of centers of competence in certain areas of activity, that combine production resources with the necessary competencies for their use is of particular importance. The presence of such centers will allow to quickly form the scientific and production structure of new projects from individual components, minimizing the dispersion of resources to duplicate units. In organizational terms, it does not have to be limited to a single organization.

Further economic analysis of the product development process of rapid development involves assessing the competitiveness of products taking into account the necessary costs for the development and transfer of competencies, the development of innovative technologies, etc. At this stage, it is necessary to determine (adjust) the cost and timing of the creation of new products.

To quantify the competitiveness of products, we can use the method proposed in Nesterov et al. [1]. The following parameters will be considered as the initial data of the algorithm for determining the competitiveness of products:

S_0 cost of financial investments made by an organization to create a product and start production for its release
S_1 unit cost of the product
S_F the price of a similar product on the market (if there are no analogues, $S_F = S_1$)
N forecast number of units of the product to be sold on the market
T time required to develop, organize the release and start of sales of the product
T_0 planned time of operation of the product
α coefficient from 0 to 1, reflecting the importance of the waiting time for the product release
Q product innovation (in comparison with the current level of scientific and technological progress)
R internal risks arising in the process of production by the enterprise
F external risks associated with the sale of the product on the market

The indicator of competitiveness of production is calculated by the following formula:

$$IQ = \left(\frac{Q \cdot S_F}{\frac{S_0}{N} + S_1} - \alpha \frac{T_0}{T} \right) \cdot (1 - R) \cdot (1 + F).$$

This formula expresses the numerical index of competitiveness of the product under consideration in the market.

The competitiveness of an organization is directly related to the competitiveness of products. Moreover, it depends on the entire range of products. If we know the integral indicators of competitiveness of all types of the products $IQ_1, IQ_2, \ldots, IQ_N$, where N is the number of types of products, then the integral indicator of the competitiveness of the entire set of products produced by the organization may be found by the following formula:

$$IQ = \sum_{i=1}^{N} \omega_i IQ_i,$$

where ω_i is the weight coefficient characterizing the contribution of each type of products to the formation of an integral indicator of the organization's competitiveness.

As the weight coefficient corresponding to a certain type of product, we will consider the share of revenue from sales on the market in the total revenue of an organization:

$$\omega_i = \frac{V_i}{V},$$

where V_i is revenue from sales of products of the type i; V is total revenue of an organization. Thus, $\sum_{i=1}^{N} \omega_i = 1$.

The competitiveness of an organization is determined not only by the competitiveness of its products, but also by the level of:

- Development of an adaptive production system (K_1)
- Competencies of an organization (K_2)
- Competences of subcontractors (K_3)
- Interaction with customers in terms of the formation of technical specifications for product development and determination of consumer preferences (K_4)

To assess the listed factors of competitiveness, we can use known methods [2]. The result of calculation by the method should be normalized to unit in order to obtain dimensionless values.

Given the known values of the indicator of competitiveness of the whole set of products $\widetilde{IQ}$ and assessment of factors K_1, K_2, K_3, K_4, the integral indicator of an organization's competitiveness is written as:

$$IQO = w_0 \cdot \widetilde{IQ} + w_1 K_1 + w_2 K_2 + w_3 K_3 + w_4 K_4,$$

where w_i are the weight coefficients of the competitiveness of the organization. Thus, $\sum_{i=0}^{4} w_i = 1.$

An organization, whose integral indicator of competitiveness is $IQO \geq 1$ should be considered competitive on the market.

If the developed product is not competitive at the price, then it is necessary to determine the design and technological solutions in order to bring the price of products to a competitive price in the framework of product design at a given cost.

For the production of products with a technical and economic image with certain technical and economic characteristics, it is necessary to have the appropriate production capabilities. The task of assessing the technical level of production is to determine the degree of an organization's ability to produce developed products. For solving such problems, for a long time the main methods used were those that rely on the assessment of the complexity of the production operations necessary for the production of products. This approach is justified if the organizations in question have approximately the same technological level, and production involves the operation of widely available technological competencies.

In modern economic conditions, which imply the rational use of financial resources, high requirements for the technological level and competitiveness, as well as orientation to the maximum possible use of the domestic component base and materials, new effective methods are needed to assess the ability of organizations to produce products in the face of high demands. Such methods can be based on a comprehensive qualitative and quantitative assessment of an organization based on a set of indicators and features.

A modern high-tech organization is a complex system that has a multilevel structure with vertical and horizontal connections between level elements. An organization has administrative, production, research, and other elements. Each unit performs its function and uses different resources (financial, labor, etc.) to conduct activities. The activities of an organization are influenced by various factors, which can be both internal and external. External factors are associated with the overall development of the economy of the state, the competitiveness of the industry, the state support of the industry, and the organization itself. Such factors cannot be managed at the organizational level. Normally, external factors are common to all organizations in the industry. In this regard, to assess the technical level of various industries for the implementation of the project, the urgent problem is the identification and evaluation of factors that characterize the internal state of an organization (financial, technical, etc.). Internal factors form the internal microenvironment of an organization that will contribute to or hinder the successful development of new products.

When considering the processes of increasing the efficiency of development and production of new products, we believe that the most important information source for managing these processes are the big data of the global information space, the knowledge and prospects for the development of fundamental science, as well as the results of the intellectual processing of this information by modern mathematical methods (artificial intelligence, machine learning, discrete mathematics). This means, in particular, that we can obtain data on certain design and technological solutions, advanced techniques and technologies from the information space. This will enable organizations to implement the processes of designing, preparing production and manufacturing new products and their components at the lowest cost by constantly monitoring the external environment: the actions of competitors, factors of macroeconomic instability, new achievements of scientific and technological progress. Therefore, as part of the digitalization process of organizations, we suggest to form a new organizational and economic environment where many aspects of activities related to the release of new types of products will be improved. Under the influence of these processes a transformation of management models of organizations will take place and there will be a need for obtaining an effective economic analysis of much more information of high accuracy that is required for the implementation of activities and management decisions in the traditional production system.

In this situation, a major role is played by the development of a new management decision-making system using modern mathematical methods based on artificial intelligence and machine learning, which includes:

- Formation and implementation of modern electronic communication systems for communication and data transfer—electronic module
- Formation of a modern organizational sphere that defines issues of reforming and optimizing organizational management structures—organizational module
- Formation of new management methods—management module
- Development and implementation of promising systems for processing, analysis, and distribution of information using unified tools for its processing and transmission formats—information module

The purpose of such a system is not to combine automated means of obtaining, processing, synchronizing, and storing information created by various methods using modern information and communication technologies, but to be able to use it for a balanced integration of existing management resources and databases with new automated analysis and processing technologies to take them to a whole new level. The creation of such a system in the future can increase efficiency and synchronize key business processes of creating new products, which, in turn, will become the basis for increasing the international competitiveness of an organization.

Next, we will show how the development of new highly competitive products increases an organization's competitiveness and creates conditions for its entry into a state of rapid development.

As an example, let us examine a high-tech organization engaged in the production of modern devices and other high-tech products. We will simulate the process of

Table 10.1 Initial data for calculations for product 1

Index	Value
Cost of financial investments made by the organization to create the product and launch production for its release	300,000
Unit cost	2200
Price of a similar product on the market	3500
Forecast number of units to be sold on the market	100
Time taken to develop, organize the release, and start selling the product	0.5
Planned time of product operation	3
Coefficient of importance of the waiting time for the release of the product	0.01
Product innovation (in comparison with the current level of scientific and technological progress)	0.85
Internal risks arising in the production process	0.015
External risks associated with the sale of the product on the market	0.5

Coefficient of competitiveness of Product 1: $IQ_1 = 0.76$

Table 10.2 Initial data for calculations for product 2

Index	Value
Cost of financial investments made by the organization to create the product and launch production for its release	150,000
Unit cost	2200
Price of a similar product on the market	3500
Forecast number of units to be sold on the market	100
Time taken to develop, organize the release, and start selling the product	0.5
Planned time of product operation	3
Coefficient of importance of the waiting time for the release of the product	0.01
Product innovation (in comparison with the current level of scientific and technological progress)	0.85
Internal risks arising in the production process	0.015
External risks associated with the sale of the product on the market	0.5

Coefficient of competitiveness of Product 2: $IQ_2 = 1.1$

assessing the competitiveness of the organization and will show the impact of the development of a new highly competitive product on the formation of competitiveness.

At the first stage, we evaluate the current competitiveness of a high-tech organization. First, according to the proposed formula, we evaluate the competitiveness of each type of products manufactured by the organization. The source data for calculating the competitiveness of products are presented in Tables 10.1, 10.2, 10.3, 10.4, and 10.5.

Product 1 On-board equipment for generation and emission of navigation radio signals, inter-satellite measurements, inter-satellite information exchange.

Table 10.3 Initial data for calculations for product 3

Index	Value
Cost of financial investments made by the organization to create the product and launch production for its release	150,000
Unit cost	1700
Price of a similar product on the market	4000
Forecast number of units to be sold on the market	1000
Time taken to develop, organize the release, and start selling the product	0.1
Planned time of product operation	4
Coefficient of importance of the waiting time for the release of the product	0.02
Product innovation (in comparison with the current level of scientific and technological progress)	0.85
Internal risks arising in the production process	0.015
External risks associated with the sale of the product on the market	0.8

Coefficient of competitiveness of Product 3: $IQ_3 = 1.073$

Table 10.4 Initial data for calculations for product 4

Index	Value
Cost of financial investments made by the organization to create the product and launch production for its release	200,000
Unit cost	1700
Price of a similar product on the market	4000
Forecast number of units to be sold on the market	1000
Time taken to develop, organize the release, and start selling the product	0.1
Planned time of product operation	4
Coefficient of importance of the waiting time for the release of the product	0.01
Product innovation (in comparison with the current level of scientific and technological progress)	0.85
Internal risks arising in the production process	0.015
External risks associated with the sale of the product on the market	0.5

Coefficient of competitiveness of Product 4: $IQ_4 = 1.43$

Product 2 On-board equipment which provides command reception, receipt transmission, generation and issuance of control commands, retransmission of signals to measure the current navigation parameters.

Product 3 On-board rescue radio complex.

Product 4 On-board information system.

Product 5 On-board radio complex of automatic identification system.

The weight coefficients of the products will be calculated as a share of sales revenue in total revenue (total revenue of the organization is 1,500,000) (Table 10.6):

According to the proposed formula, the coefficient of competitiveness of the whole set of products equals to: $IQ = 0.33 \cdot 0.76 + 0.13 \cdot 1.1 + 0.2 \cdot 1.073 + 0.23 \cdot 1.43 + 0.11 \cdot 0.75 = 1.018.$

Table 10.5 Initial data for calculations for product 5

Index	Value
Cost of financial investments made by the organization to create the product and launch production for its release	100,000
Unit cost	2000
Price of a similar product on the market	1500
Forecast number of units to be sold on the market	1000
Time taken to develop, organize the release, and start selling the product	0.1
Planned time of product operation	2
Coefficient of importance of the waiting time for the release of the product	0.01
Product innovation (in comparison with the current level of scientific and technological progress)	0.87
Internal risks arising in the production process	0.015
External risks associated with the sale of the product on the market	0.8

Coefficient of competitiveness of Product 5: $IQ_5 = 0.75$

Table 10.6 The results of calculations for products 1–5

Type of product	Product 1	Product 2	Product 3	Product 4	Product 5
Revenue	500,000	200,000	300,000	350,000	150,000
Weight coefficient w_i	0.33	0.13	0.2	0.23	0.11

Table 10.7 Weight coefficients reflecting the importance of an organization's competitiveness factors

Factor	$\widetilde{IQ}$	K_1	K_2	K_3	K_4
Weight	0.6	0.1	0.1	0.1	0.1

Weight coefficients reflecting the importance of an organization's competitiveness factors are presented in Table 10.7.

Other factors affecting the competitiveness of an organization, take the following values:

- Level of development of the adaptive production system $K_1 = 0.7$
- Level of competence of the organization $K_2 = 0.9$
- Level of competence of subcontractors $K_3 = 0.8$
- Level of interaction with customers in terms of the formation of technical specifications for product development and determination of consumer preferences $K_4 = 0.5$

According to the proposed formula, the coefficient of competitiveness of the considered high-tech organization equals to:

$$IQO = 0.9.$$

Since the index of competitiveness of the organization takes a value less than one, such an organization is not competitive in the market.

Achieving competitiveness on the market is possible due to rapid development. This development is associated with the implementation of a breakthrough project, creating a product that will form a new market or take a significant share of the existing market. At the same time, the competitive advantages of the newly created products will be based on the previously formed factors of competitiveness.

For simplicity of numerical calculations, when constructing the development trajectory, we will assume that the totality of all previously produced types of products with competitiveness Q_1 and a new type of product with competitiveness Q_2 are being considered.

Let the competitiveness of Q_1 and Q_2 change in time according to the graph presented in Fig. 10.5.

As we can see from the graph, the existing innovative potential of the organization contributes to an increase in the level of competitiveness of new products at a constant level of competitiveness of the whole set of already manufactured types of products.

Finally, let us build the organization's development trajectory on the basis of the forecasted revenue values from the sale of newly created products. We also recalculate the values of the weight coefficients of the indicators of product competitiveness and, based on the results obtained, we obtain new values of the organization's competitiveness (Tables 10.8 and 10.9).

The graph of the development trajectory is shown in Fig. 10.6.

After moment $t = 2$, the value of the integral indicator of the organization's competitiveness becomes more than 1, which indicates the achievement of priority development due to the release of a new highly competitive type of product.

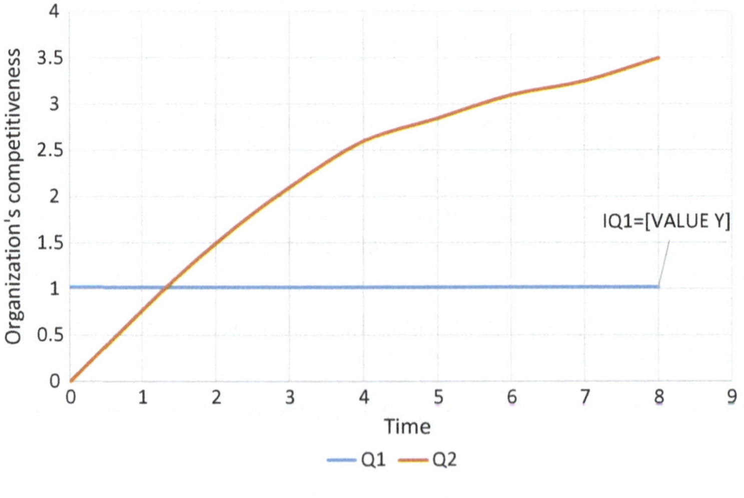

Fig. 10.5 Assessment of changes in competitiveness in time

Table 10.8 Values of competitiveness of the whole set of products taking into account the new type of products

Time	0	1	2	3	4	5	6	7	8
$\widetilde{IQ}$	1018	0.96	1.16	1.37	1.57	1.71	1.85	1.97	2.12

Table 10.9 Competitiveness of an organization on condition of invariability of assessment of factors K_1, K_2, K_3, K_4 and weight coefficients of factors

Time	0	1	2	3	4	5	6	7	8
IQO	0.9	0.87	0.99	1.11	1.23	1.31	1.4	1.47	1.56

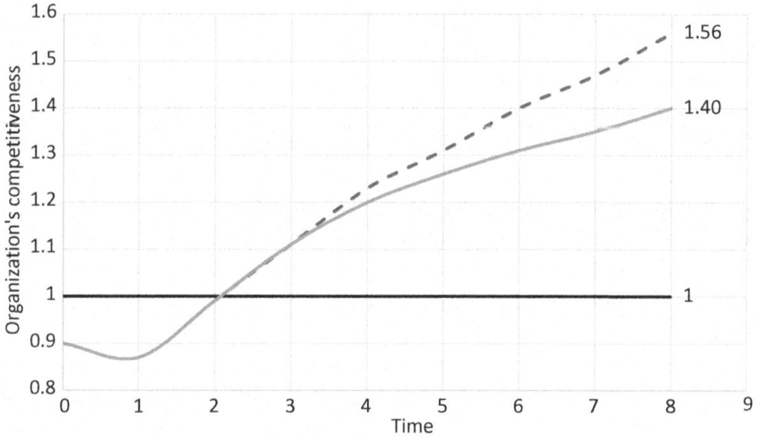

Fig. 10.6 The trajectory of the economic development of an organization, passing into rapid development (solid line—on condition of a constant cost of sales on the market; dashed line—on condition of an increase in the cost of sales at each moment of time by 5%, starting from the moment $t = 3$)

Permanent management of rapid development based on the elaboration of new highly competitive products will allow an organization to be proactive, to a certain extent anticipating the transformation of the economic environment, and to some extent even driving them by segmenting individual markets, initiating life cycles of innovations, creating a demand for high-tech products, taking preventive measures related to the rapid formation of the necessary competencies and the development of knowledge management toolkits used by the organization. The use of rapid development management tools in the complex forms an integral mechanism for the transition of organizations to the rapid development path. The innovative activity of an organization on the rapid development path, and the creation of unique products (or new markets), provided with resources, are harmonized (consistent in pace and intensity) with the development of the production system and business processes, accompanied by the economic stability of the organization.

10.2 Mechanisms for the Transition of Organizations to the Rapid Development Path

An organization's rapid development being the most important condition for dynamic development necessitates the creation of a fundamentally new innovative system for managing the processes of product development, preparation of production, direct production, and sale of finished goods. This system should be based on the extensive use of artificial intelligence and machine learning, digital economy, digital Earth. The term "rapid development" refers to the management of research, development, innovation, production, organizational, and other processes of an organization, which determines the formation of an organization's potential, sufficient to create a completely new product with high consumer properties, allowing to win a significant share of the existing market or to create a new market for high-tech products.

We realize that the developed products can be presented on the market only after their production. And the question here is that it is either necessary to create fundamentally new production facilities for the production of these products or to upgrade existing ones. It is the head of an organization who should determine the direction, based on the availability of the organization's competencies, financial resources, terms of delivery of products to the customer or access to the market, bearing in mind possible risks, including the risk that competitors might be ahead of time and might put on the market similar products with higher technical characteristics.

It should be noted that the set of the main projected long-term goals and objectives of rapid development of an organization, agreed on all types of resources and terms, should play the central role in an organization's development strategy, since today rapid development is a necessary condition for the successful development and effective operation of high-tech organizations.

The most significant indicators that characterize the basis of rapid development of an organization and determine its stable economic development are indicators that characterize both the organization's competitiveness and the ability to reformat its activities to solve a fundamentally new problem of creating products of new generation.

The key factor determining the ability of rapid development of an organization is its ability to reach new levels of industry positioning. This underlines an organization's ability (competence) to develop and implement advanced technologies for various industries. For the organization under analysis, it is necessary to determine the current point among the levels of industry positioning and form a bank of unique competencies of the organization that will allow it to reach new levels of industry positioning.

A high level of competitiveness indicators in general can provide organizations with a time-predicted appearance of products with high competitive advantages. Consequently, this opens up new opportunities both for shaping new markets, and for increasing the volume of sales of products in the existing market, which is the basis of an organization's rapid development.

The basis of an organization's transition to the rapid development path is the assessment of its competitiveness. Based on this quantitative assessment, it is concluded whether an organization is ready for rapid development or whether it is necessary to improve certain technical and economic characteristics that determine the organization's competitiveness, necessary to gain superiority. Such superiority over competitors may be due to the use of unique competencies, innovative solutions, transfer of critical technologies to obtain fundamentally new consumer properties of products and their ability to displace existing products on the market or create a new market.

Next, we will consider the process of forming the optimal rapid development path. It is based on the possibility of determining: firstly, increasing sales of products with high added value due to its intensive demand in the markets. Secondly, the innovative potential of new products should be determined, which can be used as an indicator to create sufficient conditions of an organization's readiness to release a product with new consumer characteristics based on the intensive application of advanced scientific achievements. This can be achieved by issuing private technical tasks to subcontractors with competencies in the development of unique technical and technological solutions.

The successful implementation of a project to create a highly competitive product that could displace existing products or create a new market depends on the availability of key competencies, through which unique technical, economic, and consumer characteristics of new products can be created. Thus, the base of rapid development is the competitiveness already formed by an organization. In the conceptual aspect, it is a system of knowledge, skills, methods, and tools for managing the activities of an organization, allowing to achieve strategic goals through the implementation of breakthrough projects to create highly competitive products. Organizational project management can be applied both within the entire organization and at the level of an organization that is part of it, which depends on the way the product will be created and produced (by an organization itself or by an organization involving subcontractors).

As a result of calculation according to the model for assessing the competitiveness of products proposed in Sect. 10.1, it is possible to estimate dynamics of growth of an organization's competitiveness. In doing do, the planned rapid development path (i.e., the trajectory of change in time of the integral indicator of an organization's competitiveness) can be set. Correspondence of the real values of the integral indicator of an organization's competitiveness to the values of the planned rapid development path at each moment of time indicates the adequacy of incentive measures to ensure rapid development.

As a result of using the innovative potential of an organization in the production of new types of products, the integral indicator of assessing its competitiveness will change.

The task of achieving rapid development in the context of the tools presented is to achieve such indicators of innovation and investment activity, in which the integral indicator of an organization's competitiveness will be greater than one, i.e., $IQO > 1$. The following situations are possible:

1. If an organization produces products similar to those existing on the market and its $IQO > 1$, then rapid development is expressed in gaining a significant share in the market of highly competitive products with high added value.
2. If an organization has released a new type of product, not on the market to date, indicator of the competitiveness of these products is $Q > 1$, the integral indicator of the organization's competitiveness is $IQO > 1$, then its rapid development is due to the formation of a completely new market of highly competitive products with high added value (while the organization can occupy a significant part in other markets).

Thus, if an organization's competitiveness changes in accordance with the trajectory of rapid development, then its profits will increase with the growth of sales of products in the market, since highly competitive products have high profitability and high added value.

Let us consider in dynamics the process of functioning of a high-tech organization, which includes the management of the current operational activities of its commercial and production divisions and the development determined by innovative processes and investment measures. The operational activities of a high-tech organization are related to the implementation of current contracts with customers, development of custom-made products, serial production, sales, operational logistics, procurement of materials and components, including all subsidiary and support processes. The main task is to fulfill the sales and production plan and reduce costs. Development measures are aimed at increasing the scale of an organization's activities, expanding its product line and forming markets, creating fundamentally new product lines. The three-level system of planning the activities of a high-tech organization includes strategic planning (where the prospects for the technical development of an organization in the future and updating the range of its products are determined, the planning period is usually 10 years), an activity program (where the main directions of an organization's work in the medium term are formulated, the planning period is usually 5 years), and budgeting (production and financial planning of an organization for a period of 1 and 3 years). Such a planning system is static, and any changes are difficult and require a thorough revision of planning documents and indicators. The task of an effective system of planning the activities of a high-tech organization is to ensure that it enters rapid development path by organizing the development and production of highly competitive products. Ensuring the growth of competitiveness of products will lead to an increase in an organization's competitiveness, since highly demanded products on the market create competitive advantages for an organization. This ensures the growth of revenues and profits of an organization, which leads to an increase in its economic performance and sustainability. This provides a resource reserve, which can be aimed at developing a product line, technological base, and competence component. This development of an organization leads to increased competitiveness in general and increase its ability to make the transition to the rapid development path.

The sustainability of an organization's development when various economic indicators of its activities change, also has an important effect on an organization's competitiveness and its ability to transition to rapid development through making products capable to create new markets or to expand considerably borders of existing ones ensuring competitiveness of production on quality and the price. In this regard, the development of an effective planning system for the activities of a high-tech organization is associated with the transition from a static planning system to a flexible one, in which various uncertainty factors (both positive and negative) will be taken into account in dynamics. For this, many organizations today are moving to planning cycles 5–3–3 (1), i.e., strategic planning for a 5-year period, development of a program of activities for a 3-year period, and budgeting for 1 or 3 years. Moreover, an annual assessment of the state of a high-tech organization and its sustainable development takes place. With a significant deviation from the trajectory of development (both up and down), it is necessary to adjust the planned indicators, as well as strategic and budgetary documents.

To assess the factors that affect the economic stability of an organization, a dynamic approach to the assessment of the main parameters of its activities is proposed. When modeling dynamic planning, the main problem is that the considered parameters of an organization's activity are influenced by various random factors. For such factors it is necessary to have adequate mathematical methods to assess their impact. As a basic model for estimating the influence of random factors, we use the doubly stochastic Poisson processes of so-called Cox processes. Such models describe the situation of cumulative impacts. We will also consider max-generalized doubly stochastic Poisson processes that describe models of catastrophic shock effects. The main difference between the considered methods and standard Poisson processes is that these models allow us to describe inhomogeneous chaotic flows of extreme events.

Let us give the definition of inhomogeneous Poisson process. Let us consider the function $\Lambda(t)$, with respect to which we will assume the following conditions:

$$\Lambda(t) \geq 0; \quad .$$

$$\Lambda(t_1) \leq \Lambda(t_2), t_1 \leq t_2;$$

$$\Lambda(0) = 0;$$

$$\Lambda(t) < \infty, t > 0.$$

Point process $N(t)$, we will call $t \geq 0$ an inhomogeneous Poisson process with an intensity measure $\Lambda(t)$, if:

- $N(t)$ there is a random process with independent increments
- For any $0 \leq s < t < \infty$ the increments $N(t) - N(s)$ has a Poisson distribution with parameter $\Lambda(t) - \Lambda(s)$

If the intensity measure $\Lambda(t)$ has a derivative, then the function

$$\lambda(t) = \frac{\mathrm{d}\Lambda(t)}{\mathrm{d}t}$$

then the function is the intensity of the process $N(t)$, and the function $\Lambda(t)$ is a measure of intensity, which we will call the accumulated intensity.

Let us denote a standard Poisson process by $N_1(t)$.

Let $\Lambda(t)$ be a random measure independent of $\Lambda(t)$. A random process defined by the formula:

$$N(t) = N_1(\Lambda(t)),$$

is called a doubly stochastic Poisson process or Cox process. The following relations for mathematical expectation and variance are valid for a Cox process:

$$\mathrm{EN}(t) = E\Lambda(t);$$
$$\mathrm{DN}(t) = E\Lambda(t) + D\Lambda(t).$$

Let us now turn to the application of doubly stochastic processes to build models for assessing the main parameters of a high-tech organization. The main idea of the model is that the effects of random factors on the parameters of an organization are heterogeneous in time, as it is assumed in standard Poisson processes. The economic justification is due to the fact that at the time of the development of a crisis, the probability of factors influencing the state of the parameters of an organization's activities can significantly increase. With this approach, it is necessary to assess the influence of heterogeneity of random processes, which describe the dynamics of the influence of external factors on the state of the parameters of an organization.

Let $X_1, X_2, \ldots$ be a set of random variables having the same probability distribution. Let us suppose that at each moment of time $t \geq 0$ the random variables $N(t)$, X_1, $X_2, \ldots, X_n$ are stochastically independent. The process specified by the formula

$$S(t) = \sum_{j=1}^{N(t)} X_j, \quad t \geq 0,$$

is called the generalized Cox process. This process can be considered as a natural generalization of the classical Poisson process.

For the case under consideration, we assume that the random variables $\{X_j\}_{j \geq 1}$ have the first and second moments. For certainty, we will denote these moments as follows:

$$EX_1 = a; DX_1 = \sigma^2,$$

$$0 < \sigma^2 < \infty$$

We will also assume that

$$E\Lambda(t) < \infty; E(\Lambda(t))^2 < \infty,$$

then the following relations take place:

$$\mathrm{ES}(t) = a\, E\Lambda(t),$$
$$\mathrm{DS}(t) = \left(a^2 + \sigma^2\right)E\Lambda(t) + a^2 D\Lambda(t).$$

The practical application of random Cox processes is due to the general limit theorem, which in the case under consideration can be formulated as follows.

Let $G_{\alpha,\,\theta}(x)$ be a distribution function with parameters α and θ, which corresponds to the characteristic function of the form:

$$g_{\alpha,\theta}(s) = \exp\left\{-|s|^{\alpha}\exp\left\{-i\frac{\pi\theta\alpha}{2}\mathrm{signs}\right\}\right\};$$
$$s \in \mathbb{R}, 0 < \alpha \leq 2, |\theta| \leq \theta_\alpha = \min\left(1, \frac{2}{\alpha} - 1\right).$$

Under the assumption that the control process $\Lambda(t)$ unlimitedly increases in probability with $t \to \infty$, in order to have the limiting law:

$$\lim_{t\to\infty} P\left(\frac{S(t)}{\sigma\sqrt{\mathrm{d}(t)}} < x\right) = G_{\alpha,\theta}(x), x \in \mathbb{R},$$

it is necessary and sufficient that

$$\lim_{t\to\infty} P\left(\frac{\Lambda(t)}{\mathrm{d}(t)} < x\right) = G_{\alpha/2,1}(x), x \in \mathbb{R}.$$

Let us assume that we have a deterministic estimate of a certain parameter of the activity of a high-tech organization, which we obtained without taking into account the influence of random factors. Besides, it should be noted that the resulting estimate should be adjusted due to the influence of various factors. It is well known that the influence of instability factors is heterogeneous in time. Therefore, we use the apparatus of twice stochastic Poisson processes.

Let us denote the flow of factors affecting the main parameters of the implementation of innovative projects through equally distributed values:

$$X_1, X_2, \ldots;$$

$$EX_1 = 0; 0 < \sigma^2 = DX_1 < \infty.$$

Let us introduce the notations:

$$\overline{S}(t) = \max_{0 \leq \tau \leq t} S(\tau),$$

$$\underline{S}(t) = \min_{0 \leq \tau \leq t} S(\tau).$$

We denote the distribution function of the maximum of the standard Wiener process by $G(x)$ on the segment [0, 1]. This function can be calculated by the formula

$$G(x) = 2\Phi(\max\{0, 1\}) - 1, x \in \mathbb{R}.$$

Let $d(t)$ be a positive function satisfying the condition

$$\lim_{t \to \infty} d(t) = \infty.$$

Let us assume that the following condition is met

$$\Lambda(t) \to \infty, t \to \infty.$$

Then the one-dimensional distributions of the normalized process of extremums of the generalized Cox process converge weakly to a certain distribution, which we denote by Z:

$$\frac{\overline{S}(t)}{\sigma\sqrt{d(t)}} \Rightarrow \overline{Z}, t \to \infty;$$

$$\frac{\underline{S}(t)}{\sigma\sqrt{d(t)}} \Rightarrow \underline{Z}, t \to \infty,$$

then and only then if there exists such a non-negative quantity U that the relation holds

$$\frac{\Lambda(t)}{d(t)} \Rightarrow U, \quad t \to \infty.$$

In this case, the distribution function of the random variable Z can be found from the relations

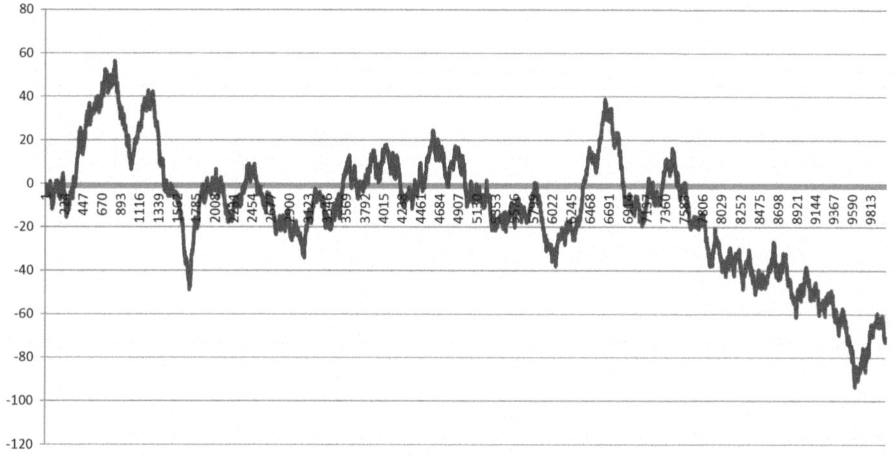

Fig. 10.7 The trajectory of Cox processes

$$P(\bar{Z} < x) = EG\left(\frac{x}{\sqrt{U}}\right);$$

$$P(\underline{Z} < x) = 1 - EG\left(-\frac{x}{\sqrt{U}}\right).$$

These relations can be used as the basis for a mathematical model for assessing the main parameters of an organization's activities under conditions of an inhomogeneous flow of factors influencing the value of parameters.

As an illustration we refer to a typical example of the implementation of the trajectories of Cox processes, with various parameters that show the effect of heterogeneity of the event flow on the implementation. Figure 10.7 presents the trajectory of the Cox processes with the following parameters: the mathematical expectation is 0, the variance is 1.

Figure 10.8 presents the graph of the probability density function.

The given graphs of computational experiments show an approximate behavior of the trajectories of doubly stochastic Poisson processes.

The application of the proposed approach allows us to refine the assessment of the parameters of the activity taking into account uncertainties and more accurately assess an organization's competitiveness. In this case, the factor of heterogeneity of economic parameters will be fully taken into account in the mathematical model. Another indicator, which should also be evaluated taking into account uncertainties, is the technical level of production, the sufficiency of which, along with competitiveness, is a key condition for an organization to enter the rapid development path. Let us describe the approach which enables to quantitatively describe the technical level of production, as well as formally describe the criterion of an organization's readiness for rapid development based on an assessment of its technical level.

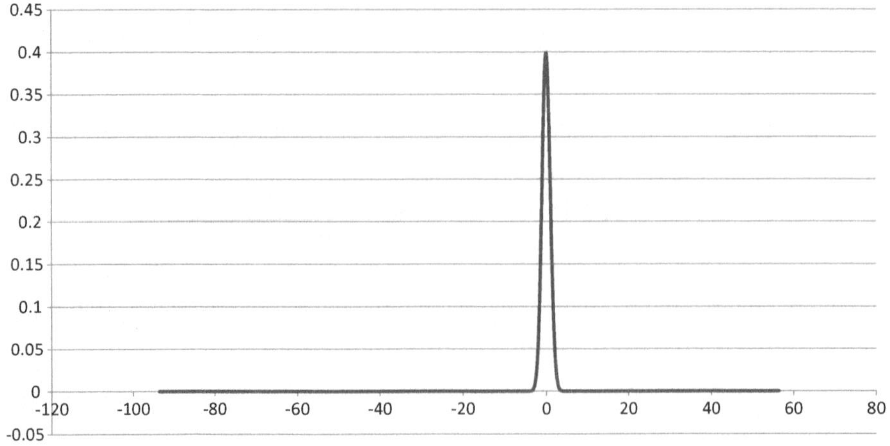

Fig. 10.8 The graph of probability density function

In modern conditions of global competition in the markets of high-tech products, innovative organizations pay great attention to the development, commercialization, and transfer of technologies. This requires the creation of an innovative infrastructure that meets the challenges of development and production of rapid development products, which is the basis for organizing effective processes for the development, preparation of production, and production itself.

Difficulties in using an organization's innovative and technological potential may be associated with the absence or poorly developed methods for assessing the technical level of production and its sufficiency for the release of new highly competitive products.

Assessment of the technical level of production helps to form a complete picture of the level of efficiency of an organization in order to make informed decisions in the field of production excellence, resource management, as well as investment and innovation.

To date, one of the problems in determining the comprehensive assessment of the technical level of production is the lack of a unified methodology for assessing the technologies underlying the creation of advanced development products. Today, the use of the Technology Readiness Level scale (TRL), proposed by NASA, is supported worldwide. The TRL approach is used by advanced industrial organizations such as "Boeing," "General Electric," "Airbus," "Rolls-Royce," and many others.

The levels of technology readiness allow us to assess how far development has progressed, from the idea of its creation to an already fully fledged industrial design.

TRL scale has nine levels, each of which has a number of criteria to help determine the exact level of technology at the moment. These readiness levels allow not only to evaluate the technology itself, but also to evaluate it relative to the market and competitors.

The key benefits of using TRL are:

- Assessment of technology in terms of its elaboration
- Monitoring the number of developments within the technology at each stage
- Technology transfer
- Ability to control risks

In turn, the TRL methodology does not cover the following issues:

- Production cost
- Sufficiency of resource provision
- Assessment of production capabilities when transferring production from the pilot site to the main production, etc.

Thus, the methods for assessing the level of technology should be supplemented by methods that evaluate the production infrastructure created in an organization.

Let us describe a mathematical model for assessing the readiness of an organization for the production of new products. Earlier we examined the formation of an integral indicator of an organization's competitiveness, which was based on such indicators as the level of competitiveness of products (IQ), the level of development of an adaptive production system (including the level of automation, the use of systems with artificial intelligence, 3D technologies) and flexible production systems (K); organization competency level (K_2); competency level of subcontractors (K_3). When assessing the readiness of an organization for the production of rapid development products, we will consider a single indicator of the intellectual potential of an organization (A), which characterizes a sufficient level of competence (of an organization and its subcontractors) and innovative potential, thereby achieving an organization's competitiveness.

The integral indicator of assessing an organization's readiness for launching production of rapid development products based on the described indicators, as well as assessing the level of applied technologies using the TRL method, is formed on the basis of the following formula (all values in the formula are normalized and dimensionless):

$$M = w_1 \cdot \mathrm{TRL} \cdot IQ + w_1 \cdot K \cdot A,$$

where TRL is the normalized score on the scale of the TRL methodology: $w_1 + w_2 = 1$.

The quantitative estimates obtained as a result of the calculation according to the proposed model have the following economic sense: we will consider the technical level of an organization as sufficient, if its integral readiness indicator is $M \geq 1$. Such an organization is ready for rapid development. Moreover, the first term in the expression characterizes the level of technological development of the organization, and the second—its production capabilities.

The competent combination of methods for assessing technological readiness and production capabilities of an organization make it possible to reach the following cumulative effect. If the current assessment of the level of production readiness is far ahead of the assessment of the level of technology readiness, it can lead to the launch of products based on insufficiently developed technology, which in turn will cause a loss of investment. If the current assessment of the level of technology readiness is far ahead of the assessment of the level of production readiness, then such a combination can lead to the launch of low-quality products at inflated prices. Therefore, a necessary condition for the transition of an organization to the rapid development path is the balanced development of an organization's technological level and its production capabilities.

The process of entering the rapid development path will be considered balanced and stable in time if the condition $(\mathrm{TRL}_1 \cdot IQ_1 - \mathrm{TRL}_0 \cdot IQ_0) \approx (K_1 \cdot A_1 - K_0 \cdot A_0)$ is met. It means that innovative activity on development and introduction of innovative technologies and creation of unique products is harmonized (consistent in pace and intensity) with the development of the production system and production capabilities of an organization through the effective use of intellectual potential.

Effective management of competitiveness and the technical level of production determine the ability of an organization to enter the rapid development path and should be a continuous process in time. This is due to the fact that the economic environment is constantly changing (market conditions are changing, new players are appearing, scientific and technological progress is occurring), and in practice it is rather difficult to obtain in time reliable information on the influence of environmental factors. In this regard, the management system for rapid development of an organization should be structured in a way that enables taking timely measures to maintain a high level of an organization's competitiveness and technical level of production based on incomplete information on environmental factors.

Early formation of the system's flexibility allows to respond to the manifestation of negative environmental factors at an early stage and with high efficiency. Well-known approaches from the practice of risk management, such as responding to weak signals may be used as an economic tool. This will allow taking into consideration the likely trends of the external economic environment in the planning process and to predict the need for evolutionary and revolutionary changes in the product strategy of an organization.

10.3 Rapid Development of an Organization: Key to Improving Sustainability and a Revolutionary Transition to a New Technological Order

The modern economy is revolutionizing toward the intellectual and digital economic environment. Moreover, the processes of formation and change of the economic environment are significantly ahead of the existing economy. The latest achievement in information technology, machine learning, and artificial intelligence presents completely different requirements for managing organizations and the economy as a whole. Therefore, it is rapid development of organizations that will allow knowledge-based organizations to take and maintain a leading position in the future economic paradigm.

The outline of the future economy will undoubtedly be associated with the use of automated intellectual solutions that will minimize human participation (especially in terms of physical labor) in the operation of production. Even today we can see that scientific achievements in the field of digitalization and intellectualization of management methods of organizations are becoming the basis for the formation of new economic realities. Of course, these changes will soon require radical changes in the development path of organizations; therefore, rapid development for high-tech organizations will be the key to their successful development in a changing economic system, with regard to the next change in technological order.

The new technological order and its impact on the development of an industrial organization is governed by the rule of reduced manual and physical labor, as well as an increase in the share of mental labor in the structure of work performed, an increase in the competencies of employees, which leads to an increase in the intellectual potential of the organization and, as a result, an increase in innovative potential. More and more employee functions are transferred to automatic and mechanized complexes. This trend leads to a change in the nature of labor and the growth of competencies of employees of an organization (Figs. 10.9 and 10.10).

Thus, each technological order, bringing new technologies and equipment, carries a natural change in the organizational structure of the organization. At the same time, technological orders become shorter in duration, which is justified by the acceleration of R&D developments and the continuous improvement of existing labor technologies and production equipment.

In the transition from one technological order to a more advanced one, a gradual transfer of employee functions to machine complexes is observed. Note that the directions for improving the technology with the onset of the next technological order are laid in the previous technological order. Employees of organizations, leaning on their own work experience and the functions they perform, transfer their specific production functions to machine complexes, freeing up their time and energy to perform more important mental work. The automation of production

Fig. 10.9 Change in the share of manual and physical labor of employees during the five technological orders (TO)

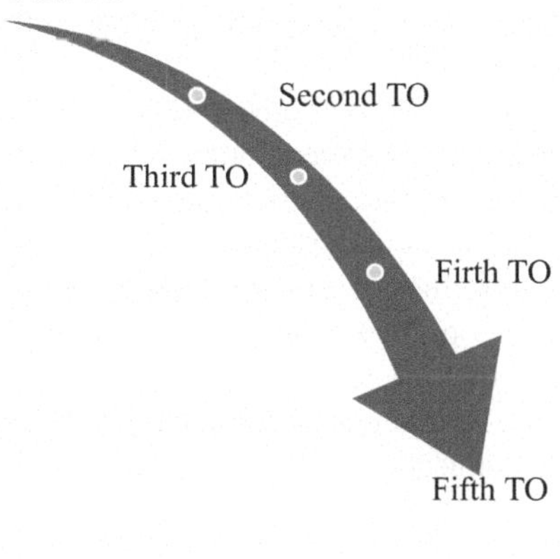

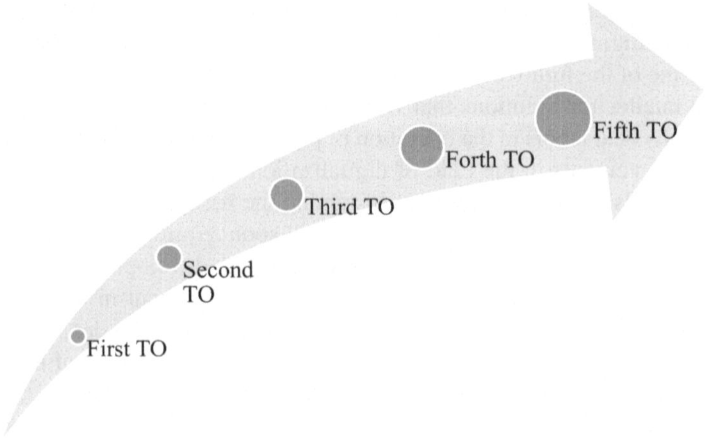

Fig. 10.10 Change in the share of mental labor and the number of competencies of employees during the five technological orders (TO)

increases, since not only production functions, but also some production management functions are transferred from employees to automated complexes.

At the same time, when transferring employee functions to an automated complex, new functions arise that require the creation of more advanced equipment and technologies for modeling. Using the word "modeling," we emphasize that automated systems currently are not able to copy the entire set of functions of an employee, so they model only their individual functions. However, for successful implementation and functioning in an organization, an automated complex should consider the nature of production and the features of its formation.

In the course of its operation, the elements of the automated complex may change, which entails a change in the entire system. In the conditions of modern technological order, an organization's employees have to monitor such changes and make the necessary adjustments to the system and some of its elements in accordance with the changing needs and functions of production and employees. However, in the near future, automated systems will acquire new functions related to the ability to self-learn and readjust their elements and their relationships for the best functioning of the system.

It should be noted that the competencies and needs of employees are growing faster than the development of the automated system used in the production, which leads to contradictions associated with the inability to transfer some functions of employees to the automated complex. So, there is a need to use the competencies and intellectual capital of employees to improve the existing automated system or develop a new system that could fulfill the necessary volume of production functions of employees.

The automated complex passes explosively from one qualitative state to another in connection with accumulation and application of necessary quantity of innovative technologies by employees for its improvement and in connection with transition of an organization to a new technological order. We see that over time, transferring their functions to automated systems, the employee ceases to be part of the technical system of an organization, acquiring new competencies, that are more relevant to the modern market. The acquisition of new competencies by an employee leads to the emergence of new professions and new products. For this reason, using a competency-based approach, a modern organization can enter the path of economically sustainable development.

The increase in the level of automation of production has already made it possible to create fully automated ("unmanned") organizations corresponding to the sixth technological order, also called the digital economy. The digital economy triggers another change in the requirements for the competencies of employees of existing organizations: in particular, today thanks to digital economy technologies not only development operations, but also product testing and even production processes can be transferred to computer systems. Thus, we will come to a production form in which machines produce machines and there will be no need for the physical labor of the employee directly associated with the production processes.

Bearing in mind the previously discussed experience of automation of production in Japan and Germany, we note that production systems of the sixth technological order are able to independently produce new production systems: machines are already able to produce machines.

Naturally, the sixth technological order will be replaced by the seventh already in the middle of the twenty-first century. This view is based on observations confirming the creation of innovative potential in organizations, which can become the basis for another "explosion".

According to the assumptions of some scientists, the new seventh order will bring innovations that surpass in their level the innovations of previous ways. For example, Bill Zebuhr writes the following: "I am proposing that the seventh wave is the wave of innovation and development that will make use of new science and the technology resulting from it. I am also proposing that the six waves are components of a super wave and that the seventh will be the first of a series of waves all based on the new science but with different technologies dominating as the science is better understood and utilized."

Based on the history of the development of technological orders and taking into account the view of modern economists we can predict the shape of the organization in the future—the one that will function in conditions of the seventh technological order.

One of the main changes brought by the seventh TO will be a new approach to production, associated with the next wave of growth in employee mental work. In this regard, a new stage of development of scientific approaches in the field of management of intellectual potential of an organization is forecasted, corresponding to the current state of research in the field of intellectual capital, especially taking into account the latest scientific developments in the context of ecosystems, at the national and regional levels. Modern research reemphasizes the approach to understanding the driving forces of creating material values based on the totality and

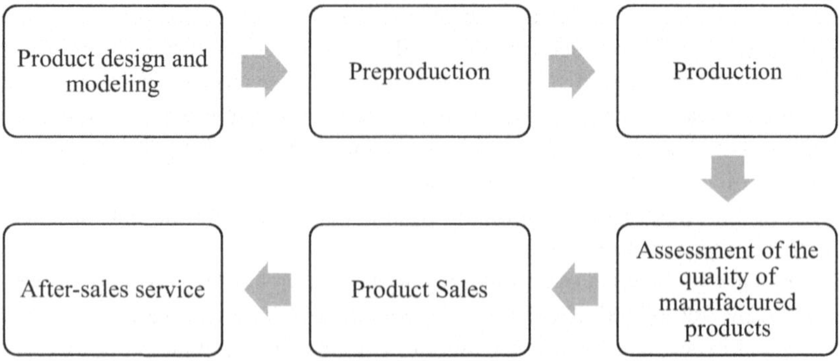

Fig. 10.11 Organization of production activities in an organization of the future

balance of organization's intellectual and financial resources in order to create a more holistic vision of its innovative potential. Most current scientific research mainly deals with the organizational level of the relationship between the intellectual potential of an organization and its productivity.

Accordingly, it is fair to assume that in the short term the creation of an organization of the future will be closely connected with the creation and implementation of robotic systems based on artificial intelligence (AI). So, an organizational structure will be achieved in which the employee creates artificial intelligence, and AI, in turn, will be able to independently create mechanical and automated systems without the intervention of employees.

The algorithm of production activities of an organization of the future is presented in Fig. 10.11.

The stage of product design begins with a preliminary design of the product, which determines its technical and economic appearance. Organizations with rapid development should use modern digital intelligent methods for product design and modeling, based on technologies for constructing digital doubles of products and full mathematical modeling with imitation of various scenarios of product functioning to design its appearance that most fully corresponds to the given technical characteristics of the product.

For successful design and modeling of high-tech products in organizations with rapid development, the technology of machine learning and artificial intelligence should be fully used, which, through the intelligent analysis, will allow to find the most effective design solutions.

To prepare the production of complex products, a digital double of products and a digital production plan are used, which create a single virtual environment for modeling production processes taking into account various factors and variations in different scenarios. As a result of the preproduction stage in organizations with rapid development and the creation of a digital product double, the preproduction stage begins, including its robotization—equipping the production with intelligent robotic systems that are able to quickly respond to changes in the work area and adapt their activities to such changes.

When organizing production processes in an organization of the future, it is necessary to use cybernetic systems consisting of natural and artificially created objects, subsystems, and controllers that form a functionally improved automated control system. The fundamental difference of such a system from the systems used at present is a qualitatively new executive mechanism capable of perceiving changes in the environment and internal environment of the organization, responding to those changes and learning independently.

At the stage of production in organizations of the future, automated and robotic complexes will play a significant role, which will not only be able to perform mechanical work, but also independently plan and manage other robotic complexes without human intervention. Building on the latest achievements of artificial intelligence, such complexes can be delegated not only production processes, but also the processes of "machine management," which will be able to predict and solve "production issues" more accurately and better than humans.

At the same time, it is possible to use Brain-computer interface technology at a number of critical production sites, which will allow for effective interaction between humans and control complexes to adjust production processes in real time. In addition, augmented reality technologies are promising for human participation in production in organizations of the future, when a highly qualified employee can perceive the process in the workplace using digital and virtual technologies.

The creation of smart production facilities with rapid development should lean on open production technology based on the principles of open source (open design, etc.). This approach is well established in the field of software development and for modern digital production it will be a very effective method of creating and developing an organization of the future.

Speaking of the direct production and equipment for their production in an organization of the future, it is necessary to note the additive technology, which consists in creating products through the consecutive layering. Models made by additive method can be used both at the stage of pilot production and mass production. Additive manufacturing techniques have been used for a long time (welding, for example). However, the additive methods used in organizations of the future should lean on the most innovative three-dimensional methods of creating products. For example, 3D printing, in which an object consists of layers superimposed on each other with a 3D printer.

It is also proposed to use mobile applications and devices (mobile technologies) to solve problems of assessing the quality of manufactured products and for their after-sales service. For example, mobile applications can monitor the status of a product and inform owners about its breakdowns and the need for a scheduled inspection by specialists, offering a list of organizations or checkup points closest to the owner (in terms of geographic location).

Thus, we see how production on the basis of new technological solutions will be organized. The integrated use of the abovementioned technologies in the organizations of the future will allow to create the most efficient system for product development and creation and will be able to carry out the mass transition of production organizations to functioning in the sixth technological order, which will help achieve faster development.

Clearly, not every organization has the potential to achieve rapid development with proper use. To assess the possibility of achieving rapid development by a high-tech organization, we will consider the "boundaries" of rapid development, in which the indicators of the economic condition of an organization (determining its competitiveness) demonstrate the achievement of balanced sustainable economic development. This means that innovation and the creation of unique products (or new markets), provided with resources, are harmonized (consistent in pace and intensity) with the development of the production system and business processes, accompanied by economic sustainability of the organization.

Sustainable development of a high-tech organization, turning into rapid development is similar to a situation in which there will be not only an increase in an organization's competitiveness indicators, but also the vector indicators under Pareto optimality condition, which directly determine an organization's competitiveness. Development management within the framework of a three-level planning system of a high-tech organization is Pareto optimal if the condition for Pareto optimality of the vector of indicators forming the organization's competitiveness *IQO* is met. This condition requires improvement of at least one component of the vector while preserving the values of the rest.

In such a situation, for an organization being on the path of rapid development (expressed in the framework of the model by the transition to a new Pareto optimal vector *X*) is possible only as a result of the effective implementation of operational and investment-innovative activities within the framework of a three-level planning system.

Let us consider an organization's competitiveness management model to ensure its rapid development. The mathematical model of an organization's competitiveness management is described by the formula:

$$I\dot{Q}O = D(t) \cdot IQO(t) + M(t, IQO(t)),$$

where *IQO* is a vector whose components form an integral indicator of an organization's competitiveness (see Sect. 10.1); $D(t)$ is a matrix describing the mutual influence of the components of the *IQO* vector.

At each moment of time t, the elements of the matrix take on new values. The economic meaning of the matrix element is to describe the tendency to change the components j of the vector *IQO* under the influence of components i; $M(t, IQO(t))$ are the control actions on the operational activities and development program of an organization.

The model mathematically describes the economic processes of the influence of all of the above parameters on the final result—the integrated competitiveness of an organization and entering the rapid development path, as well as clearly demonstrates how it is possible, by managing operational activities and a development program, to enter and take hold in the regime of rapid development of an organization.

The most important element of the model is the matrix $D(t)$ of the relationship of elements of the vector *IQO*. As a result of the management of operational activities by an organization's development program, the matrix should be changed in accordance with the following economic principles. Thus, ensuring the growth of competitiveness of products will lead to an increase in the organization's competitiveness as a whole, since highly demanded products in the market create competitive advantages of the organization. This ensures the growth of the organization's income and profits, which triggers an increase in its economic indicators and sustainability, while creating a resource reserve that can be used to develop a product line, technological base, and competency component, which eventually also leads to an increase in the organization's competitiveness as a whole and increase its capabilities for the transition to the rapid development path.

Thus, the solution of the proposed equation is the trajectory of rapid development of a high-tech organization. This is due to the fact that at certain points in time an organization may experience the effect of uncertainty factors (or the result of the effect of the negative impact of such factors accumulated over time), which may not allow it to achieve the necessary values of the parameters of rapid development. In this case, it is necessary to take into account the value of the integral indicator of competitiveness *IQO* of a high-tech organization, which should remain greater than one (Fig. 10.12).

If the parameters of competitiveness of a high-tech organization do not go beyond the established boundaries and the value of the integral indicator of competitiveness is $IQO \geq 1$, it means that innovation and the creation of unique products (or new markets), provided with resources, is harmonized (consistent in pace and intensity) with the development of the production system and business processes, accompanied by rapid development of the organization. The redistribution of resources or other measures (for example, the creation of a risk fund) neutralizes the negative impact of uncertainty factors on the process of economic development of the organization. Any case out of boundary requires a review of all planning documents, including the strategy. At the same time, the market opportunities of an economic entity are determined by the quantitative and qualitative characteristics of its

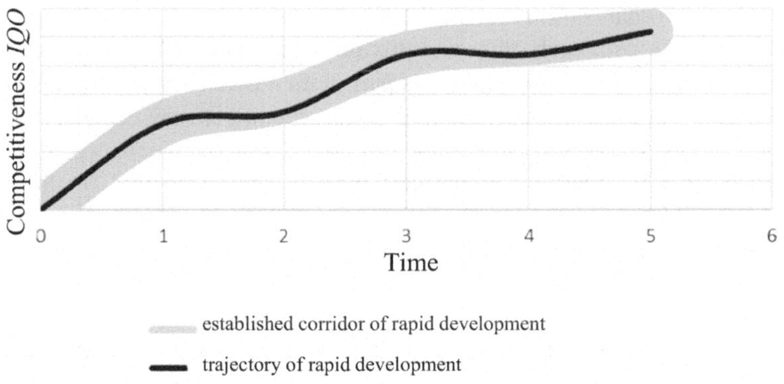

Fig. 10.12 The boundaries of rapid development of high-tech organizations

resources and the efficiency of their distribution, the possibility of additional attraction on favorable terms, the ability to implement stable economic development through the use of innovative technologies, taking into account the actions of competitors to increase their competitive advantages.

Through the redistribution of resources, an organization can form the direction of its strategic development. For example, it could be:

- Creation of competitive advantages that will strengthen positions in a certain market segment or let enter new markets
- Organization of new activities capable of generating significant income
- Creation of new markets based on the development of previously not marketed goods

Practical approaches to managing the operational activities and development program of an organization to ensure its rapid development can be implemented in two development scenarios: moderate market and progressive market.

Scenario I "Progressive market" (high level of consolidation of high-tech industry assets and high degree of market orientation).

All high-tech organizations position their assets as centers of competence in their field. The interaction model of all key organizations and private organizations is of a partnership nature. Both on the global and Russian markets, they work with clearly defined and strategically approved product and technological profiles and competencies. In this scenario, the most extreme solution would be to combine all organizations of the same industry into a single structure, managed from a single center. Holdings may consist of divisions—independent divisions (within the framework of the holding's general strategy), market organizations for specific, narrower market niches. Such organizations are active in the M&A market, acquiring key competencies and buying organizations, startups in all technological spheres.

In this case, competition among high-tech organizations will rapidly decline.

Scenario II "Moderate market" (average level of consolidation of organizations and assets of high-tech industry and high degree of market orientation).

All high-tech organizations retain their independence and structure, while continuing to compete with each other and private organizations. But at the same time, each organization creates independent (as part of their own strategy) market-oriented structures and develops production and product competence centers. As in the progressive market scenario, the model of interaction between organizations is transformed from competitive to partner: both in the world and in the Russian markets, the holdings strengthen other holdings with their competencies as part of cooperation. Organizations actively acquire key competencies by buying private startups, small innovative organizations and serial competitor plants across the entire spectrum of technologies.

There is internal competition only at the first stages of the scenario, gradually decreasing as the centers of competence are built and their competitive advantages are increased. International competition will increase more intensively than in the progressive market scenario, organizations will become full-fledged world-class players, competing with leading organizations, as well as experiencing pressure from other domestic organizations which are entering the world market. While having the state order as their priority, organizations should meet market key performance indicators (capitalization, market share, performance, profit, etc.).

"Progressive market scenario" implies a more efficient structure, production system and management system than "Moderate market." Besides, "Progressive Market Scenario" will require less resources and will provide greater competitiveness by consolidating the industrial assets of organizations and eliminating internal competition when working in the Russian and international markets.

Choosing one of the two scenarios under consideration, a high-tech organization can choose the pace of increasing its presence in many niches, focusing and developing only promising and critical products and technologies, rather than "blurring" resources into over-sized and underutilized assets. For global leadership in the market, high-tech organizations need to focus today on the development of the sixth technological order. Only in this case projects can be launched (on a new technological basis) in 2018–2023 to create equipment and systems by 2025–2030 surpassing world analogues in their characteristics. Technology should be developed at a faster pace than design and product challenges. Otherwise, organizations will always be catching up-players—both in traditional markets and in new ones.

Thanks to the integration of information on the current level of competitiveness of the entire range of products manufactured by an organization, the existing innovative potential and the level of unique competencies of both the organization and its subcontractors, the level of development of the adaptive production system, the level of interaction with customers in terms of determining the appearance of new products and accounting market consumer preferences, it is possible to form a trajectory of economic development of a high-tech organization and determine the conditions under which the organization is in a state of rapid development.

The proposed criterion for rapid development is to assess the competitiveness of high-tech organizations. The condition for achieving rapid development is the excess of the competitiveness index of an organization of a single value of the integral indicator. An organization will remain competitive as long as its products, which have unique technical and economic characteristics, will remain in demand in the market.

References

1. Nesterov, E. A., Pankov, A. A., & Yudin, A. V. (2019). Economic-mathematical model for assessing the competitiveness of a high-tech corporation. *Economics and Management: Problems, Solutions, 1*(86), 70–75.
2. Chursin A. & Makarov Y. (2015), *Management of competitiveness. Theory and practice* (521 p.). Cham: Springer.

Chapter 11
Conclusions

As part of this monograph, new approaches to creating a theory of the product life cycle have been considered. On its basis, a strategy for rapid development of organizations can be built. The idea of achieving rapid development is based on existing economic theories: the competitive management theory (in particular, in the context of digital transformation); the theory of corporate governance; the theory of innovation management; as well as fundamental economic laws.

An algorithm for creating a new theory of the product life cycle in the context of achieving rapid development has been proposed, and the main areas of research necessary to implement this algorithm in practical activities of industrial companies in the face of dynamically changing factors and risks have been identified.

The conditions necessary at the initial stage of designing products of rapid development have been formulated. Models for assessing an organization's capacity to achieve rapid development in selected priority areas have been built. In the context of creating products of rapid development, issues of modernization to increase the life cycle have been considered and the criteria for the sustainable economic development of an organization are determined. The authors have outlined the directions of development of competencies, information solutions, digital technologies for creating a fundamentally new product with unique consumer properties, and high competitiveness. The approaches to the implementation of advanced methods (primarily digital) of product design, preproduction, and production have been analyzed.

It is suggested to ensure the market efficiency and competitiveness of products at all life cycle stages using the developed effective tools and mechanisms and, as a consequence, the methodology for calculating the cost of the product life cycle, taking into account the risks and uncertainties when creating its competitive advantages at the main stages of the life cycle, ensuring the extension of the presence of products on the market.

Following the study of the problems of managing the product life cycle and the need for its update, it has been concluded that the observed processes have certain regularities, which summary allowed the authors to formulate and mathematically

A. Tyulin, A. Chursin, *The New Economy of the Product Life Cycle*,
https://doi.org/10.1007/978-3-030-37814-1_11

justify the law of the manufacturer's rapid development. The law of rapid development shows that the evolution of technologies, development of competencies and innovations, which are mutually influencing, lead to the creation of new promising products of rapid development.

Personification of needs is considered as a reference point in the creation of promising products, while it is established that the emerging compliance with the level of technology development and consumer expectations determines the creation of promising products. Accordingly it is proposed to identify such correspondences on the basis of a digital product life cycle management infrastructure, which, on the one hand, will enable the analysis of the main consumer expectations for the promising technical and economic image of new products, and, on the other hand, the determination of the possibilities of creating such products by an enterprise given the identified consumer expectations based on an analysis of the current development of equipment and technologies.

The effective processes of development and marketing of products of rapid development are based on the proposed economic and mathematical models of a balanced process for determining their technical and economic image. These models are based on the analysis of consumer expectations and an organization's capacity to meet them. It is suggested to conduct the analysis of the market and future needs when shaping the product image using modern intelligent technologies to support the adoption of effective management decisions.

Models for assessing the innovative potential and unique competencies of an organization for the production of goods under specified parameters have been developed. However, it is proven that the most effective organizational and information basis for the involvement of innovative potential and unique competencies in the production process is the transformation of an organization to a digital enterprise and the placement of relevant methodological tools on its platform. A single knowledge base, methods for analyzing the global information space to identify new areas for the development of engineering and technology, as well as mechanisms for replenishing and updating the knowledge base form the core of intelligent methods for supporting the adoption of effective managerial decisions.

Use of digital infrastructure for product development management is proposed as an organizational and information basis for managing the competitiveness of products at the stage of shaping their technical and economic image. To create competitive products in accordance with the developed technical and economic profile, the process of creating the corresponding production capabilities has been described. By means of modeling, approaches to determining resources for preparing production capacities for production with a given technical and economic image, cost, and timing have been proposed. Models for managing the competitiveness of a product of the future have been developed, taking into account various organizational and economic aspects and risks and allowing to manage all the processes from creating the image of future products to the moment of their replacement with a product with higher consumer value with higher consumer properties.

To comply with the established boundaries, it has been proposed to design products using applied intelligent automated systems (software solutions) that can build digital twins of the future product and calculate its approximate cost based on variations in the parameters of its individual components, subject to the required technical characteristics.

All these processes involve Big data analytics, collected in enterprises, holdings, industries, as well as the global information space. The collected Big data form a data bank in the areas that provide product design management and allow to run different options of economic development of an organization depending on various product parameters of products and volumes of their sale in the market by means of the created simulation mathematical models of process of designing and production.

An economic basis for preproduction of new products with high technical and economic characteristics has been formed and modern methods for managing the production process, as well as methods and tools for managing the life cycle of a product of the future have been proposed.

Models for determining the cost and complexity of manufacturing parts and assemblies based on information about the factors affecting the cost are built, including data on materials and own production capacity. By varying these factors, provided that the part meets the specified technical characteristics, it is possible to obtain the optimal value of its cost and complexity of production. The creation of an automated intelligent decision support system for the formation of new products is proposed, which is the evolutionary development of simulation methods. For its implementation, it is proposed to use new innovative developments based on the knowledge of fundamental science and the capabilities of artificial intelligence methods, machine learning, and decision-making on the basis of the analysis of large amounts of data on factors affecting cost.

The monograph describes the basics of building a digital enterprise, which are primarily associated with the effective organization of production based on the organization of flexible automated production taking into account the advantages of digital technologies that allow the production of various positions of a wide diversified product range of the enterprise. To determine the optimal ratio of the work of an organization and the work of subcontractors, a system has been developed for the distribution of costs for production between the producers of the main products and subcontractors.

The basic methods for constructing large organizational systems based on artificial intelligence, neurosystems, and the main approaches to their creation are described with recommendations for shaping the foundations of building a large management system of production organization.

It has been proved that the development and production of products of the future ensures the advance development of an organization in the event that an effective process of managing rapid development is built on the basis of using a whole range of developed tools which allow to assess the dynamics of changes in the competitiveness of products depending on the use of certain technologies and competencies in the process of their production; to select the most innovative technical solutions and evaluate an organization's capacity for their implementation and production of products of the future that can lead to the rapid development path; to assess the economic sustainability and sufficiency of resource support necessary to enter the rapid development path. The use of rapid development management tools in the complex forms a holistic mechanism for the transition of enterprises to the rapid development path.

Models describing the impact of instruments on the final result—reaching the rapid development path and achieving a balanced sustainable economic development are constructed. This means that innovation and the creation of unique products (or new markets), provided with resources, are harmonized (consistent in pace and intensity) with the development of the production system and business processes, accompanied by economic sustainability of the corporation.

As a whole, the scientific and practical significance of this monograph is that it contains:

- Practical recommendations on the organization of optimal product life cycle management, which are of interest to industrial enterprises, especially high-tech companies, individual entrepreneurs, and economic analysts.
- Management framework of all stages of the production cycle of new products with high technical and economic characteristics, which will increase the economic stability and positioning of high-tech companies in the world market.
- Modern models and methods of production process management taking into account the advantages of digital technologies.
- Mechanism of transition of high-tech enterprises to the rapid development path, which can be used in practice in order to achieve global competitive leadership.

Bibliography

1. Angang, H. (2015). Embracing China's "new normal". *Foreign Affairs, 3*, 5–11.
2. Baden-Fuller, C., & Morgan, M. S. (2010). Business models as models. *Long Range Planning, 43*, 154–156.
3. Boisoit, M. (1998). *Knowledge assets: Securing competitive advantage in the information economy*. Oxford: Oxford University Press.
4. Chen, H. C., & Dahlman C. J. (2004, August). *Knowledge and development: A cross-section approach.* World Bank Policy. Research Working Paper, vol. 3366.
5. Chursin, A., & Makarov, Y. (2015). *Management of competitiveness. Theory and practice* (378 p.). Cham: Springer.
6. Chursin, A., Tyulin, A., & Yudin, A. (2016). The model of risk assessment in the management of company's competitiveness. *Journal of Applied Economic Sciences, 11*(8), 1781–1790.
7. Chursin, A., Semenov, A., & Danilchanka, A. (2016). Analysis of innovation development in the economy with exhaustible resource sector by first order dynamical systems application. *Nonlinear Phenomena in Complex Systems, 19*(3), 254–270.
8. Chursin, A., Vlasov, Y., & Makarov, Y. (2016). *Innovation as a basis for competitiveness: Theory and practice* (336 p.). Cham: Springer.
9. Collis, D., & Montgomery, C. A. (1995). Competing on resources: Strategy in the 1990s. *Harvard Business Review, 73*, 118–128.
10. Conte, T. (2008, September 8). *A framework for business models in business value networks* (pp. 5–7). (White paper No 001-08). London: Institute of Information Systems and Management (IISM).
11. Crosby, P. (1979). *Quality is free: The art of making quality certain: How to manage quality – So that it becomes a source of profit for your business* (309 p.). New York: McGraw-Hill.
12. Cukier, K. (2005, October 22). Survey: Patents and technology. *The Economist*, 3–20.
13. Davenport, T., De Long, D. W., & Beers, M. C. (1998). Successful knowledge management projects. *Sloan Management Review, 17*(8), 43–57.
14. Davenport, T., & Prusak, L. (2000). *Working knowledge*. Boston, MA: Harvard Business School Press.
15. Davenport, T. H., Jarvenpaa, S. L., & Beers, M. S. (1996). Improving knowledge work processes. *MIT Sloan Management Review, 1*(8), 53–65.
16. Deming, W.E. (1982). *Quality productivity and competitive position* (373 p.). Cambridge, MA: Massachusetts Institute of Technology.
17. Department of Defense Fiscal Year. (2016). *President's budget submission, 2015.*
18. Dornbusch, R. Fischer, S., & Startz, R. (2013). *Macroeconomics* (672 p.). New York: McGraw-Hill Education.

A. Tyulin, A. Chursin, *The New Economy of the Product Life Cycle*,
https://doi.org/10.1007/978-3-030-37814-1

19. Drucker, P. (1993). *Post capitalist society* (232 p.). Oxford: Butterworth & Heinemann.
20. Drucker, P. (1969). *The age of discontinuity* (436 p.). Piscataway, NJ: Transaction Publishers.
21. Drucker, P. (1998). The next information revolution. *Forbes ASAP, 24*, 57–69.
22. Edvinsson, L., & Malone, M. (1997). *Intellectual capital: Realizing your company's true value by finding its hidden roots*. New York: HarperCollins Publishers.
23. Feigenbaum, A. (1951). *Quality control: Principles, practice, and administration* (443 p.). New York: McGraw-Hill.
24. Firer, S., & Williams, M. (2001). Intellectual capital and traditional measures of corporate performance. *Journal of Intellectual Capital, 4*, 348–360.
25. Flanagan, R.J. (2006). *Globalization and labor conditions: Working conditions and worker rights in a global economy* (272 p.). Oxford: Oxford University Press.
26. Frascati Manual. (2002). *Proposed standard practice for surveys on research and experimental development*. Paris: OECD.
27. Futron Corporation. (2014). *Futron's 2014 Space Competitiveness Index* (180 p.). Washington, DC: Futron Corporation.
28. Garicano, L., & Rossi-Hansberg, E. (2005, June). *Organization and inequality in a knowledge economy* (NBER working paper No. 11458, pp. 16–25).
29. Garvin, D. A. (1993). Building a learning organization. *Harvard Business Review, 18*(7), 78–91.
30. Gibbons, M., Limoges, C., Nowotny, H., Schwartzman, S., Scott, P., & Trow, M. (1994). *The new production of knowledge: The dynamics of science and research in contemporary societies*. Thousand Oaks, CA: Sage.
31. Granovetter, M. (1982). *Social structure and network analysis*. Thousand Oaks, CA: Sage.
32. Grant, R. (1996). Prospering in dynamically-competitive environments: Organizational capability as knowledge integration. *Organization Science, 4*(7), 375–387.
33. Hamel, G., & Prahalad, C. K. (1994). *Competing for the future*. Boston, MA: Harvard Business School Press.
34. Hammer, M. & Champy, J. (1993). *Reengineering the corporation: A manifesto for business revolution* (272 p.). New York: Harper Business.
35. Ishikawa, K. (1988). *What is total quality control? The Japanese way* (240 p.). J. Lu David (Trans.). Englewood Cliffs, NJ: Prentice Hall.
36. Kay, J. (1999). Strategy and the illusions of grand designs, mastering strategy. *Financial Times, 15*, 2–4.
37. Kim, W. C., & Mauborgne, R. (2005). *Blue Ocean strategy*. Harvard: Harvard Business School Press.
38. Luthy, D. (1998). *Intellectual capital and its measurement* (pp. 115–119). Osaka, Japan: Asian Pacific Interdisciplinary Research in Accounting Conference (APIRA).
39. Machlup, F. (1984). The economics of information and human capital. In *Knowledge, its creation, distribution, and economic significance* (Vol. 3, pp. 67–78). Princeton, NJ: Princeton University Press.
40. Osel, R. R., & Wright Robert, V. L. (1980). Allocating resources: How to do it in multi-industry corporations. In *Handbook of business problem solving*. New York: McGraw-Hill.
41. Osterwalder, A., & Pigneur, Y. (2010). *Business model generation: A handbook for visionaries, game changers and challengers*. Hoboken, NJ: Wiley.
42. Ouchi, W.G. (1981). *Theory Z: How American business can meet the Japanese challenge* (192 p.). Boston, MA: Addison-Wesley.
43. Penrose, E. (1980). *The theory of growth of the firm* (p. 21). New York: M. E. Sharpe. 265 p.
44. Penrose, E. (1959). *The theory of growth of the firm*. Oxford: Blackwell.
45. Porter, M.E. (1980). *Competitive strategy: Techniques for analyzing industries and competitors* (397 p.). New York: The Free Press.
46. Porter, M. E., & Stern, S. (2000). *Measuring the ideas production function: Evidence from international patent output. NBER Working Paper* (Vol. 7891, pp. 46–53). Cambridge, MA: National Bureau of Economic Research.

47. Schein, E. H. (1996). Culture: The missing concept in organizational studies. *Administrative Science Quarterly, 41*(2), 229–240.
48. Selznick, P. (1957). *Leadership in administration*. New York: Harper.
49. Senge, P. M. (1994). *The fifth discipline fieldbook: Strategies and tools for building a learning organization paperback* (593 p.). New York: Doubleday.
50. Shewhart, W. A. (1931). *Economic control of quality of manufactured product*. New York: D. Van Nostrand Company.
51. Sloan, A. (1964). *My years with general motors* (472 p.). New York: Doubleday.
52. Smith, A. (2014). *The wealth of nations* (524 p.). North Charleston: Create Space Independent Publishing Platform.
53. Solow, R. (1957). Technical change and the aggregate production function. *The Review of Economics and Statistics, 39*(3), 312–320.
54. Tallman, S. (2014). Business models and multinational firm. *Multidisciplinary Insights from New AIB Fellows Research in Global Strategic Management, 16*, 115–138.
55. Teece, D. J. (2010). Business models, business strategy and innovation. *Long Range Planning, 43*, 172–194.
56. The Seventh Framework Programme (FP7). (2007). Brussel.
57. The Sixth Framework Programme in brief. (2002, December). Brussel.
58. The State of Human Capital. (2012). *False summit*. Why the human capital function still has far to go. A Report by McKinsey & Company and The Conference Board, New York.
59. Unleashing America's Research & Innovation Enterprise. (2013). Cambridge, MA: American Academy of Arts and Sciences.
60. Wernerfelt, B. (1984). A resource-based view of the firm. *Strategic Management Journal, 5* (2), 171–180.
61. Akerman Yelena, N., & Burets Yulia, S. (2014). Transformation of models of innovative development on the way to openness of innovative systems. *Tomsk State University Journal, 378*, 178–183.
62. Asaul, A.N., Karpov B.M., Perevyazkin V.B., & Starovoytov M.K (2008). *The modernization of the economy based on technological innovation* (606 p.). St. Petersburg: ANO IPEV.
63. Bely, E. M., Rozhkova, E. V., & Tyulin, A. E. (2013). Integrated structures in the modern economy: The essence and trends of development. *Fundamental Research, 6*, 1482–1484.
64. Bely, E. M., & Tyulin, A. E. (2013) *Statistical tools to solve the problems of integration of industrial enterprises*. In Materials of the 1st International Conference on the Formation of the Main Directions of Development of Modern Statistics and Econometrics, Orenburg, 26–-28 September 2013, vol. 1, pp. 204–211.
65. The Open Future. (2015). (pp. 13–35) Thomson Reuters.
66. Vanyurikhin, G., Paison, D. B., & Makarov Y. N. (2013). *The economy of space activities: Monograph* (600 p.). Under the scientific. ed. of the Doctor of technical sciences, prof. G.G. Raikunova. Moscow: FIZMATLIT.
67. Vikhansky, O., & Naumov, A. I. (2006). *Management: The textbook* (4th Revised and Enlarged ed., p. 670). Moscow: Economist.
68. Womack, J., & Jones, D. (2004). *Lean production*. Moscow: Alpina-Publisher.
69. Gates, B. (2001). *Business @ the speed of thought: English translation* (480 p.). Moscow: Eksmo-press.
70. Gitelman, L. D., & Kozhevnikov, M. V. (2013). Centers of competencies – The progressive form of organization of innovative activity. *Innovations, 10*(180), 92–98.
71. Ivanter, V. (2005). *The innovative and technological development of economy of Russia* (592 p.). Ivanter: V. V. Max Press.
72. Ilyin, V. A., Gulin, K. A., & Uskova, T. V. (2010). Intellectual resources as an innovative development factor. *Economic and Social Changes: Facts, Trends, Forecast, 3*(11), 14–25.
73. Castells, M. (2000). *Information age: Economy, society and culture: English translation* (608 p.). Moscow: HSE.

74. Kleiner, G. B. (2011). The resource theory of the system organization of economics. *Russian Management Journal, 3*, 3–28.
75. Maurik, J. (2002). *Effective strategist: English translation* (208 p.). Moscow: INFRA-M.
76. Mintzberg, H., Ahlstrand, B., & Lampel, J. (2001). Школы стратегий (*Schools of strategies*). Trans. from English. Under the editorship of Yu. Kapturevsky. St. Petersburg: Peter.
77. Prahalad, C. K., & Hamel, Г. (2001). Core competence of the corporation. In H. Mintzberg, J. Quinn, & S. Ghoshal (Eds.), *Strategic process*. Trans. from English. Under the editorship of Yu. Kapturevsky. St. Petersburg: Peter,
78. Kondratiev, V. (2008). *Seven notes of management. Manager's handbook* (7th edn., revised and extras, 976 p.). Moscow: Eksmo.
79. Soifer, V., & Shakhmatov, E. (2007). The development of a competence centre in the field of aerospace and geo-information technologies. *Herald of the Nizhny Novgorod University After N.I.Lobachevsky, 2*, 41–48.
80. Stewart, T. A. (2007). *Intellectual capital. A new source of wealth for the organization*. Trans. from English by V. Nozdrina (368 p.). Moscow: Pokoleniye.
81. Suprun, V. (2006). *Intellectual capital: The main factor of competitiveness of economy in the XXI century* (192 p.). Moscow: Komkniga.
82. Taylor, F. W. (2008). *The principles of scientific management*. Moscow: Dashkov.
83. Teece, D. J., Pisano, G., & Shuen, A. (2003). Dynamic capabilities and strategic management. *Herald of St. Petersburg State University Management, 4*, 133–171.
84. Thompson, A.A., & Strickland, A.J. (2006). *Strategic management: Concepts and situations for analysis* (12th ed, 928 p.). Moscow: Williams Ed. House.
85. Tyulin, A. (2013). *Public-private partnership: The essence and problems of formation* (pp. 65–68). Collection of materials of II International Scientific-Practical Conference "Scientific Aspects of Innovative Research." (Samara, June 13, 2013).
86. Tyulin, A. (2013). *Staffing corporate network of competencies centers in aircraft instrument making* (pp. 218–222). Scientific results of 2013: Achievements, projects, hypothesis: Compilation of the III International Scientific and Practical Conference materials (Moscow, December 27, 2013).
87. Tyulin, A. (2013). The core competency of the organizations participating in the integrated structures. *Problems of Economics and Management, 6*(22), 62–65.
88. Tyulin, A. (2013). The selection criteria for the optimal integrated structure of the production. *Quality, Innovation, Education, 9*(100), 66–71.
89. Tyulin, A. (2012). *Marketing analysis of the global market of avionics* (pp. 72–74). Collection of articles of the International Scientifically-Practical Conference on Marketing and Mass Communication in Sustainable Development of the Territory and of the Enterprise (Penza, June 2012-Penza).
90. Tyulin, A., Ozhiganov, E., & Korneenko, V. (2014). The technique of a rating estimation of efficiency of use of human capital. *Economics and Entrepreneurship, 12*(3), 183–191.
91. Tyulin, A. (2013). *Mechanisms of public-private partnership* (pp. 90–91). Collection of scientific materials of International Scientifically-Practical Conference on the Science and Education in the Modern World, Vol. II (May 31, 2013 in 4 volumes, Moscow).
92. Tyulin, A. (2013). Organizations participating in integrated structures as competencies centres. *Academic notes: Economic Sciences/Ulyanovsk State University, 30*, 44–47.
93. Tyulin, A. (2013). Organizational-economic mechanism of functioning of the industry based on *competencies* centres. *Quality, Innovation, Education, 5*(96), 65–69.
94. Tyulin, A. (2013). *Main types of public-private partnership* (pp 206–208). Materials of the fourth all-Russia Scientific–Practical Conference with International Participation "The regional innovation economy: The nature, the elements, the problems of formation" (Ulyanovsk, May 2013 Ulyanovsk).
95. Tyulin, A. (2013). Basic principles for the establishment of sectoral centres of competence. *Resources Information Supply Competition, 1*, 140–143.

96. Tyulin, A., & Chursin, A. (2016). Fundamentals of management of innovation processes in knowledge-intensive industries (practice), Moscow.
97. Tyulin, A., Ostrovskaya, A., & Chursin, A. (2015). Fundamentals of management of innovation processes in knowledge-intensive industries (theory), Moscow (290 p.).
98. Tyulin, A., & Barashkov, S. (2014). *Improving the competitiveness of the industry on the basis of a network of competence centres* (Vol. 4, pp. 160–168). Science and technology: Materials of XXXIV conference devoted to the 90-th anniversary of academician V. P Makeev (Moscow, 10–12 June 2010–2014).
99. Tyulin, A., & Rusinov, A. (2015). Approaches to measure competitiveness of rocket and space equipment. *Business in Law, 1*, 179–182.
100. Tyulin, A. (2016). Proposals for the improvement of the innovative development program of the corporation based on the selection of the strategy to create new competitive advantages. *Business in Law, 3*, 40–44.
101. Tyulin, A. (2016). Benchmarking in the creation of new competences й. *Economics and Entrepreneurship, 6*, 491–497.
102. Tyulin, A. (2013). *Principles for the establishment of sectoral centres of competence* (pp. 298–302). Collection of materials of III International Scientific-Practical Conference "Economics and Management: Analysis of Tendencies and Prospects of Development" (Novosibirsk, November 15, 2013).
103. Tyulin, A., & Rozhkova, E. (2013). Problems of conceptual competence approach in management: Structuring competences. *Economics and Entrepreneurship, 4*(33), 368–372.
104. Tyulin, A. (2013). *Promotion of aircraft instruments making enterprises to the world market* (pp. 152–153). Collection of scientific papers on materials of International Scientifically-Practical Conference "Modern Issues of Education and Science" (Moscow, December 30, 2012).
105. Tyulin, A. (2012). Product and technological specifics of aircraft instruments making. *Academic Notes: Economic Sciences, 28*(2), 31–42. Ulyanovsk State University.
106. Tyulin, A. (2013). *The project management process within the corporate network of competence centres* (pp. 177–180). Collection of materials of International Scientific-Practical Conference "Modern Tendencies of Development of Economy, Management and Law" (Moscow, 22 December 2013).
107. Tyulin, A. (2014). The development and operation of a branch network of competence centers. *Economic Science: Scientific and Information Journal, 2*(111), 95–98.
108. Tyulin, A. (2016). Recommendations on the practical implementation of open innovation by high-tech Russian companies. *Microeconomics, 3*, 55–59.
109. Tyulin, A. (2013). *The role of technological and economic factors in the development of sectoral integration* (pp. 149–154). Collection of materials of the X International Scientific-Practical Conference "Economics and Management in XXI Century: Trends of Development" (Novosibirsk, November 2013).
110. Tyulin, A. (2013). *Network interaction of sectoral competence centres in instrumentation: Main elements* (pp. 221–223). Collection of materials of II International Scientific-Practical Conference "Fundamental and Applied Sciences Today" (North Charleston, June 2013).
111. Tyulin, A. (2015). *Modern development paradigm of rocket and space technology* (pp. 33–40). Technologies and services: International forum "Russian innovative technologies and global market" (Moscow, November 27, 2015).
112. Tyulin, A., & Bely, E. (2012). Strategic analysis in aviation engineering. *Contemporary Problems of Science and Education, 3*, 25–32.
113. Tyulin, A. (2015). *Theory and practice of creating and managing competences to enhance the competitiveness of integrated structures* (312 p.).
114. Tyulin, A. (2014). Formation of the conceptual model of sectoral management based on competence centres. *Russian Entrepreneurship, 9*(255), 4–11.

115. Tyulin, A. (2013). Formation of the mechanism of public-private partnership in integrated business structures. *In the World of Scientific Discoveries (Economy and Innovative Education), 8.1*(44), 199–221.
116. Tyulin, A. (2011). *The goals and objectives of the innovation development of domestic aircraft engineering* (pp. 74–75). Collection of articles of the III International Scientific-Practical Conference "Problems of the Innovation Economy, Modernization and Technological Development". (Penza, March 2011).
117. Tyulin, A., & Yudin, A. (2015). Methodical approach to assessment of influence of innovative technologies on the competitiveness. *Microeconomics, 6*, 59–64.
118. Tyulin, A., & Yudin, A. (2015). Methodical approach to assessment and ranking of unique technological competence. *Economics and Entrepreneurship, 12*(2), 681–685.
119. Harrington, J., & Vole, F. (2008). *Knowledge management excellence*. Moscow: Standards and Quality.
120. Hodgson, G. (2001). Socio-economic consequences of the advance of complexity and knowledge. *Questions of Economy, 8*, 32–46.
121. Hodgson, G. (2003). *Economic theory and institutions: The manifesto of modern institutional economics* (464 p.). Moscow: Delo.
122. Tsvetkov, V., & Pushkareva, K. (2010). Competencies and competitiveness of staff. *International Journal of Applied and Fundamental Research, 1*, 85–86.
123. Chursin, A., & Tyulin, A. (2018). *Competence management and competitive product development: Concept and implications for practice* (pp. 1–241). Heidelberg: Springer. https://doi.org/10.1007/978-3-319-75085-9
124. Chursin, A., Drogovoz, P., Sadovskaya, T., & Shiboldenkov, V. (2017). The dynamic model of elements' interaction within system of science-intensive production under unstable macroeconomic conditions. *Journal of Applied Economic Sciences, 12*(5), 1520–1530.
125. Chursin, A., Drogovoz, P., Sadovskaya, T., & Shiboldenkov, V. (2017). A linear model of economic and technological shocks in science-intensive industries. *Journal of Applied Economic Sciences, 12*(6), 1567–1577.
126. Chursin, A. A., Shamin, R. V., & Fedorova, L. A. (2017). The mathematical model of the law on the correlation of unique competencies with the emergence of new consumer markets. *European Research Studies Journal, 20*(3), 39–56.
127. Chursin, A. A., Kashirin, A. I., Strenalyuk, V. V., Ostrovskaya, A. A., & Kokuytseva, T. V. (2018). The approach to detection and application of the company's technological competences to form a business-model. *IOP Conference Series: Materials Science and Engineering, 312*(1), 012003. https://doi.org/10.1088/1757-899X/312/1/012003
128. Chursin, A.A., & Shevchenko, V.V. (2017). *About the possibilities of operational gaming scenario modeling activities of enterprises and corporations*. In Proceedings of 2017, 10th International Conference Management of Large-Scale System Development, MLSD. doi: https://doi.org/10.1109/MLSD.2017.8109609
129. Chursin, A. A., Semenov, A. S., & Danilchankay, A. V. (2016). Analysis of innovation development in the economy with exhaustible resource sector by first order dynamical systems application. *Nonlinear Phenomena in Complex Systems, 19*(3), 254–270.
130. Batkovskiy, A., Leonov, A., Pronin, A., Chursin, A., & Nesterov, E. (2018). Regulation of the dynamics of creating high-tech products. *International Journal of Engineering and Technology (UAE), 7*(3), 261–270. Special Issue 14, 2018.
131. Vlasov, Y. V., & Chursin, A. A. (2016). Management of diversification system in aerospace industry. *Economy of Region, 12*(4), 1205–1217. https://doi.org/10.17059/2016-4-21
132. Tyulin, A., Chursin, A., & Yudin, A. (2017). Production capacity optimization in cases of a new business line launching in a company. *Espacios, 38*(62).
133. Ermakov, V. A., Burmistrova, E. M., Bodin, N. B., Chursin, A. A., & Shevereva, E. A. (2018). A letter of credit as an instrument to mitigate risks and improve the efficiency of foreign trade transaction. *Espacios, 7*(3), 261–270. Special Issue 14, 2018.

134. Chursin, A. A., & Strenalyuk, V. V. (2018). Synergy effect in innovative activities and its accounting in the technological competencies of an Enterprise. *European Research Studies Journal, XXI*(4), 151–161.
135. Chursin, A. (2013). *Problems of competitiveness management in the aerospace industry in modern economic conditions*. Defense tech. Moscow: FSUE Scientific Technical Center "Informtehnika" No. 1.
136. Chursin, A., & Afanasyev, M. (2014). *Reform and development of rocket-space industry of Russia (methods, concepts and models): Monograph* (451 p.). Moscow: Publishing House "Spectrum".
137. Chursin, A., & Baymuratov, U. (2010). Global economy and problems of competitiveness of developing economies. *Reports of the National Academy of Sciences of the Republic of Kazakhstan, 1*, 73–80.
138. Chursin, A., & Vasilyev, S. (2011). *Competition, innovation and investment (non-linear synthesis)* (480 p.). Moscow: Engineering.
139. Chursin, A., & Volkov, V. (2012). *Some theoretical approaches to evaluating the competitiveness of space-rocket industry when implementing innovative technologies. System analysis, management and navigation: Book of abstracts* (133 p.). Moscow: MAI Publishing House.
140. Chursin A. A., Davydov V. A., & Ozhiganov E. N. (2012). *Factors and indicators of investment attractiveness of enterprises of space-rocket complex in the current economic conditions* (No. 6–7, pp. 36–44). Defense tech. Moscow: FSUE "Scientific—Technical Center informtekhnika".
141. Chursin, A., Danilyuk, R., & Ostrovskaya, A. (2014). *Evaluation of the effectiveness of projects in knowledge-intensive industries* (Vol. 4, pp. 148–151). Business in law, Publishing house "Yur – VAK".
142. Chursin, A., & Dranayeva, A. (2010). Quantitative evaluation of the competitiveness of the organization. *Defense Complex to Scientific and Technical Progress in Russia, 2*, 95–100.
143. Chursin, A., & Ivanov, A. (2010). Guidelines for creating a concept of sustainable innovative development of structures with the consolidated capital. *Defensive Complex to Scientific and Technical Progress of Russia, 3*, 110–117.
144. Chursin, A., Kovkov, D., & Shamin, R. (2013). Approaches to assess the impact of external and internal factors on competitiveness of products of rocket and space industry. Publishing house "Yur-VAK". *The Journal "Business in Law", 1*, 127–131.
145. Chursin, A., & Kokuytseva, T. (2011). The law of competitiveness. *Problems of Modern Economics, 1*, 43–45.
146. Chursin, A., & Kokuytseva, T. (2010). Innovative economy as the strategic goal of development in crisis and post-crisis conditions. *Bulletin of the National Academy of Sciences of the Republic of Kazakhstan*. Almaty: The National Academy of Sciences, *4*, 80–88.
147. Chursin, A., Makarov, Y., & Baymuratov, U. (2011). *Investment with innovation: Synergy in the competitiveness of the economy. Under the scientific editorship of A. Chursin* (496 p.). Moscow :Publishing house "MAKD": Engineering.
148. Chursin, A., & Milkovsky, A. (2014). *The role of information and communication technologies in the management of enterprises* (Vol. 4, pp. 123–127). Business in Law, Publishing house "Yur–VAK".
149. Chursin, A., & Okatyev, N. (2008). Approaches to optimization of resources on creation and production of competitive goods. *Engineering Journal, 5*, 35–39.
150. Chursin, A., & Sergeev, S. (2009). *To the question of the mechanism of management of large Russian corporations on the example of the aviation company of holding type, the article of defense equipment* (No 4–5, pp. 59–68).
151. Chursin, A., & Solovyev, V. (2013). The impact of innovation on mechanisms of competitiveness. *Innovations, 3*(173), 60–66.
152. Chursin, A., & Shamin, R. (2011). *Investments and innovations and their role in enhancing the competitiveness of the organization. Defensive complex – to scientific and technical progress of Russia* (No. 2. pp. 83–87).

153. Chursin, A., & Shmakov, E. (2014). Economic-mathematical model of optimal allocation of investments in the modernization of technology-intensive enterprises. *Business in Law, Publishing House "Yur-Vak", 3*, 239–243.
154. Chursin, A., & Vasiliev, S. (2011). *Competition, innovation and investment (nonlinear synthesis): Monograph* (478 p.). Moscow: Engineering.
155. Chursin, A. (2012). *Theoretical foundations of competitiveness management. Theory and practice* (590 p.). Moscow: Spektr.
156. Shinkevich, A., Sultanova, D., & Burganov, R. (2013). Managerial innovations as a factor of productivity growth. *Bulletin of the Kazan University, 24*(16), 217–220.
157. Shiryaeva, K., & Anishchenko, Y. (2014). Current state of space-rocket industry of Russia. *Actual Problems of Aviation and Cosmonautics, 10*(T. 2), 48–49.
158. Shifrin, M. (2006). *Strategic management.* St. Petersburg: Peter.
159. Schoenberg, R. (1988). *Japanese methods of business management* (251 p.). Moscow.: Economica
160. Schumpeter, J. A. (1988). *Capitalism, socialism and democracy* (trans. from English) Foreword and the general editorship of V. Avtonomov (540 p.). Moscow: Economica.

The manufacturer's authorised representative in the EU is Springer Nature Customer Service Centre GmbH, Europaplatz 3, 69115 Heidelberg, Germany. If you have any concerns regarding our products, please contact ProductSafety@springernature.com

Printed and bound by CPI Group (UK) Ltd, Croydon, CR0 4YY
30/09/2026
02225591-0016